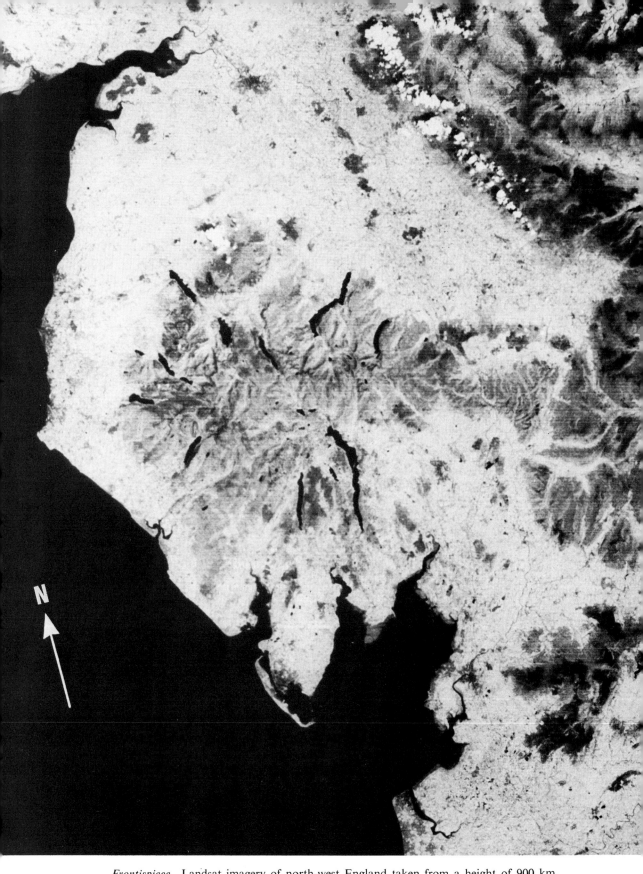

N

Frontispiece Landsat imagery of north-west England taken from a height of 900 km.
Photograph: National Aeronautical Space Administration, U.S.A.

THE GEOLOGY

OF

THE LAKE DISTRICT

Edited by

F. MOSELEY

YORKSHIRE GEOLOGICAL SOCIETY

(Occasional Publication No. 3)

1978

ISBN 0 9501656 2 X

PRINTED IN ENGLAND BY W. S. MANEY AND SON LIMITED   LEEDS

CONTENTS

LIST OF PLATES

Frontispiece Satellite photograph of the Lake District and surrounding areas

FOREWORD

The Lake District has long been a classical area for geologists. Since the original pioneer studies of Jonathan Otley and Adam Sedgwick in the early years of the last century the area has been known to show many features of major geological interest. The fact that there has been no comprehensive account of the geology of the area since that of J. E. Marr in 1916 encouraged the Yorkshire Geological Society to attempt a synthesis of modern knowledge of the district. Even after more than a hundred and fifty years of study there are still several major problems unsolved. This book attempts to provide a synopsis of current views — a sort of interim report, and not the final answers which, geology being a rapidly evolving science, may well be many years in coming.

The volume has been prepared under the direction of a committee appointed late in 1976 under the chairmanship of Dr W. H. C. Ramsbottom, then President of the Society, Dr F. Moseley, who most kindly offered to edit the volume, Dr N. J. Soper, and Messrs A. J. Wadge, R. S. Arthurton and J. G. O. Smart (Treasurer of the Society). The Society is much indebted to the specialists who have so generously agreed to contribute to the volume. Together they have brought a wealth of knowledge and experience to the task. Mr R. S. Arthurton has been responsible for the compilation of the coloured geological map, which has been skilfully drawn by Mr I. J. Wilkinson, and Miss Sonia Hodges has drawn more than half the text figures in the Drawing Office at Birmingham University.

The book is dedicated to the memory of the late Dr G. H. Mitchell, a former President of the Society, so well known for his studies of the volcanic rocks of the Lake District which he summarized in two Presidential Addresses to the Society in 1955 and 1956. It is a pleasure to so acknowledge the debt geology owes to him, and to remember the ready help and encouragement he gave, especially to younger colleagues. We are sad that he is no longer with us.

The Society is much indebted for financial help to those listed on an adjacent page. It is through the generosity of these firms that the book can be issued at so low a price in these hard financial times.

We have been fortunate in our printers, W. S. Maney and Son Ltd, and would like to thank in particular Mr B. G. Maney for his unfailing patience and technical guidance. We also wish to thank the Geological Society of London, *Geological Magazine* and the *Geological Journal* for permission to reproduce diagrams, either in their original or redrawn form.

JOHN DOSSOR

President
Yorkshire Geological Society

DONORS

The Society gratefully acknowledges generous financial assistance towards the cost of this book from the following organizations:

Selection Trust Limited

Shell U.K. Limited

Cluff Oil Limited

H. V. Dunnington and Associates

Mobil North Sea Limited

Home Oil (U.K.) Limited

Robertson Research International Limited

The British National Oil Corporation

British Nuclear Fuels Limited

Terresearch Limited

The Society is grateful to those members who contribute regular subscriptions to the Special Publications Fund and acknowledges the covenanted donations from Imperial Chemical Industries Limited and Associated Portland Cement Manufacturers Limited.

Dedicated
to the memory of
G. H. Mitchell

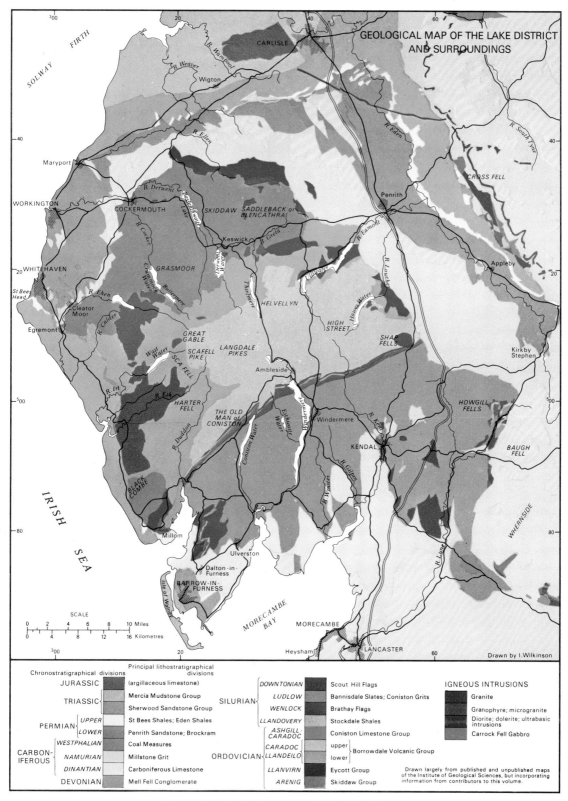

GEOLOGICAL MAP OF THE LAKE DISTRICT
AND SURROUNDINGS

Drawn by I. Wilkinson

Chronostratigraphical divisions	Principal lithostratigraphical divisions
JURASSIC	(argillaceous limestone)
TRIASSIC	Mercia Mudstone Group
	Sherwood Sandstone Group
PERMIAN UPPER	St Bees Shales; Eden Shales
PERMIAN LOWER	Penrith Sandstone; Brockram
CARBON-IFEROUS WESTPHALIAN	Coal Measures
CARBON-IFEROUS NAMURIAN	Millstone Grit
CARBON-IFEROUS DINANTIAN	Carboniferous Limestone
DEVONIAN	Mell Fell Conglomerate

SILURIAN DOWNTONIAN	Scout Hill Flags
SILURIAN LUDLOW	Bannisdale Slates; Coniston Grits
SILURIAN WENLOCK	Brathay Flags
SILURIAN LLANDOVERY	Stockdale Shales
ORDOVICIAN ASHGILL-CARADOC	Coniston Limestone Group
ORDOVICIAN CARADOC	upper Borrowdale Volcanic Group
ORDOVICIAN LLANDEILO	lower Borrowdale Volcanic Group
ORDOVICIAN LLANVIRN	Eycott Group
ORDOVICIAN ARENIG	Skiddaw Group

IGNEOUS INTRUSIONS

Granite

Granophyre; microgranite

Diorite; dolerite; ultrabasic intrusions

Carrock Fell Gabbro

Drawn largely from published and unpublished maps of the Institute of Geological Sciences, but incorporating information from contributors to this volume.

PLATE 1. Geological map of the Lake District

I

The Geology of the English Lake District
An Introductory Review

F. MOSELEY

A general perspective

It is scarcely 50 km from one side of the Lake District to the other and yet in that small space there is a variety of scenery and geology almost unequalled in Britain. It is not surprising therefore that this is a region which has inspired some of Britains most admired poets, as well as attracting outstanding geologists. Jonathan Otley was the first, followed by Adam Sedgwick whose memorial stone of Shap Granite stands in the village of Dent. Then came a whole string of great figures; Harkness, Nicholson, Aveline, Ward, Goodchild and Harker, culminating in the numerous contributions of J. E. Marr, whose publications span 50 years from 1880 to 1930. There were many others and between them all they established the fundamentals of Lakeland geology and revealed the mysteries of the older slate belt to the north, the craggy volcanic peaks of central Lakeland, the lower lying, younger, slates and sandstones of the south, and the surrounding rim of limestone, forming the often described Lake District dome. Since those days large numbers of geologists have flocked to the region, for the problems were by no means solved by those pioneers, they simply indicated the way ahead. There can be no doubt in the minds of any who know Lake District geology that this latest phase of activity has been dominated by G. H. Mitchell, especially for his work on the central core of volcanic rocks, and it is remarkable that the greatest proportion of this work has been a spare time activity. This greatly respected geologist had the closest association with the Yorkshire Geological Society, and it is most appropriate that this volume should be dedicated to him. But others have been stimulated on an accelerating scale and Smith's (1974) bibliography lists more than one hundred works on Lakeland geology published during the previous ten years. It is significant that this unending research has created more problems than it has solved so there is no shortage of geological interest in the Lake District for the forseeable future.

For the visitor a first introduction is likely to be via the motorway from either Kendal in the south or Penrith in the north. In each case the first outcrops to be seen will be Carboniferous Limestone, the younger rock surrounding the older core. From the south the more interesting approach is from Levens and along the picturesque Lythe Valley. Here the estuarine sands and clays of Morecambe Bay and the peat bogs of Levens are dominated by the well wooded scars and pavements of white limestone which form Whitbarrow. Further north these regions give way to the knobbly partially wooded hills of Silurian slates surrounding Windermere, whilst

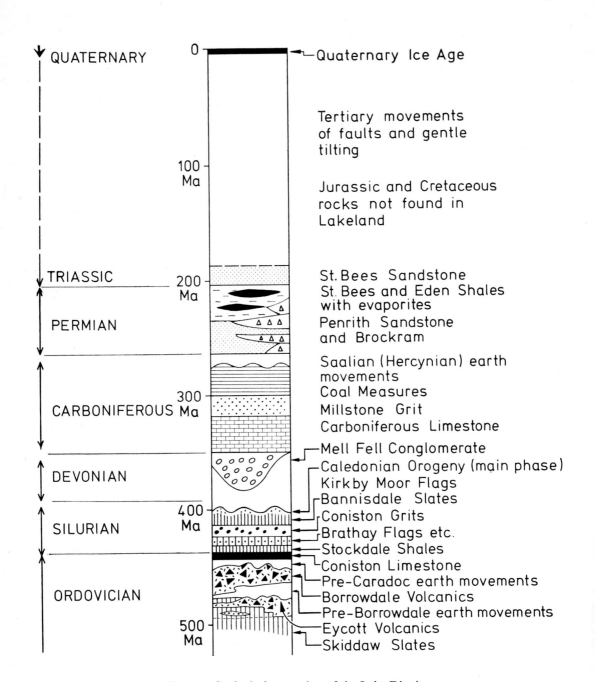

FIG. 1. Geological succession of the Lake District.

further north still there are the more precipitous rough volcanic crags of central Lakeland, deeply carved and scoured by the glaciers of the Quaternary Ice Age. From the north the rim of limestone is low lying and less obvious but there is no difficulty in recognizing the clear rounded silhouettes of the Mell Fells, composed of conglomerates of Devonian age, or in identifying the volcanic crags around Ullswater, and the softer outlines of the Skiddaw Slates of Blencathra and Skiddaw.

Most of these rocks and the unseen basement beneath have complex origins, and for full understanding many of the diverse disciplines of the geological sciences have to be consulted. Recent advances in geophysics, geochemistry, structural geology, palaeontology, hydrogeology and Quaternary biology require and have received the attention of specialists in those fields, so that the chapters which follow will bring the reader to the forefront of present knowledge of Lake District geology. In this introduction the overall geological history is outlined for the benefit of those not familiar with the Lake District and Lakeland geology.

Caledonian plate tectonics

The most important period in the geological history of the Lake District is that extending from early Ordovician to the end of the Silurian, and it is the events of this first 100 of the 500 Ma of Lakeland history which must now command our attention. It was a time when a major Late Proterozoic–Lower Palaeozoic ocean (Iapetus) was steadily destroyed, leading to continental collision and the climax of the Caledonian Orogeny.

Since Wilson (1966) suggested the existence of a proto-Atlantic Ocean, later named Iapetus by Harland and Gayer (1972) to distinguish it from the Mesozoic opening of our present Atlantic, there have been many publications referring to the supposed sutures between ancient America and ancient Europe, and to the sub-duction zones and plate movements which eventually destroyed this former ocean. Most of the earlier interpretations (Dewey 1969, 1971; Fitton and Hughes 1970; Church and Gayer 1973) placed the greater part of Scotland in 'America', England and Wales in 'Europe', and regarded the Southern Uplands as a relic of the former ocean Iapetus. The Girvan–Ballantrae ophiolite was taken to be a relic of a subduction zone descending beneath the Scottish–American continent, with another zone, perhaps centred on or near to the Solway, descending beneath England and Wales. Fitton and Hughes (op. cit.) also noted that the progressive north–south variation in the Ordovician volcanic rocks from the tholeiitic tendencies of the Eycotts in the north of the Lake District, to the calc-alkali Borrowdale Volcanics, and the predominantly alkali volcanics of Wales, could be paralleled by similar variations between continental and oceanic sides of island arcs, as described by Kuno (1966). Indeed there seemed little doubt that the volcanic rocks of the Lake District were formed in an island arc or continental margin environment and related to a south or south-easterly inclined subduction zone.

However further research in other fields resulted in modifications to the initial schemes. The first was when Powell (1971) demonstrated on geophysical grounds that the Southern Uplands must be underlain by continental crust. This led Gunn (1973) and then Jeans (1973) to suggest that perhaps the Midland Valley of Scotland was the Iapetus relic, with the Girvan ophiolite subducted not beneath Scotland, but southwards beneath the Lake District, where it was a source for the calc-alkali vol-canics of England and Wales. This idea did not survive long since Williams (1975)

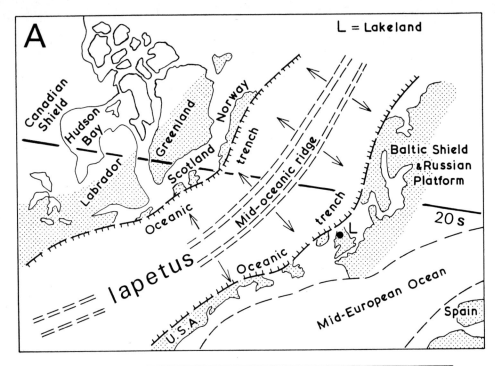

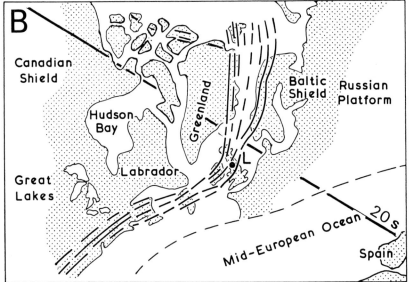

FIG. 2.   Lower Palaeozoic palaeogeographical reconstructions
       A — Lower Ordovician,
       B — End Silurian continental collision and the main phase of the
           Caledonian Orogeny (thick lines).
       Latitude from Faller and Briden (Chapter 2).

reiterated his earlier declarations that, on palaeontological grounds, the Southern Uplands belonged to ancient America. He showed conclusively that when brachiopod species were subjected to cluster analysis, Scottish–American faunas, including those of the Southern Uplands, were quite distinct from Anglo–Welsh faunas until Onnian times, after which they were indistinguishable, the faunas having intermingled. The same seemed to be true even for the pelagic graptolites (Skevington 1974). It was therefore clear that the Midland Valley could not be a relic of Iapetus and another revised interpretation became necessary. During the greater part of the Ordovician the Scottish–American and Anglo–Welsh provinces must have been separated by a barrier to faunal migration, but this did not exist during late Ordovician and Silurian times. The ocean Iapetus would have formed such a barrier with faunal mixing becoming possible upon near closure of the ocean. Palaeomagnetic evidence confirms the existance of the ocean, but suggests that by early Ordovician times it had narrowed to less than 1000 km (Faller and Briden Chapter 2). Subsequent geophysical work, the Lithospheric Seismic Profile in Britain (Bamford et al. 1976), has assisted inter-pretation still further. This profile extends from southern England to northern Scotland and has revealed that the whole of Britain is underlain by continental crust with the Moho at an average depth of 35 km.

In summary the conclusions now emerging are that an ocean (Iapetus) separated the continents for the greater part of the Ordovician, virtually closing towards the end of this period, but with true continental collision and orogeny delayed until the end of the Silurian, when practically all traces of Iapetus were obliterated. The most probable line for the continental suture is along the Solway where, according to Phillips et al. (1976), there was also more than 1000 km of dextral displacement, essentially taking place during Silurian times.

Lower Palaeozoic palaeogeography

The earliest of the Lake District rocks are the Skiddaw Slates, more than 3000 m of graptolitic mudstones with subsidiary greywacke sandstones, the latter especially with a great variety of sedimentary structures. For the most part they are considered to be distal turbidites deposited on the continental margin of ancient Europe (Jackson Chapter 7). Until recently it was believed that there was little volcanic activity asso-ciated with the Skiddaw Slates, but work by Downie and Soper (1972) has shown that the Eycott Volcanics, formerly equated with the Borrowdale Volcanics, in fact belong to this earlier episode. The lowest of these volcanic rocks are interbedded with pelites of Arenig age and there follows more than 2500 m of basalts, andesites and subsidiary dacites of transitional tholeiitic–calc-alkaline composition (Fitton and Hughes 1970), all now regarded as earlier than the main Borrowdale Volcanic episode. These rocks and the associated pelites have now been removed from the Skiddaw Group to form an entirely separate Eycott Group (Wadge Chapter 6). This volcanic episode is strong evidence for an active subduction zone, and the tholeiitic tendencies, early date and geographical location nearest to the continental margin all fit into an acceptable environment of continental margin or island arc underlain by continental crust (Bamford et al. 1976). It is likely also that some of the basic intrusions, some of which have a tholeiitic trend, belong to this Eycott episode, in particular the Carrock Complex, and minor intrusions such as those of Castle Head, Keswick and Embleton (Firman Chapter 10).

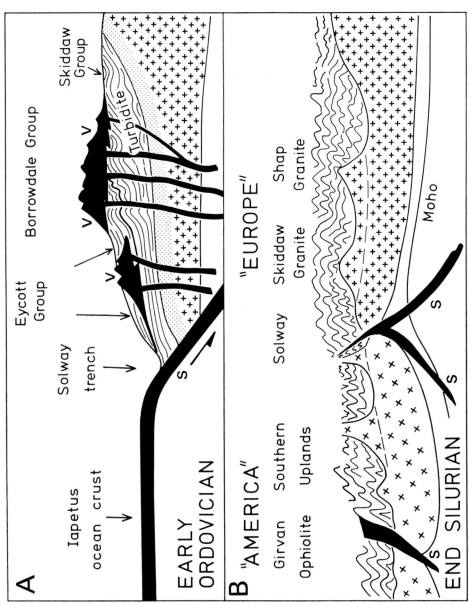

Fig. 3.   Diagrammatic sections to show the Lower Palaeozoic evolution of Lakeland.
A — Lower Ordovician. Skiddaw and Eycott sediments and Eycott
and Borrowdale volcanoes (V) resting on continental basement.
B — End Silurian continental collision and the main Caledonian Orogeny.
S — Subduction and former subduction zones.

The Skiddaw Slates are separated from the Borrowdale Volcanics by a noticeable unconformity (Wadge 1972, Jeans 1972), although it is not regarded here as of orogenic proportions as suggested by Simpson (1967) (see Moseley 1972). The pre-Borrowdale folds have north–south orientations, nearly at right angles to the continental margin, and difficult to account for. They could be related posthumously to north–south structures in a Pre-Cambrian basement similar to the north–south trends seen elsewhere in England (Moseley and Ahmed 1973).

The next important occurrence was the eruption of the calc-alkaline Borrowdale Volcanic rocks, partly as subaqueous, and partly as subaerial lavas and tuffs. The whole structure can be referred to as the Borrowdale volcano, and visualized as one of a string of volcanoes at a latitude of about 20°S (Faller and Briden Chapter 2) with neighbouring volcanoes in North Wales, Pembrokeshire and Southern Ireland, in fact midway between an island arc situation and a continental margin such as the North American Cascades of today. Precise dating of these essentially unfossiliferous rocks is difficult but they seem to belong to the period between *D. murchisoni* (Wadge et al. 1972) and the Ashgillian (Ingham and Rickards Chapter 9). The total thickness exceeds 5000 m and includes basaltic, andesitic and dacitic lavas and tuffs, with the acidic tuffs usually ignimbritic. The field relations are complex, but nevertheless it is possible to interpret the sequence in terms of three basic to acid magmatic cycles, and to conclude that several major centres were operating. The latter are not immediately obvious, but the Haweswater Complex would appear to be one (Nutt 1968) and there are other intrusives and vent agglomerates which can be interpreted either as major or as parasitic vents. There seems little doubt that these eruptions, like those of the Eycotts, were related to a SE dipping subduction zone and that the whole process was part of the gradual closure of Iapetus.

The next event also provides important evidence for the ocean closure. The volcanic rocks are followed unconformably by the Coniston Limestone Group, which, from NE to SW, oversteps the whole 5000 metres of the volcanic sequence to rest on Skiddaw Slates; a considerable degree of uplift and erosion. The pre-Ashgillian, or possibly pre-Caradocian, earth movements responsible for this also resulted in open ENE trending folds (Soper and Numan 1974), and were perhaps generated as the two continents came into close proximity. At approximately the same time vulcanicity and presumably subduction, largely ceased, and by the beginning of the Ashgillian the previously separate faunas of both continents were able to mingle.

The Coniston Limestone Group of Ashgillian age is widespread but is not usually more than 150 m thick in the Lake District. However it exceeds 600 m further east (Ingham and Rickards 1974). The greater part consists of calcareous mudstone, grading locally into nodular limestones, with the tuffaceous sandstones of the Stile End beds at the base. Most of the sequence is interpreted as a benthos rich shelf sea facies.

The Ashgill Shales at the top of the Coniston Limestone Group are followed by Llandovery (Skelgill) Shales at the base of the Silurian. They are black carbonaceous shales, very rich in graptolites and probably indicate anaerobic conditions and a deepening sea (Ingham et al. Chapter 9). There is then transition to the greenish Browgill Beds and to the 3000 m of striped mudstones and greywackes which constitute the Wenlock and Ludlow Series. It is probable that by this time all that remained of Iapetus was a narrow geosynclinal trough with turbidity currents bringing in floods of sediment, the greywackes being proximal turbidite influxes into a predominantly mudstone environment (Norman 1963). Evidence for current directions comes mainly from turbidity current structures (Norman op. cit.) and partly from fossil orientations (Furness 1965), and suggests flow mostly from the NW down the flanks of the geosyncline, but with some ENE–WSW axial flow, much of the latter from fossil orientations. In this connexion Loubere (1977) has shown that elongate fossils (Brathay Flags orthocones) were probably deposited at right angles to currents and some of the ENE–WSW current directions derived on the assumption of alignments parallel to currents may be suspect. The top of this thick Silurian sequence, now dated as Downtonian, shows some reddening, perhaps a forerunner of the Old Red Sandstone conditions to come.

The end-Silurian climax of the Caledonian Orogeny

From the Late Pre-Cambrian until the closing stages of the Ordovician the ancient continents of America and Europe, separated by Iapetus, had developed independent structural histories (Wright 1976). After the near closure of Iapetus by Onnian times (Williams 1975) there was little tectonic activity until the end of the Silurian, when renewed continental movement brought about the long imminent continental collision. The remaining relics of Iapetus were destroyed as the two continents became firmly welded together, and the rocks of both continental margins became strongly deformed (Phillips et al. 1976, Moseley 1977).

The most complex deformation in the Lake District is found in the weaker, incompetent rocks, especially the Skiddaw and Bannisdale Slates, in which there are tight folds and moderately strong cleavage. Polyphase deformation of the Skiddaw Slates with new structures superimposed upon those of earlier episodes makes them particularly difficult to study. More competent rocks such as the Loweswater Flags, Borrowdale Volcanics and Coniston Grits were less effected, the folds being more open and minor folds less common. Cleavage also is unimportant in most of these competent rocks, the exceptions being fine grained and chlorite rich tuffs within the volcanics which provide the best slates of the region. Faulting was also an important part of the tectonic activity with NW dextral and N sinistral wrench faults prominent. Strike slip displacements of up to a mile have been recognized.

The end of the orogeny was marked by the intrusion of the granite batholith which, geophysical investigations have shown, underlies the north and central parts of the Lake District, extending east into the Weardale Granite beneath the Alston Block and probably west to the Isle of Man (Bott 1974 and Chapter 3). The high points of the batholith are now exposed as the Skiddaw and Shap Granites (Fig. 4), radiometric dating placing them in the early Devonian (Brown et al. 1964). All these exposed granites are similar chemically, although there are local differences such as the late potash metasomatism of the Shap Granite (Firman Chapter 10). It is also probable that the batholith was important as a source of the epigenetic mineralization for which the Lake District is so well known, numerous veins having yielded tungsten, lead, zinc, copper, manganese, graphite and other important minerals. Although working has practically ceased it has been estimated that reserves are still considerable (Firman Chapter 15).

The Old Red Sandstone Continent

A principle result of the Lower Palaeozoic plate movements was the formation of an entirely new desert continent extending from Russia to western Canada. The Caledonian Mountains were on its SE margin, bounded by a coastal plain which

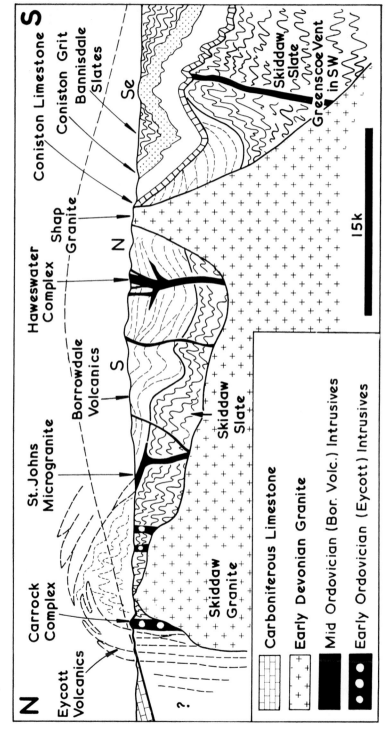

FIG. 4. Diagrammatic profile across the Lake District showing the early Devonian granite batholith intruded into Skiddaw Slates, Borrowdale Volcanics and post volcanic sediments.

varied in width as the sea advanced and retreated during different stages of the
Devonian Period (House 1968) whilst the palaeomagnetic evidence indicates an
equator not far away (Faller and Briden Chapter 2 and Fig. 5).

The rocks of the Lake District believed to be of this age are the Mell Fell Conglomerates which outcrop near Ullswater, and the Polygenetic Conglomerate of the
Cross Fell Inlier. The former is probably about 300 m thick and was formed as
coalescing fans descended from surrounding high ground, rather like some alluvial
fans in parts of South Arabia today. The abundance of Silurian greywacke phenoclasts suggests that these rocks were much more widely exposed during the Devonian
than now. The polygenetic conglomerate is similar and was probably derived from an
east facing escarpment (Wadge Chapter 6).

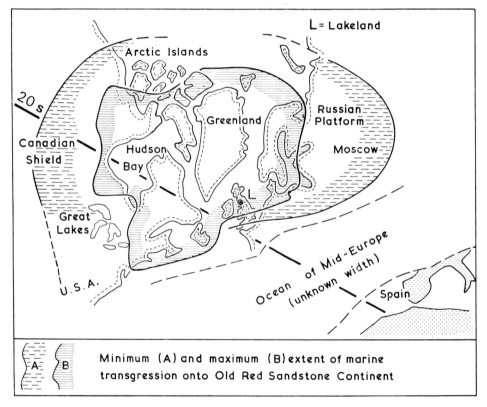

FIG. 5. The Old Red Sandstone Continent, modified from House 1968.
Latitude from Faller and Briden Chapter 2.

Further north in the Cheviots and Scotland there are thick accummulations of
tholeiitic to calc-alkali volcanics (G. B. Slater in press) and Slater (op. cit.) suggests
that subduction, now inclined northwards, continued in a modified way through
Silurian and early Devonian times until the continents finally became stable.

The Carboniferous Period

The mountains of the Old Red Sandstone Continent were eventually reduced to a landscape of low relief, and by the beginning of the Carboniferous epicontinental seas of the mid-European Ocean transgressed across much of the British area. Palaeomagnetic latitude determinations show that the region was still tropical (Faller and Briden Chapter 2) and this is supported by the Carboniferous faunas and floras. There is also no doubt that the old Caledonian structures still exerted important influences, affecting Carboniferous sedimentation and the orientations and styles of the Hercynian structures. There were stable massifs which either remained above the sea for the whole of the Carboniferous or received limited thicknesses of sediment, and intervening basins characterized by great thicknesses of Carboniferous Limestone, Millstone Grit and Coal Measures (Fig. 6). The Lake District was one of the massifs made resistant by the lenticular Eycott and Borrowdale Volcanoes and the intrusion of the northern England Batholith, whilst to the north and south there were the subsiding gulfs of Solway–Northumbria and the Central Pennines.

Carboniferous rocks are, in fact, little noticed within the Lake District itself, but are of major economic importance in adjacent parts of Cumbria and in offshore areas, where the tropical Carboniferous Limestone seas were followed by deltaic, lagoonal and estuarine deposits of the Millstone Grit and Coal Measures, the latter characterized by rich floras of equatorial swamp forests. Carboniferous Limestones, peripheral to the Lower Palaeozoic core of the Lake District do deserve further mention however, since they form extremely attractive scenery, especially in the south near Morecambe Bay. The limestone pavements and escarpments of Whitbarrow and Underbarrow are particularly impressive, and form strong scenic contrasts to the older rocks.

The Hercynian Orogeny

During the Upper Palaeozoic the region to the south of the Old Red Sandstone Continent was part of the wide Mid-European Ocean which closed towards the end of the Carboniferous (Fig. 5), a situation comparable with the Lower Palaeozoic closure of Iapetus. Most of Britain was too remote from these new plate movements to feel severe effects, although the rocks of SW England were strongly deformed. In the Lake District the only tectonic phase of importance was the Saalian at the end of the Carboniferous. At this time there was severe folding and faulting in some areas, the vertical strata of the Silverdale disturbance and the complex tectonics of the Cross Fell Inlier being good examples, but elsewhere the rocks were merely tilted to moderate angles, so that the Carboniferous rocks generally dip outwards at less than 10° from the central core of Lower Palaeozoic rocks.

The New Red Sandstone Deserts

It has long been known from the nature of the sediments that by Permian times Britain had moved into a desert environment; now confirmed by Palaeomagnetic studies, which have revealed that between the Lower Carboniferous and Triassic Britain drifted 30°N into Sahara-like latitudes (Faller and Briden Chapter 2). The landscape of this time is believed to have been low lying with a badly defined coastal plain, wide desert basins and low escarpments, all indirectly related to underlying Caledonian and Hercynian structures.

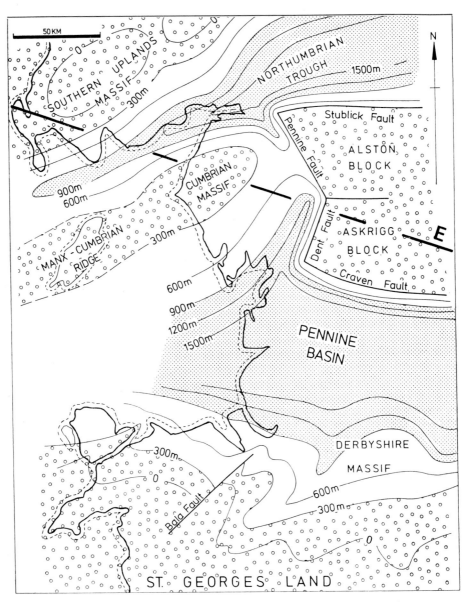

FIG. 6. Lower Carboniferous isopachs illustrating the relationships between massifs and basins in Northern England. Modified from George 1958. Latitude from Faller and Briden Chapter 2 (E — Equator).

During the Lower Permian the barchan sand seas of the Penrith Sandstone spread across the whole region, orientations of dune bedding suggesting an easterly derivation (Shotton 1956, Waugh 1970). Interbedded lenses of Brockram were formerly alluvial fans of angular gravel brought in from the higher ground by flash floods (Arthurton et al. Chapter 13). These gravels and breccias are commonly calcreted and the whole complex is closely reminiscent of deposits of modern hot deserts. By upper Permian times the sand seas had given way to coastal sabkahs upon which there was extensive evaporite accumulation (gypsum and anhydrite), still with occasional Brockram influxes (Arthurton et al., op. cit.). These are the St Bees and Eden Shales and Evaporites. The succeeding Triassic Period is largely represented by the thick accumulation of the St Bees Sandstone, believed to be alluvial plain deposits laid down by braided streams, although there was some local dune formation and there are also red mudstones with intercalations of salt which, like those of the Upper Permian, seem likely to have been formed on coastal flats.

None of these Permo–Triassic sediments occur within the Lake District proper, and although at one time they must have crossed the entire area, erosion has stripped them away from central Lakeland so that they are only to be found in the low lying border areas of the Vale of Eden, the north and west Cumbrian plains and parts of Morecambe Bay.

Later Mesozoic and Tertiary Times

The Lake District provides little evidence for geological events between the Triassic and the Quaternary, but enquiries elsewhere have revealed two important and progressive changes in palaeogeography. The first disrupted the Eur–American (Old Red Sandstone) Continent as new episodes of plate movement resulted in the opening of our modern Atlantic, with Europe and America now drifting apart, and the second saw Britain and Europe moving steadily northwards from tropical to temporate latitudes (Smith and Briden 1977).

Although there are no Jurassic, Cretaceous or Tertiary rocks within the immediate area of Lakeland it is a treasured belief of many that there were occasions when the whole area was beneath the sea, and that Mesozoic and possibly Tertiary sediments were laid down. In particular there is the long accepted concept of superimposed drainage from a former chalk dome, proposed by Marr (1916). This hypothesis required Lakeland to have been covered by the Chalk seas, with the resulting formations domed upwards during post-Triassic (presumed Tertiary) earth movements. Erosion followed and a consequent radial drainage system developed upon the dome (or 'inverted caddy spoon') with the rivers eventually carving their way through the younger sediments to become superimposed upon the Lower Palaeozoic rocks. The idea was brilliant and almost certainly contains much truth, but the details are not so convincing and require closer scrutiny. For example, there are numerous post-Triassic faults in northern England many of which cross the Lake District and must have affected any former chalk cover (Moseley 1972). It is also noticeable that most of the radial drainage lines follow important Caledonian structures in the Lower Palaeozoic rocks. Much of Ullswater is aligned along the Ullswater Anticline, which brings to the surface the easily eroded Skiddaw Slates. Windermere, Coniston Water and Thirlmere follow the directions of pronounced northerly Caledonian wrench faults, and other examples are not difficult to find. However, there still remains a residue of unexplained alignments (Bassenthwaite, Crummock, Ennerdale, etc.) which suggests that there may be something in the old idea after all.

Post-Triassic faulting has already been mentioned. It is noticeable that all faults which moved during these times, so judged by their effects on the Permo-Triassic rocks of the surrounding regions, followed exactly the lines of much older faults, and are almost certainly posthumous structures with Caledonian, Hercynian and Mesozoic to Tertiary phases of movement (Moseley and Ahmed 1967). In the past these faults have been seen as expressions of the distant Alpine Orogeny, but this connexion is an unlikely one since northern Britain was at the time within the tensional regime of the ever widening Atlantic Ocean, and the pattern of normal faults in the Lake District and surrounding areas fits well with this type of plate margin. For example, to the south of the Lake District the Askrigg Block is separated from the Irish Sea Basin by a series of normal faults most of which step down to the south-west with extension in this direction (Moseley 1972).

The Quaternary Ice Age

The latest, shortest and possibly most dramatic event so far as it affects today's scenery has been the extensive glaciation of Lakeland. The spectacular scenery of sharp peaks, deep corries and impressive ribbon lakes is immediately obvious, and the morainic hummocks left behind by melting ice are almost equally well known, resulting in a landscape of imposing aspect, even though the highest mountain does not reach 1000 m.

The Quaternary Ice Age is of course by no means unique to earth history and certainly the last 1000 Ma has been punctuated by periodic glaciations. The last event of this kind prior to the present ice age, was however, a long time ago at the end of the Carboniferous, and it did not affect Britain, which at that time was basking in the tropical glories of Coal Measure swamps. Since then, during the Mesozoic and Tertiary epochs, there is nothing to suggest the existence even of polar ice caps, and the earth appears to have been warm and ice-free until the beginning of the present ice age, an event which is by no means complete (Wright and Moseley 1975).

The first signs of our present ice age are Antarctic records of glacial deposits interbedded with basalt lavas which have been radiometrically dated at 8.7 Ma. Since then the ice has advanced and retreated many times, gradually extending into latitudes far from the poles. The beginning of the Quaternary Period is placed, not at the first record of glaciation, but much later (1.8 Ma) when the cold conditions had become well established in non-polar latitudes, although even then Britain remained unglaciated. Eventually the glaciers reached Britain, and during a span of close on a million years advanced and receded on several occasions. Each of the glaciations scoured and carved the landscape, and left behind an extensive detritus of moraine, till and outwash, and each of the interglacials, as warm or warmer than the present day, softened the scenery, and allowed rivers to redistribute the glacial sediments. Nor is the Quaternary Ice Age necessarily at an end since most of the evidence suggests that we are now merely indulging in an interglacial or interstadial pause. Indeed it is alarming to be told of the rapidity with which climates can change. Studies of fossil coleoptera (Coope 1975) have shown that on two occasions during the latest Devensian Glaciation, chilling has been so rapid that should it happen now there would be substantial glaciers in Britain within the span of one lifetime.

It will be understood that details of Quaternary history are derived from the stratigraphy of interglacial and interstadial deposits rather than observations of glacial erosion, and in this respect it is lowland rather than mountain areas which yield most information. In the mountains of the Lake District records of early glaciations appear to have been largely swept away by the latest Devensian (Weichselian) glaciers. Most of the erosional land-forms almost certainly belong to the main episode of this glaciation, but they were emphasized no doubt by the last phase of the Younger Dryas corrie glaciation (Walker 1966) not much more than 10,000 years ago. Interglacial records are insignificant and it is not until the Late Glacial Interstadial and Post Glacial times that detailed stratigraphical sequences are preserved. They are best seen in the sediments of lake basins and have been studied in great detail (Pennington Chapter 14).

REFERENCES

BAMFORD, D., FABER, S., JACOB, B., KAMINSKI, W., NUNN, K., PRODEHL, C., FUCHS, K., KING, R. and WILMORE, P. 1976. A Lithospheric Seismic Profile in Britain, I. Preliminary results. *Geophys. J. R. astr. Soc.* **44**, 145–60.

BOTT, M. H. P. 1974. The geological interpretation of a gravity survey of the English Lake District and the Vale of Eden. *Jl geol. Soc. Lond.* **130**, 309–31.

BROWN, P. E., MILLER, J. A. and SOPER, N. J. 1964. Age of the principal intrusions of the Lake District. *Proc. Yorks. geol. Soc.* **34**, 331–42.

COOPE, G. R. 1975. Climatic fluctuations in northwest Europe since the Last Interglacial, indicated by fossil assemblages of Coleoptera, *in* Wright, A. E. and Moseley, F. (Eds). Ice ages ancient and modern, *Geol. J. spec. issue*, No. 6, 153–68.

CHURCH, W. R. and GAYER, R. A. 1973. The Ballantrae ophiolite. *Geol. Mag.* **110**, 497–510.

DEWEY, J. F. 1969. Evolution of the Appalachian/Caledonian orogen. *Nature, Lond.* **222**, 124–29.

—— 1971. A model for the lower Palaeozoic evolution of the southern margin of the early Caledonides of Scotland and Ireland: *Scottish Jour. Geology*, **7**, 219–40.

DOWNIE, C. and SOPER, N. J. 1972. Age of the Eycott Volcanic Group and its conformable relationship to the Skiddaw Slates in the English Lake District. *Geol. Mag.* **109**, 259–68.

FALLER, A. M. and BRIDEN, J. C. 1977. Palaeomagnetic results from the Borrowdale Volcanic Group, English Lake District. *Geophys. J.R. astr. Soc.* **48**, 111–21.

FITTON, J. G. and HUGHES, D. J. 1970. Volcanism and plate tectonics in the British Ordovician: *Earth planet Sci. Lett.* **8**, 223–28.

FURNESS, R. R. 1965. The petrography and provenance of the Coniston Grits east of the Lune Valley Westmorland. *Geol. Mag.* **102**, 252–60.

GEORGE, T. N. 1958. Lower Carboniferous palaeogeography of the British Isles *Proc. Yorks geol. Soc.* **31**, 227–318.

HOUSE, M. R. 1968. Continental drift and the Devonian System. *Inaugural lecture, University of Hull.* 1–25.

INGHAM, J. K. and RICKARDS, R. B. 1974. Lower Palaeozoic Rocks. *In* D. H. Rayner and J. E. Hemingway (Eds). The Geology and Mineral Resources of Yorkshire, *Yorkshire geol. Soc.* 29–44.

JEANS, P. J. F. 1972. The junction between the Skiddaw Slates and Borrowdale Volcanics in Newlands Beck, Cumberland. *Geol. Mag.* **109**, 25–8.

—— 1973. Plate tectonic reconstruction of the southern Caledonides of Great Britain. *Nature Phys. Sci.* **245**, 120–22.

KUNO, H. 1966. Lateral variation in basalt magma type across continental margins and island arcs. *Can. Geol. Surv.*, Paper 66–15, 317–36.

LOUBERE, P. 1977. Orientation of orthocones in the English Lake District based on field observations and experimental work in a flume. *Jour. Sed. Petrol.* **47**, 419–27.

MARR, J. E. 1916. The Geology of the Lake District. Cambridge Univ. Press. xii + 220 pp.

MOSELEY, F. 1972. A tectonic history of north-west England. *Jl geol. Soc. Lond.* **128**, 561–98.

—— 1977. Caledonian plate tectonics and the place of the English Lake District. *Bull. geol. Soc. Am.* **88**, 764–68.

—— and AHMED, S. M. 1967. Carboniferous joints in the north of England and their relation to earlier and later structures. *Proc. Yorks. geol. Soc.* **36**, 61–90.

—— and AHMED S. M. 1973. Relationship between joints in Pre-Cambrian, Lower Palaeozoic and Carboniferous rocks in the West Midlands of England. *Proc. Yorks. geol. Soc.* **39**, 295–314.

NORMAN, T. N. 1963. Silurian (Ludlovian) Palaeo-current directions in the Lake District area of England. *Bull. geol. Soc. Turkey*, **8**, 27–54.

NUTT, M. J. C. 1968. Borrowdale Volcanic Series and associated rocks around Haweswater, Westmorland. *Proc. geol. Soc. Lond.* No. 1649, 112–13.

PHILLIPS, W. E. A., STILLMAN, C. J. and MURPHY, T. 1976. A Caledonian plate tectonic model. *Jl geol. Soc. Lond.* **132**, 579–609.

POWELL, D. W. 1971. A model for the Lower Palaeozoic evolution of the southern margin of the early Caledonides of Scotland and Ireland. *Scottish Jour. Geology*, **7**, 369–72.

SHOTTON, F. W. 1956. Some aspects of the New Red Desert in Britain. Liverpool/Manchester *Geol. J.*, **1**, 450–65.

SIMPSON, A. 1967. The stratigraphy and tectonics of the Skiddaw Slates and the relationship of the overlying Borrowdale Volcanic Series in part of the Lake District. *Geol. J.* **5**, 391–418.

SKEVINGTON, D. 1974. Controls influencing the composition and distribution of Ordovician graptolite faunal provinces. *In* Rickards, R. B., Jackson, D. E. and Hughes, C. P. (Eds). *Graptolite studies in honour of O. M. B. Bulman Spec. Pap. Palaeont.* **13**, 59–73.

SMITH, A. G., BRIDEN, J. C. and DREWRY, G. E. 1973. Phanerozoic world maps. *In* Hughes, N. F. (Ed.). *Special Papers in Palaeontology. Pal. Assoc. Lond.* 1–42.

—— and BRIDEN, J. C. 1977. Mesozoic and Cenozoic paleocontinental maps. *Cambridge Earth Sciences Series. Camb. Univ. Press.* 63 pp.

SMITH, R. A. 1974. A bibliography of the geology and geomorphology of Cumbria. *Cumberland geol. Soc.* 1–32.

SOPER, N. J. and NUMAN, N. M. S. 1974. Structure and stratigraphy of the Borrowdale Volcanic rocks of the Kentmere area, English Lake District. *Geol. J.* **9**, 147–66.

WADGE, A. J. 1972. Sections through the Skiddaw–Borrowdale unconformity in Eastern Lakeland. *Proc. Yorks. geol. Soc.* **39**, 179–98.

—— NUTT, M. J. C. and SKEVINGTON, D. 1972. Geology of the Tarn Moor Tunnel in the Lake District. *Bull. geol. Surv. Gt. Br.* No. 41, 55–62.

WALKER, D. 1966. The Glaciation of the Langdale Fells. *Geol. J.* **5**, 208–15.

WAUGH, B. 1970. Petrology, provenance and silica diagenesis of the Penrith Sandstone (Lower Permian) of north-west England. *J. Sedim. Petrol.* **40**, 1226–40.

WILLIAMS, A. 1975. Plate tectonics and biofacies evolution as factors in Ordovician correlation. *In* Bassett, M. G. (Ed.). *The Ordovician System proceedings of a Palaeontological Association symposium.* Univ. of Wales Press and National Museum of Wales, Cardiff, 18–53.

WILSON, J. T. 1966. Did the Atlantic close and then re-open? *Nature, Lond.* **211**, 676–81.

WRIGHT, A. E. 1976. Alternating subduction direction and the evolution of the Atlantic Caledonides. *Nature, Lond.* **264**, 156–60.

—— and MOSELEY, F. 1975. Ice ages ancient and Modern, a discussion. *In* Wright, A. E. and Moseley, F. (Eds). *Ice ages ancient and modern, Geol. J. special issue.* No. 6, 301–12.

F. MOSELEY, D.SC., PH.D.
Department of Geological Sciences, University, Birmingham

2

Palaeomagnetism of Lake District Rocks

ANGELA M. FALLER and JAMES C. BRIDEN

In principle, identification of the direction of original remanent magnetization in a suite of rocks can be translated into an inference of its palaeolatitude and palaeo-meridional orientation at the time of its formation. Sequences of data with good age-control and covering a large span of time may be used to construct a standard record. This standard may be expressed (a) as time-sequences of directions of magnetization (which are identified with ancient magnetic field directions), (b) as palaeolatitudes and meridional orientations, or (c) by the positions of the palaeomagnetic poles which the data imply, on the assumption that the ancient magnetic field has been dipolar, much as it is now. This last method, the contruction of apparent polar wander paths, is the most widely used, because it eliminates the effects of geographic variation of the geomagnetic field and enables data to be summarized from areas as large as an entire lithospheric plate. Figure 7 shows the standard apparent polar wander path of the south pole relative to Britain through Palaeozoic and early Mesozoic time, and figure 8 is composed of maps showing some corresponding palaeolatitudes and palaeomeridional orientations.

Amongst the standard data used to compile figures 7 and 8 are the two most important and securely based palaeomagnetic results from the Lake District. They are from the Eycott Group (Briden and Morris 1973) and the Borrowdale Volcanic Group (Faller et al. 1977). A further feature, not illustrated herein, which is conveniently thought of as standard in the present context, is the discovery from palaeomagnetic studies in North Wales (Thomas and Briden 1976) that the palaeomagnetic field may have been anomalous and steeply inclined in Britain for at least one brief spell in Caradocian time.

Palaeomagnetic studies of rock units whose age or tectonic situation is inadequately known may, by comparison with the standard record, assist in deducing age or tectonic relationships. The study of the Carrock Fell Complex is the best example in this category, and the method has also been applied to the interpretation of smaller quantities of data from various minor intrusions throughout the Lake District.

The Eycott Group

The first palaeomagnetic study in the Lake District was by Nesbitt (1967) who sampled seven flows from Eycott Hill and found a stable remanence with a direction similar to that in other British Palaeozoic rocks. A more extensive survey was carried out by Briden and Morris (1973) with the objectives of defining the local Ordovician geomagnetic field and obtaining a point on the apparent polar wander path for Britain, comparing the palaeomagnetic directions with those from the Borrowdale

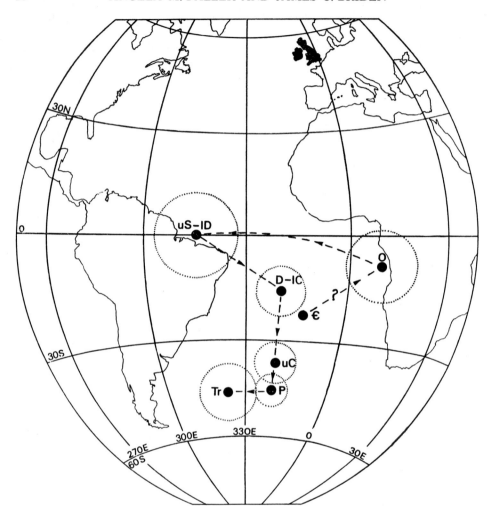

Fɪɢ. 7.   Palaeozoic south polar path for the British Isles, based on Ordovician
data summarized by Faller et al. (1977) and other Phanerozoic data summarized
by McElhinny (1973). Mean poles are shown for each period designated,
together with circles of 95 per cent confidence. Cambrian denoted by Ꞓ,
Ordovician O, Silurian S, Devonian D, Carboniferous C, Permian P, Triassic
Tr (u denotes upper, 1 denotes lower). Lambert equal area projection.

Volcanic Group and the Carrock Fell Complex, and contributing to a palaeomagnetic
study of the British Caledonides in the context of plate tectonics. In this study many
flows in both the Binsey and High Ireby Formations were sampled in detail at Binsey
and Eycott Hill. Alternating field (AF) demagnetization was used to remove any low
stability secondary magnetization and the *in situ* palaeomagnetic directions were
corrected for the measured dip of the strata by rotation about the present strike.
The appreciable difference in attitude between the lavas at Binsey and at Eycott Hill

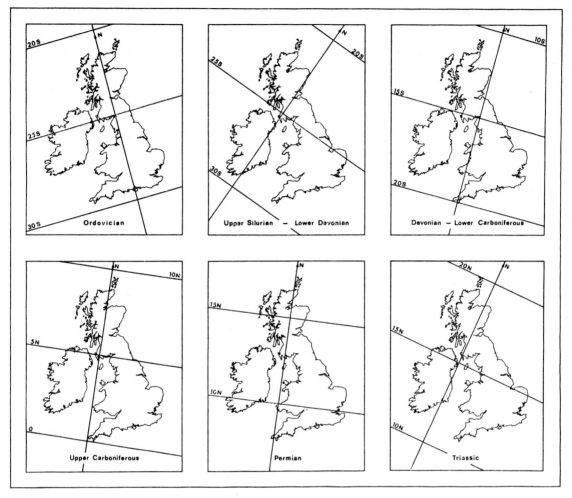

FIG. 8. Palaeolatitudes and palaeomeridional orientations for the British Isles derived from the Phanerozoic poles shown in Figure 7, from Ordovician onwards.

permits a fold test of the age of remanence. After tilt correction the between-site precision was greatly improved; this implies that the remanence predates the folding, and also lends support to the interpretation of Eastwood et al. (1968, Fig. 8) that the upper part of the Eycott sequence is inverted to the north of Binsey. Briden and Morris also demonstrated that any errors due to oversimplifying the fold-history of the rocks, as though they had been simply rotated about the present strike, were likely to be small.

Fourteen of fifteen sites sampled in the Binsey Formation gave statistically significant mean directions which were combined to give an overall result of Declination D = 5°, Inclination I = −50° (circle of 95 per cent confidence α_{95} = 8°). The corresponding result for the High Ireby Formation, derived from thirteen sites, was D = 356°, I = −41° (α_{95} = 7°). Although the results from the two formations can be shown to be significantly different, because of the possibility that secular variation may not have been adequately averaged, Briden and Morris combined them to give for the Eycott Group $D = 0$, $I = -46°$ ($\alpha_{95} = 6°$), which corresponds to a palaeomagnetic pole at 7°N, 177°E. All the samples had the same polarity, suggesting that no geomagnetic reversal occurred during the eruption of the Eycott Group. The main magnetic constituent seen in polished specimens from the lavas is fine-grained titanomagnetite, generally in a low oxidation state.

The Borrowdale Volcanic Group

Morris (1973) sampled the Borrowdale Volcanic Group in the Haweswater–Ullswater area and Faller et al. (1977) extended the sampling to the south-west. It was hoped that small changes in palaeomagnetic directions could be detected and that these would be of correlative value. However, this was found not to be so for several reasons, some connected with sampling. There is limited outcrop of suitable uncleaved flows and it is not known what time-span a flow represents and therefore whether secular variation has been averaged. Accurate tilt-correction cannot always be made. Massive, uncleaved andesites were sampled but the local dip had to be measured on the nearest bedded tuffs; the assumption was made that the lavas were always tilted by the same amount as the tuffs. It was assumed that the lavas were erupted on to a horizontal surface and, in correcting for tectonic dip, no allowance could be made for possible depositional dip. All these factors limit the potential precision of the results.

Twenty-eight sites were sampled at eastern localities, to the north-west of Ullswater, round Haweswater, in the Kentmere Valley and at the head of Longsleddale. The Haweswater sites lie on the opposite (northern) limb of the Nan Bield Anticline from those in Kentmere and Longsleddale; the dip at the Ullswater sites is determined by the Ullswater anticline. In the west nineteen sites were sampled at localities in Dunnerdale, on the Ulpha fells and in the Coppermines Valley near Coniston, where the dips are determined mainly by the Ulpha syncline.

After AF demagnetization and correction for tilt seventeen of the western sites were combined to give $D = 322°$, $I = -55°$ ($\alpha_{95} = 11°$). Twenty-two of the eastern sites had significant mean directions after AF treatment. They fell into a principal group and a smaller number of anomalous sites. The principal group of fourteen sites (after tilt correction) is centred on $D = 336°$, $I = -43°$ ($\alpha_{95} = 9°$). Eight of the eleven anomalous sites were from the Haweswater Dolerite or the volcanics intruded by it; they gave steep inclinations, both positive and negative, and their possible interpretation will be discussed later.

There is no significant difference between these two principal results from eastern and western areas; they may therefore be usefully combined to give an overall mean direction of $D = 329°$, $I = -49°$ ($\alpha_{95} = 7°$) for the Borrowdale Volcanic Group, which corresponds to a palaeomagnetic pole at 0°, 203°E. All the sites which contribute to this result have the same polarity as the Eycott Group. The significant improvement of precision on using tilt corrected site means is again convincing evidence that the stable remanence pre-dates the regional folding.

The Carrock Fell Complex

Because the age and attitude of emplacement of the Carrock Fell Complex have been matters of controversy, palaeomagnetic directions from it cannot be used in the same way as those from the well dated volcanic rocks described above to add to our knowledge of the palaeomagnetic field and its evolution. Briden and Morris (1973) reversed the argument in order to infer the age and attitude of the intrusion by comparing palaeomagnetic directions from the Carrock Fell Complex with the standard record from the adjacent volcanics and elsewhere in Britain. Twenty-one sites were drilled from the northern and southern margins of the complex, the Round Knott Diabase and the metamorphic aureole of the Skiddaw Granite. The variety of rock types was reflected in the wide range of total NRM intensities (0.3×10^{-3} G to 7.0×10^{-6} G) and there were three main types of behaviour in response to AF demagnetization. Only eleven of these sites had significantly grouped remanence after AF treatment. None from the metamorphic aureole of the Skiddaw Granite had any systematic remanence. The *in situ* mean direction for the Carrock Fell Complex was $D = 353°$, $I = -18°$ ($\alpha_{95} = 14°$), which is similar to the tilt-corrected results from the Eycott and Borrowdale Volcanic Groups, though with shallower inclination. Correcting the *in situ* result for the regional dip of the Lower Carboniferous gave $D = 350°$, $I = -22°$. Morris (1973) added results from a further six sites and made an alternative tilt correction, a 15° northward rotation. This rotates the axial plane of the overlying Drygill Shales to the vertical and brings the mineral lamination of the complex much closer to the vertical. Morris considered that this correction gave the best estimate of the primary remanence direction, $D = 351°$, $I = -29°$ ($\alpha_{95} = 10°$) corresponding to a pole at 19°N, 184°E.

Because of the similarity of this palaeomagnetic result to others of Ordovician age from the Lake District and elsewhere in Britain, and because the characteristic Ordovician field direction persisted no later than the earliest Silurian, the palaeomagnetic evidence strongly favours the emplacement of the complex as a dyke-like body (i.e. roughly in its present orientation) in Ordovician time (Briden et al. 1973). This is an advance on the geological constraint on the minimum age of the complex which is indicated by the intrusion into it of the Skiddaw Granite. It may follow that the Eycott Group, the margins of which the complex intrudes, was folded to much its present geometry within Ordovician time and that any later Caledonian deformation had little large scale effect on it. Alternatively, since the complex was emplaced as a E–W striking dyke-like body, it lay normal to the principal compression direction of the end-Silurian deformation and therefore may not have been significantly tilted at that period.

Minor intrusions into the Ordovician rocks

Among the anomalous results, which include steep positive and negative inclinations, from the eastern Borrowdale Volcanic Group were some from the Haweswater Dolerite and the volcanics adjacent to its contact (Faller et al. 1977). As Piper et al. (in progress) have pointed out following further sampling, the palaeomagnetic results do not resolve whether the Haweswater Dolerite was emplaced before, during or after deformation of the adjacent volcanics, because the *in situ* remanence in the dolerite is not greatly different from the 'typical' Ordovician field, while on tilt correction the remanence direction is comparable to other steep anomalous directions of uncertain Upper Ordovician age. It is not even clear whether any or all of these anomalous results, and other steep downward magnetizations discovered by Morris (1973) in E–W dykes cutting the Eycott Group, and by Piper et al. (in progress) in the Threlkeld Microgranite, are to be correlated with the same anomalous event in the Ordovician geomagnetic field discovered by Thomas and Briden (1976) in North

Wales. This is because only negative inclinations were found in Wales, while the anomalous directions of the Lake District intrusions, although they are all steep, are by no means consistent; nor are perfectly antiparallel polarity groups discernible.

There are two other instances where dykes do appear to yield typical (or not grossly anomalous) Ordovician remanence directions. In the Kentmere Valley Morris (1973) sampled three sites in a NE–SW trending dyke near Stile End which intrudes the base of the Stockdale rhyolite and is therefore post-Caradocian. After AF demagnetization two of these sites have significantly grouped remanence, the *in situ* mean of which cannot be interpreted in terms of any Palaeozoic geomagnetic field direction. However after correction for the tilt of the overlying Coniston Limestone the mean direction becomes D = 5°, I = − 53°. Piper et al. (in progress) sampled three of the Embleton diorite dykes which cut the Skiddaw Slates. One of these was magnetized in a reverse direction comparable to the Borrowdale Volcanics, but the other two yielded incomprehensible results and it may be that no palaeomagnetic significance should be attached to any of them.

The Shap Granite and dykes in the Cross Fell inlier

Piper et al. (in progress) report that the Shap Granite in the Red Shap quarry is typically magnetized with shallow inclinations to the SSW which is generally (but not closely) similar to other British data from rocks of comparable (late Silurian — early Devonian) age. A similar result was obtained from minette dykes in the Cross Fell inlier (Piper et al. in progress) and a Siluro–Devonian age may reasonably be inferred; however the petrographically distinct kersantite dykes in the same vicinity are magnetized almost vertically. The latter present the same problems of interpretation as the other steep magnetizations mentioned previously. At present we lean toward the view that the anomalous results all derive from genuine geomagnetic field abnormalities during short time intervals within the late Ordovician, but we also regard such interpretation as extremely hazardous in the absence of either rigorous thermal demagnetization analysis of an anomalous suite, or a stratigraphically well controlled example.

The Cockermouth lavas

Five sites were sampled in a section which outcrops near Wood Hall, Cockermouth (Morris 1973). All had significantly grouped NRMs both before and after AF demagnetization but the overall mean of both the *in situ* and tilt corrected mean directions was not significant. The lavas appear to record different magnetic field directions at three different stratigraphic levels. Only the intermediate level gave a result (D = 21°, I = − 23°) similar to Lower Carboniferous geomagnetic field directions from elsewhere in Britain.

Discussion

The mean remanence directions for the Eycott Group, the Borrowdale Volcanic Group and the Carrock Fell Complex (whichever tilt correction is applied) all differ at the 95% confidence level using the test of Watson (1956) (Fig. 4 of Faller et al. 1977). This difference could conceivably represent polar movement relative to the Lake District from late Arenig to Caradocian time. On the other hand, when the palaeomagnetic poles from these results are compared with other reliable poles from the British Isles (Fig. 5 of Faller et al. 1977) considerable overlap between many of the ovals of 95% confidence is seen. If the correct ages of magnetization have been assigned to each group, it is impossible to construct a simple polar path for the Middle to Upper Ordovician. It would seem reasonable to attribute these apparent differences of pole position to the limitations in the sampling and statistical treatment of each study. Within the limits of the error, then, the pole appears to have remained stationary relative to the British Isles as a whole from late Arenig to Ashgillian time.

Briden et al. (1973) used the palaeomagnetic results then available from the British Caledonides to examine whether a single polar path was appropriate for the whole of the British Isles or whether closure took place across the Caledonide fold belt since the Ordovician. They compared the mean directions from the Aberdeenshire Gabbros to the north of the fold belt with those from the Eycott Group and the Builth Volcanics of Wales to the south. There was less than 5% probability that the results from the north and south margins were identical. The difference between them could be explained by approximately 10° (i.e. 1000 km) separation across the intervening Caledonides. Since the uncertainties in the mean palaeomagnetic directions are around 8°, this implies a subsequent closure across the Caledonides of 1000 ± 800 km. When the direction for the Borrowdale Volcanic Group is incorporated in the overall mean for south of the fold belt, the difference between that mean and the Aberdeen-shire Gabbros direction is very slightly reduced. Obviously the conclusion depends heavily on the Aberdeenshire result as a valid estimate of the Ordovician pole relative to the N margin of the Caledonides, but the implication must be that post-early Ordovician closure across the Caledonide fold belt, if it occurred, was comparable to the uncertainties in the palaeomagnetic method. Briden et al. (1973) pointed out that very oblique convergence of the orogenic forelands would not be in conflict with the palaeomagnetic evidence. Morris (1976) has further developed this argument by analysing all Lower Palaeozoic palaeomagnetic data from Europe and North America. He finds evidence for oblique movement between the two margins continuing into Devonian time. This is consistent with the previous conclusion, based on the British results alone, that the Iapetus Ocean on the site of the Caledonides must have been already narrow by Ordovician time.

Acknowledgements

We are indebted to Dr W. A. Morris, and to Dr J. D. A. Piper and his colleagues at the University of Liverpool, for permission to quote and discuss unpublished information. Our thanks are also due to Dr N. J. Soper for much valuable discussion on the geological implications of the results.

REFERENCES

BRIDEN, J. C. and MORRIS, W. A. 1973. Palaeomagnetic studies in the British Caledonides — III. Igneous rocks of the Northern Lake District, England, *Geophys. J. R. astr. Soc.* **34**, 27–46.
——, MORRIS, W. A. and PIPER, J. D. A. 1973. Palaeomagnetic studies in the British Caledonides — VI. Regional and global implications, *Geophys. J. R. astr. Soc.* **34**, 107–34.
EASTWOOD, S. E., HOLLINGWORTH, M. A., ROSE, W. C. C. and TROTTER, F. M. 1968. Geology of the Country around Cockermouth and Caldbeck, *Mem. geol. Surv. of Gt. Brit.* x + 298 pp.
FALLER, A. M., BRIDEN, J. C. and MORRIS, W. A. 1977. Palaeomagnetic results from the Borrowdale Volcanic Group, English Lake District, *Geophys. J. R. astr. Soc.* **48**, 111–21.
McELHINNY, M. W. 1973. Palaeomagnetism and plate tectonics, Cambridge University Press, 358 pp.
MORRIS, W. A. 1973. *Palaeomagnetic studies in the British Caledonides*, Ph.D. thesis, The Open University, Milton Keynes.

3

Deep Structure

M. H. P. BOTT

Knowledge of the deep structure of the Lake District is mainly dependent on geophysics, particularly gravity studies (Bott 1974). The region is associated with a belt of relatively low Bouguer anomalies despite the surface outcrop of dense Lower Palaeozoic rocks. This is taken to indicate an underlying composite granite batholith beneath the central and northern parts of the Lake District, to which the later vertical tectonic history of the region may be related. Seismic investigations have not revealed much information in the Lake District itself, but have been useful in the adjacent sediment-covered regions, whilst an interpretation of the aero-magnetic data is given in Chapter 4.

This chapter is mainly concerned with the gravity anomalies. In order to highlight the geological interpretation of the anomalies, technical details of the survey and its reduction and of the interpretation methods have been omitted, but can be found in Bott (1974).

MAIN SOURCES OF THE BOUGUER ANOMALY VARIATIONS

The Bouguer anomaly map of the Lake District is shown in figure 9 and Table 1 gives rock densities. The map reflects the underlying lateral density variations below O.D. The observed anomaly variations over the Lake District and the surrounding regions can be explained by four main types of structure involving lateral density variation, one of these being deep-seated and three being related to outcrops.

The Irish Sea regional high

The Bouguer anomaly level over most of the Irish Sea, after correction for the effect of sediments, is about 20 to 30 mgal higher than the comparable anomalies on the adjacent land areas. At sea to the west of the Lake District, this high is masked by the presence of the thick sediments of the East Irish Sea basin (Bott 1964, Bott and Young 1971). However, the positive gradient towards the regional high affects the western part of the Lake District. It is observed as a south-westward rise in Bouguer anomaly along the strike of the Silurian rocks towards Furness and causes the west-ward rise in Bouguer anomaly across the Cumberland Coalfield. Bott (1964) interpreted the Irish Sea regional high as caused by either thinner or denser crust beneath the Irish Sea region in relation to the adjacent areas. This crustal transition occurs near the western margin of the Lake District.

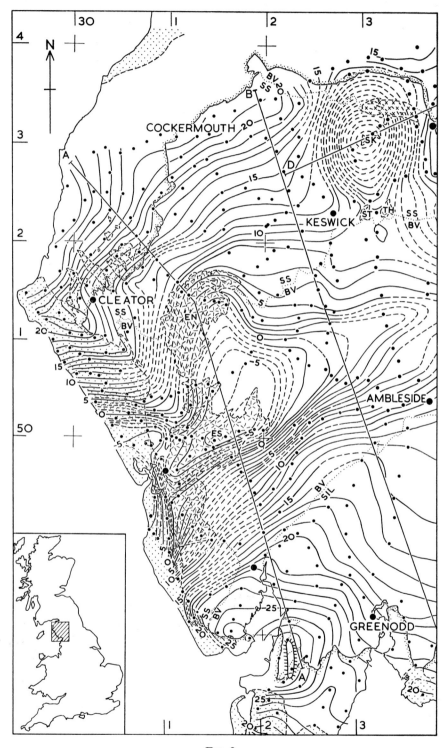

Fig. 9a

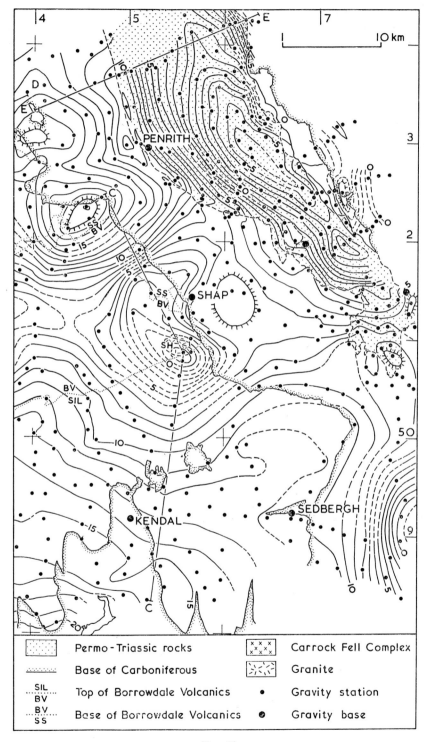

FIG. 9. Bouguer anomaly map of the Lake District and the Vale of Eden with contours at one milligal interval. Positive closures are marked by hachures and negative closures are unmarked. Formation densities for the Bouguer correction are as follows: 2.75 g/cm³ (Skiddaw Slates and Borrowdale Volcanics), 2.72 (Silurian), 2.65 (granite and Carboniferous below Coal Measures), 2.50 (Coal Measures), 2.33 (Permo-Triassic), 2.00 (thick alluvium). After Bott (1974).

Permo-Triassic rocks

Base of Carboniferous

SIL
BV   Top of Borrowdale Volcanics

BV
SS   Base of Borrowdale Volcanics

Carrock Fell Complex

Granite

•   Gravity station

⊘   Gravity base

FIG. 9b

TABLE 1

Sample density measurements (Summarized from Bott 1974)

Formation	Number of samples	Number of localities	Saturated density (g/cm³)
Skiddaw Slates	53	4	2.77 ± 0.01
Borrowdale Volcanics	247	12	2.74 ± 0.04
Brathay Flags	41	2	2.74 ± 0.01
Bannisdale Slates	21	1	2.72 ± 0.01*
Kirkby Moor Flags	20	1	2.69 ± 0.02*
Penrith Sandstone	34	2	2.42 ± 0.02
St Bees Shale	17	2	2.46 ± 0.01
St Bees Sandstone	57	5	2.22 ± 0.07
Eskdale Granite			
muscovite granite	44	7	2.62 ± 0.01
biotite granodiorite	15	2	2.71 ± 0.02
Ennerdale Granophyre	14	2	2.62 ± 0.01
Shap Granite	9	1	2.66 ± 0.01*
Skiddaw Granite	8	1	2.58 ± 0.02*
Threlkeld Microgranite	17	1	2.67 ± 0.03*

Standard deviation of localities shown, except where marked by asterisk where standard deviation of the samples from the single locality is given.

The Lake District granites

A belt of relatively low Bouguer anomalies dominates the central and northern parts of the Lake District. Individual regions of minimum Bouguer anomaly coincide with the Eskdale, Shap and Skiddaw Granites. Steep Bouguer anomaly gradients are observed and these indicate that the low density rocks causing the anomaly must locally reach within one or two kilometres of the surface. Sample density measurements (Table 1, Fig. 10) show that the exposed granites have substantially lower densities than the Skiddaw Slates and Borrowdale Volcanics, although the densities vary within and between the granites. Taken together, these observations convincingly show that the granites themselves are the main cause of the low gravity anomalies of the Lake District, and the negative anomaly belt shows that granite is much more extensive at depth than at the surface. A detailed series of interpretations is presented later in the chapter.

The Silurian rocks

The thick Ludlow succession of the southern part of the Lake District causes a local negative gravity anomaly because it is about 0.04 g/cm³ less dense than the thick underlying and surrounding Ordovician succession. This produces a southward drop in anomaly over the northern part of the Silurian outcrop, which has the effect of reducing the steep gradient of opposite sense caused by the southern flank of the granite belt. The effect can be seen on the Bouguer anomaly map (Fig. 9) by the abrupt shallowing of the steep gradients associated with the granite belt at about the position of the Ordovician/Silurian boundary. The amplitude of the anomaly caused by the Silurian rocks cannot be accurately estimated but is about -10 mgal representing a structural trough about 10 to 15 km across and about 6 to 7 km deep. The thick Silurian trough needs to be taken into account in interpreting some anomaly profiles over the granites (Figs. 12 and 14).

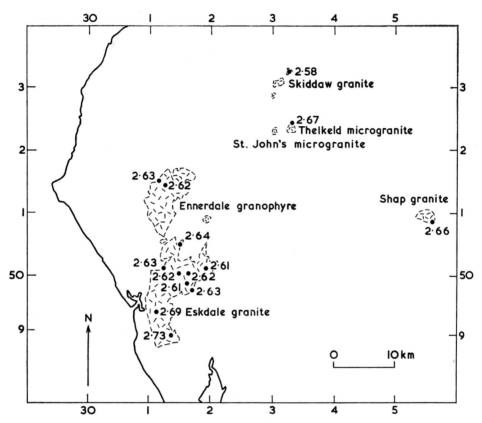

FIG. 10. Map showing sample density measurements (g/cm³) for the Lake District granites. Note the pattern of density variation of the Eskdale and Ennerdale Granites which is most consistent with the model of Fig. 11(b).

Carboniferous and Permo-Triassic sediments

Carboniferous and Permo-Triassic sediments around the Lake District cause local negative gravity anomalies. The Carboniferous strata are on average about 0.15 g/cm³ lower in density than the Lower Palaeozoic basement, and the Permo-Triassic rocks are about 0.3 to 0.5 g/cm³ less dense than basement. Thick successions of these sediments cause conspicuous local negative anomalies with steep gradients over the Vale of Eden, and along the Permo-Triassic rocks of the Cumberland and Furness coastal belt which forms the eastern edge of the East Irish Sea basin. The small negative gravity effect expected for the seaward thickening wedge of Carboniferous rocks of the Cumberland Coalfield is masked by stronger gravity gradients of opposite sign caused by the adjacent Irish Sea high and Lake District granites. Small local anomalies are also related to varying thicknesses of drift, but these are not considered here.

GRANITE BENEATH THE LAKE DISTRICT

As shown above, the main belt of low gravity anomaly over the Lake District can be attributed to low density granite emplaced within Ordovician country rock. Profiles across the anomaly belt can be used to investigate the sub-surface shape of the granite body provided that the density contrast is known. In practice, the interpretations shown below are not unique, mainly because of uncertainty in the background anomaly level and in the density contrast at depth. The broad features of the models presented are probably trustworthy, but the detail such as the exact depth to the floor of the granite is not well-established.

Eskdale Granite and Ennerdale Granophyre

The lowest observed Bouguer anomaly value of the Lake District occurs over the Eskdale Granite. The Eskdale negative anomaly also encompasses the Ennerdale Granophyre on its northern flank. This local negative anomaly is estimated to have an amplitude of about -38 mgal below background and it displays a strongly asymmetrical profile (Fig. 11) with the steeper gradients to the south. Three possible mass distributions which can satisfy to acceptable accuracy a two-dimensional interpretation of profile AA are shown in figure 11. The asymmetry of the profile can be explained either by lateral variation of the density within the granite as in models (a) and (b), or by shallowing of the floor of the granite towards the north-west as in model (c). Model (b) is regarded as the most plausible interpretation as it is most consistent with the surface geological structure and the density sampling (Fig. 10).

The gravity models do not distinguish clearly between the Eskdale Granite and Ennerdale Granophyre which appear as a single composite intrusion underlying the western Lake District. The Ennerdale Granophyre appears in model (b) to be a thin, high level phase in contrast to the more deep-seated main Eskdale body which extends north beneath it. The contacts slope outwards and the gravity anomalies suggest strong lateral variation of density within the main intrusion which is to some extent borne out by surface density measurements (Fig. 10). The depth to the base appears to be at least 8 km and may be greater if density contrast decreases with depth. A minor but significant feature is the granite ridge occurring about 3 km to the NW of the Ennerdale Granophyre (Fig. 15); this has been previously and independently inferred from the occurrence hereabouts of spotted slates (Rose 1954, Jackson 1961).

The Shap Granite

The Shap Granite corresponds to a local minimum Bouguer anomaly at the east end of the minimum anomaly belt overlying the axis of the Lake District. The amplitude of the Shap negative anomaly is about -24 mgal and it covers a much larger area of about 10×12 km^2 than does the granite outcrop itself which covers only about 8 km^2. The steepest Bouguer gradients occur 10 km NNW of the granite outcrop.

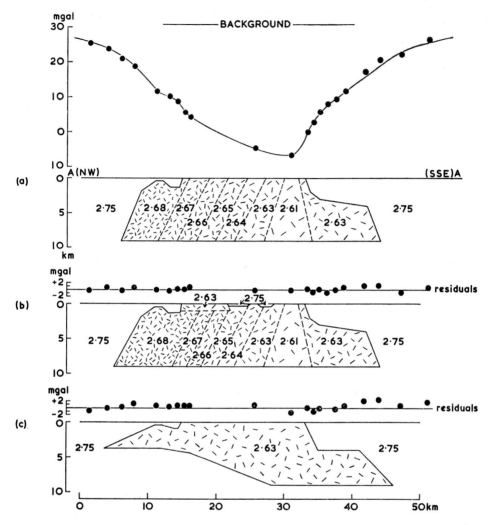

FIG. 11. Two-dimensional interpretation of the Bouguer anomaly along line AA (Fig. 9) across the Eskdale Granite and Ennerdale Granophyre, showing three possible subsurface models of granite structure. Solid circles represent observations and the continuous line is the computed anomaly for model (a). Residuals are shown for models (b) and (c). Rock densities are shown in g/cm³. After Bott (1974).

Two possible interpretations of the structure beneath profile CC across the Shap Granite, consistent with surface density observations, are shown in figure 12. A thick Silurian trough needs to be incorporated to satisfy the gravity profile. The gravity shows that the roof region of the Shap intrusion is of much larger extent than the outcrop, particularly extending at shallow depth for some 8 km further north. The granite contacts slope outwards, and the depth to the base is inferred to be about

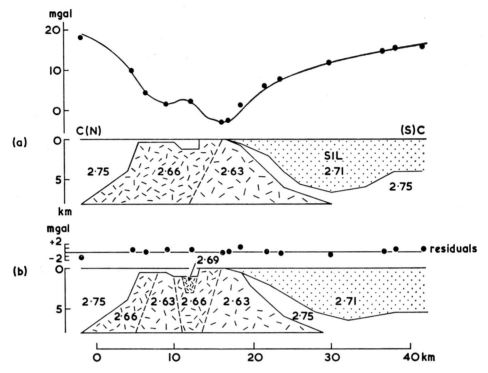

Fig. 12. Two-dimensional interpretation of the Bouguer anomaly across the Shap Granite along CC (Fig. 9) showing two possible models. After Bott (1974). A background level of 25.6 mgal has been assumed.

8 km depending on density. Lateral density variation within the intrusion is needed to satisfy the gravity profile, with an unexposed region of low density forming the southern part.

Skiddaw Granite

A conspicuous nearly circular region of low gravity of about − 19 mgal amplitude and 4.5 km radius overlies the Skiddaw Granite and its aureole. This anomaly is offset to the north of the main negative anomaly belt but joins on to it in the Threlkeld region where the St John's and Threlkeld Microgranites occur, although these intrusions are mid-Ordovician and presumably unrelated to the Skiddaw Granite.

An interpretation along profile DD, using three dimensional gravity computation methods, is shown in figure 13. This assumes a uniform density contrast of −0.15 g/cm³ based on surface sample measurements. The interpretation shows the Skiddaw Granite to be a steep-walled stock-like body embedded in the north wall of the main granite batholith. The depth to the base is inferred to be 6 km but as density relationships at depth are uncertain this should not necessarily be regarded as significantly shallower than the rest of the batholith. Whatever the detail at depth, the gravity interpretation of the sub-surface shape of the Skiddaw Granite is in excellent agreement with earlier geological inferences based on the extent of the thermal aureole

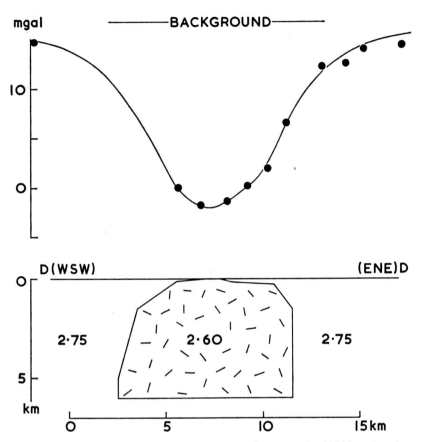

FIG. 13. Interpretation of the Bouguer anomaly across the Skiddaw Granite along DD (Fig. 9), using a three-dimensional model of approximately circular plan. After Bott (1974).

Hidden granite beneath the Lake District

The Bouguer anomaly map shows that granite underlies most of the central and northern parts of the Lake District, joining the exposed Eskdale, Shap and Skiddaw Granites. The depth to the roof of the unexposed granite cannot be estimated accurately because of the lateral variations in density, but the steep Bouguer gradients locally encountered indicate that it is probably generally less than 2 to 3 km deep and locally less than 1 km deep.

An interpretation of the structure along profile BB across the hidden part of the Lake District granite (Fig. 14) shows a shallow roof and steep walls dipping outwards. As with the Eskdale and Shap interpretations (Figs. 11 and 12). the density of the granite varies across the batholith, being lower in the south and higher in the north. The interpretation shows two prominent granite 'ridges'. The southern ridge occurs above the lowest density granite, and is visible on the map as the belt of minimum Bouguer anomaly joining the Eskdale and the Shap Granites. It underlies the main

topographical axis of the Lake District. The northern composite ridge is more difficult
to identify on the map, but extends eastwards from the Ennerdale Granophyre to the
St John's and Threlkeld Microgranites and the Skiddaw Granite. A further ridge
noted above (Fig. 11) occurs to the north-west of the Ennerdale granophyre and may
extend north-eastwards towards the Skiddaw region. Figure 15 shows the inferred
positions of the roof and wall regions of the composite batholith and the interpreted
positions of the ridges.

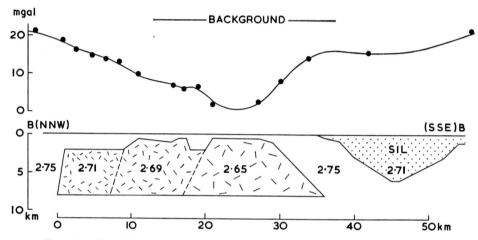

Fig. 14.   Two-dimensional interpretation of the Bouguer anomaly across the
subsurface Lake District Batholith along line BB (Fig. 9). After Bott (1974).

Connexion between the Lake District and Weardale granite

After allowing for the effect of the sediments of the Vale of Eden, a belt of
residual low gravity anomaly extends eastwards from the Shap region to the Alston
Block where it joins the negative anomaly associated with the Weardale Granite
(Bott 1974). This indicates that a deep-seated granite ridge joins the Lake District
granites to the Weardale Granite (Fig. 15), passing beneath the Vale of Eden between
Appleby and Penrith.

VALE OF EDEN

The local negative gravity anomaly following the Vale of Eden is predominantly
caused by thick, low density Permo-Triassic and Carboniferous sediments. Between
Penrith and Appleby, the postulated underlying granite ridge contributes to the
negative anomaly in the region where the lowest values occur. Profile EE across the
sediment trough is well removed from the granite ridge although the gravity effect of
the Weardale Granite affects the eastern end (Fig. 16). Two possible interpretations of
EE are shown in figure 16 based on sample density values for the sediments. The
maximum possible eastward thinning of the Permo-Triassic succession towards the
Pennine Fault is incorporated in model (a) and maximum Carboniferous thinning is
incorporated in model (b), yielding two extreme types of model.

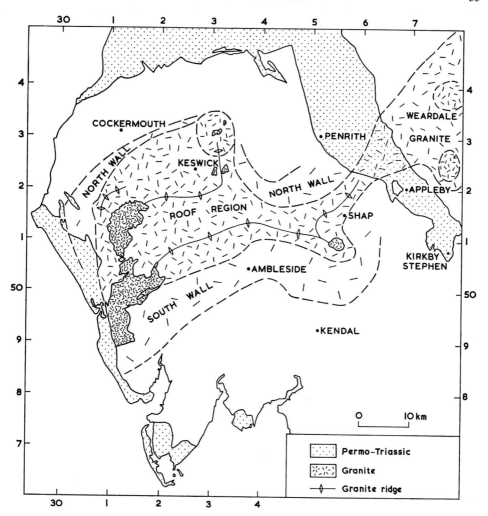

FIG. 15. Sketch map showing the roof and wall regions of the postulated
Lake District Batholith and its connexion to the Weardale Granite. After Bott
(1974).

The interpretations show the Permo-Triassic strata reaching an estimated maximum thickness of about 1 km beneath the centre of the trough. Two points of particular
significance in the interpretation arise. Firstly the Penrith Sandstone is required to
thin substantially from the centre of the trough towards the Pennine Fault belt to
satisfy the anomaly profile; it is presumed that this was caused by differential subsidence during deposition of the Penrith Sandstone. Secondly the Carboniferous
strata also give evidence of thinning towards the Pennine Fault line, which can be
attributed to pre-Permian (Hercynian) events.

The above inferences led Bott (1974) to suggest a new structural history for the
Pennine Fault belt and the Vale of Eden (Fig. 17). The two main cornerstones of this

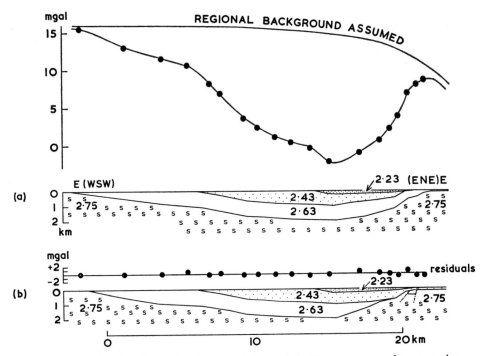

FIG. 16.   Two-dimensional interpretation of the Bouguer anomaly across the
Vale of Eden along line EE (Fig. 9), showing two alternative models. Permo-
Triassic is stippled and Carboniferous unshaded. After Bott (1974).

interpretation are that the Hercynian movements produced a faulted monocline in
the Carboniferous rather than a simple fault belt, and that the Permo-Triassic basin
was formed by contemporaneous subsidence with its eastern margin near the Pennine
fault belt. This interpretation reconciles the occurrence of Whin Sill pebbles in the
Upper Brockram with the downthrow of the Alston Block along the Inner Pennine
Faults. It also implies that the post-Triassic normal throw on the Outer Pennine
Faults may be much less than previously suspected because of the eastward thinning
of the Permo-Triassic and Carboniferous towards the fault belt. It should be noted
that this interpretation specifically refers to the profile north of the Cross Fell Inlier,
where the Hercynian Inner Pennine Faults are interpreted as lying to the west of the
post-Triassic Outer Pennine Fault.

UPLIFT OF THE LAKE DISTRICT

The Lake District now stands elevated above the surrounding regions, the geological history showing that it has tended to be a positive region since the Devonian.
A possible explanation is that the low density granitic rocks beneath the Lake District
have given the region buoyancy, so that it rises to approximate isostatic equilibrium
when stress conditions allow. This hypothesis was tested by Bott (1974), who showed

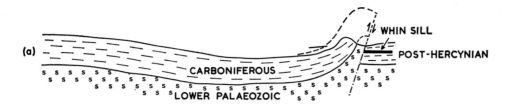

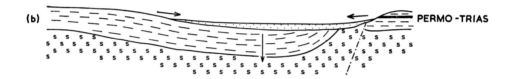

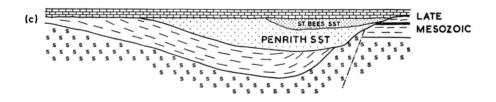

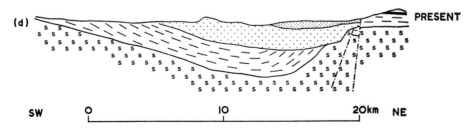

SW   0   10   20km   NE

FIG. 17.   Interpretation of the structural history of the Pennine fault line and
the Vale of Eden (vertical exaggeration × 2) After Bott (1974).

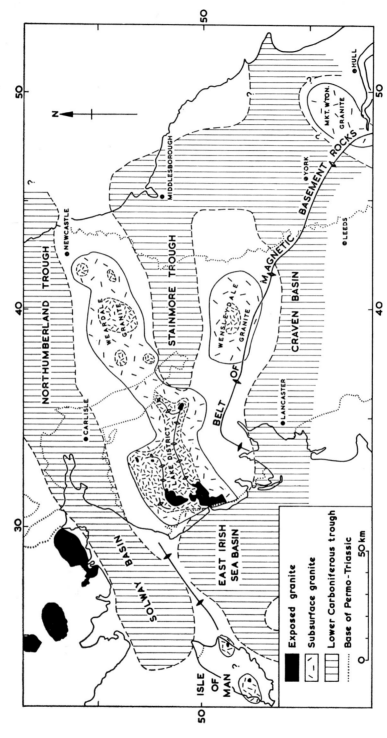

FIG. 18. The tectonic setting of the Lake District within North England, compiled with some modifications and additions from Bott (1961, 1964, 1967, 1974), Bott, Robinson and Kohnstamm (1978), Cornwell (1972), Kent (1966), and Ramsbottom et al. (1974).

that the granite mass deficiency beneath the Lake District of about 1.1×10^{18} g is about equal to the excess surface load of the Lake District topography above a 270 ft (80 m) datum. This makes it plausible to regard the present topographic elevation of the Lake District as isostatically compensated by the underlying body of low density granite. A similar equivalence between topographical load and granite mass deficiency applies to the Weardale (Bott and Masson-Smith 1957) and Cheviot Granite regions. It should be pointed out that this relationship between low density granite and elevation does not apply universally. It is not characteristic in Precambrian shield areas, but occurs in semi-mobile regions such as Britain when they undergo periods of strong differential vertical movement. Northern England appears to be an excellent example of such a region, where several (but not all) of the elevated topographical regions can be related to underlying granite.

REGIONAL SETTING

Figure 18 shows the Lake District in its regional setting. The Lake District and the Alston Block can be seen as parts of the same deep structure of ENE trend dominated by the extensive Lake District–Weardale Caledonian granite mass beneath. It is the result of accidents of later geological history that the two regions differ now in their structure and topography. This same Caledonian belt may extend WSW beneath the Ramsay–Whitehaven Ridge to the Isle of Man to join with the Dhoon and Foxdale Granites. The Lake District also forms the western termination of a belt of magnetic basement rocks of SE trend which underlies the southern part of the Askrigg Block and extends to the Wash. This belt is penetrated by the Caledonian Wensleydale Granite which almost joins on to the Shap Granite, and by another newly interpreted granite beneath the Market Weighton region (Bott, Robinson and Kohnstamm 1978). Although this belt is probably of Precambrian age, with a Charnian trend, a Caledonian age for it cannot be ruled out on present evidence. Figure 18 also emphasizes the close relationship between the basement granites of the Lake District and the Northern Pennines both of which formed positive regions during the Carboniferous, and the adjacent troughs of thick Lower Carboniferous strata.

REFERENCES

BOTT, M. H. P. 1961. A gravity survey off the coast of north-east England. *Proc. Yorks. geol. Soc.* **33**, 1–20.
—— 1964. Gravity measurements in the north-eastern part of the Irish Sea. *Jl geol. Soc. Lond.* **120**, 369–96.
—— 1967. Geophysical investigations of the northern Pennine basement rocks. *Proc. Yorks. geol. Soc.* **36**, 139–68.
—— 1974. The geological interpretation of a gravity survey of the English Lake District and the Vale of Eden. *Jl geol. Soc. Lond.* **130**, 309–31.
—— and MASSON-SMITH, D. 1957. The geological interpretation of a gravity survey of the Alston Block and the Durham Coalfield. *Jl geol. Soc. Lond.* **113**, 93–117.
—— and YOUNG, D. G. G. 1971. Gravity measurements in the north Irish Sea. *Jl geol. Soc. Lond.* **126**, 413–34.
—— ROBINSON, J. and KOHNSTAMM, M. A. 1978. Granite beneath Market Weighton, east Yorkshire. *Jl. geol. Soc. Lond.* (in press).
CORNWELL, J. D. 1972. A gravity survey of the Isle of Man. *Proc. Yorks. geol. Soc.* **39**, 93–106.
JACKSON, D. E. 1961. Stratigraphy of the Skiddaw Group between Buttermere and Mungrisdale, Cumberland. *Geol. Mag.* **98**, 515–28.

4

Aeromagnetic Survey

F. A. COLLAR and D. J. PATRICK

The aeromagnetic survey of the Lake District was flown during 1959. The flight lines were orientated north-south with an average separation of 2 km and a mean ground clearance of 300 m, although over hilly terrain the actual clearance varied considerably. Variations of amplitude of the total magnetic field were monitored with a fluxgate magnetometer and have been expressed as anomalies upon a datum which has a value of 47033 gammas at the National Grid origin and increases by 2.1728 gammas/km northwards and 0.259 gammas/km westwards. This information has been contoured at 10 gamma intervals and is shown in figure 19 reproduced from the 1.625,000 Aeromagnetic Map of Great Britain. (Institute of Geological Sciences 1965, 1972).

The largest anomaly strikes around the northern perimeter of the Lower Palaeozoic rocks, passing through Bothel, and extends with displacements as far as Penrith. It follows the outcrop of the Eycott Group lavas although west of Bothel, where they are obscured by Lower Carboniferous limestones and Cockermouth lavas, the anomaly parallels the Gilcrux Fault. Here it reaches its maximum total amplitude of 540 gammas and its typical dipolar shape is well defined. The Eycott Group lavas, particularly those in the High Ireby Formation are considered to be the principal source of this anomaly mainly because of the widespread geographical coincidence. Because of its low curvature the anomaly west of Bothel is unlikely to be caused by the shallow, gently-dipping Cockermouth lavas.

Interpretations have been made on profiles AA and BB (Fig. 20) to investigate the sub-surface structure of the Eycott Group lavas. Residual anomalies have been obtained by removing a regional anomaly which, decreasing linearly northward by 0.7 gammas/km, is asymptotic to the flanks of the observed anomaly. Interpretations of the residual anomalies have been restricted to simple, single-body, two-dimensional models, each of which incorporates the induced and remanent magnetic fields. Data for the remanent magnetism of the lavas have been taken from Faller and Briden (Chapter 2), Briden and Morris (1973), and Morris (1973). Primary remanence has been considered in preference to total natural remanent magnetism (NRM) because its mean in-situ direction at depth can be predicted with greater confidence. The mean direction parameters for the primary remanence, Declination D $= 0°$; Inclination I $= -46°$, have been adopted and re-orientated appropriately for the dip and strike of the lavas that are implied by each magnetic model. The intensities of remanence include a wide range of values (Briden and Morris 1973) and therefore an average value of 50nT has been estimated. Comparatively few data on the susceptibility of the lavas are available, so to estimate the intensity of the induced magnetic field, a minimum Koenigsberger ratio of 1 has been taken from Morris (1973).

Three types of model are shown in figure 20, each giving an anomaly almost identical to that on profile BB. By slight adjustment of shape and depths, these models also give anomalies similar to that on profile AA, although the sharper positive peak on AA cannot be satisfactorily reproduced from these simple structures, suggesting that more complicated structures which bring magnetic rock closer to surface are involved.

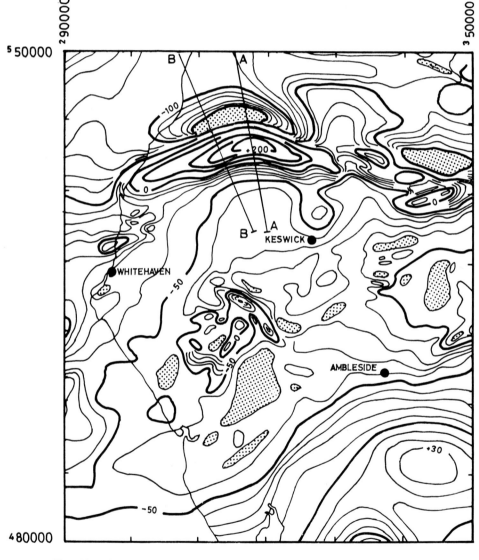

FIG. 19.   Aeromagnetic map of the Lake District showing line of profiles
A-A and B-B.

The first and simplest model (Fig. 20a) attributes the anomaly to lavas dipping south at 60° and striking perpendicular to the profile direction. Surface outcrops however show that beds dip generally northwards on Binsey Hill. The second (Fig. 20) includes a simple structure to represent the Gilcrux fault, with a normal throw of 400 m down to the north, and allows for the gradual northward thickening of the Carboniferous cover. This model however also requires the lavas to dip steeply southwards at 75°. The third model (Fig. 20c), taken from Bush (1971), shows the lavas steeply inclined northwards to a depth of several kilometres, and then extending indefinitely to the north at a shallower angle. The advantage of this model is its compatibility with the observed dips of the lavas, although the shape of its upper surface is not realistic. The most likely geological explanation for this model (Bush 1971) requires that the Gilcrux fault is a late structure developed on an established E.–W. fold. None of these simple models appears to be adequate geologically, which means that more complex types incorporating further information should be sought.

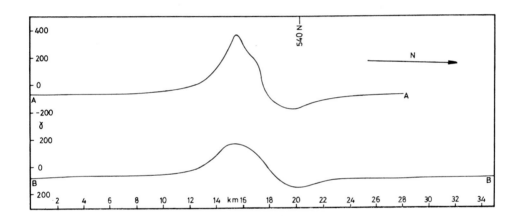

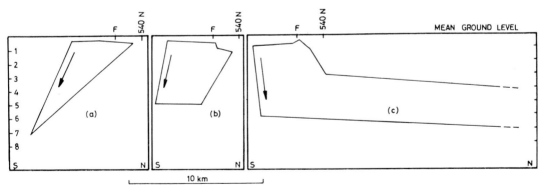

F – GILCRUX FAULT POSITION INDICATED BY 1" GEOLOGICAL MAP

IMPLIED DIP OF EYCOTT GROUP

FIG. 20.   Models of the Eycott Group lavas interpreted from the magnetic anomaly on profile B-B above.

In the southern part of the Lake District the magnetic anomaly rises gradually across the Coniston Limestone outcrop onto the Silurian rocks. This gradient defines the southern limit of a low and generally uniform background anomaly covering the outcrops of the Skiddaw, non-volcanic Eycott and Borrowdale sequences. This wide area of weaker magnetic field also coincides with the large negative Bouguer gravity anomaly attributed (Bott 1974) to a granitic batholith. If the granite can be assumed to have an average susceptibility marginally less than that of the country rock, possibly coupled with a weak remanence, then it may also contribute to the low magnetic anomaly; indeed the batholith closely coincides with the magnetic low.

A small group of dipolar anomalies occurs within this area near Ennerdale Water having a maximum total amplitude of 200 gammas. Although these anomalies are geographically associated with the Ennerdale Granophyre, the sources have not been identified. However, the recorded anomalies are not wider than 2 kms and locally the flight line separation increased to 3 km. Interpolation between such widely-spaced flight lines is unreliable and therefore the pattern of these anomalies (Fig. 19) is not wholly certain.

Acknowledgement

We are grateful to Mr M. D. Bush for permission to quote his unpublished M.Sc. thesis at Durham University. This account is published by permission of the Director of the Institute of Geological Sciences.

REFERENCES

BOTT, M. H. P. 1974. The geological interpretation of a gravity survey of the English Lake District and the Vale of Eden. *Jl geol. Soc. Lond.* **130**, 309–31.
BRIDEN, J. C. and MORRIS, W. A. 1973. Palaeomagnetic studies in the British Caledonides — III, igneous rocks of the Lake District, England. *Geophys. J. R. astr. Soc.* **34**, 27–46.
BUSH, M. D. 1971. *A magnetic survey of the northern edge of the English Lake District.* Unpublished M.Sc. thesis University of Durham.
INSTITUTE OF GEOLOGICAL SCIENCES. Sheet 2, 1975, Sheet 1, 1972. Aeromag. map of Great Britain.
MORRIS, W. A. 1973. *Palaeomagnetic studies in the British Caledonides,* Unpublished Ph.D. thesis, Open University, Milton Keynes.

F. A. COLLAR, B.SC., and D. J. PATRICK, PH.D.
Institute of Geological Sciences, Princes Gate, London, S.W.7.

5

Structure

N. J. SOPER and F. MOSELEY

In the area now known as the Lake District the earliest crustal deformations of which evidence is preserved are those associated with the Caledonian Orogeny. In Ordovician time the area lay close to a destructive plate boundary to the north and was subjected to episodes of mild compression and uplift associated with the closing of the Iapetus ocean and the attendant volcanic activity. Collision-type orogeny at the end of the Silurian resulted in polyphase deformation, mild metamorphism and granite emplacement, followed by uplift and deep erosion in Devonian time.

Much less important in the Lake District were the Hercynian movements, as the area occupied an intra-plate position in Permo–Carboniferous times. In the late Mesozoic and early Tertiary the region once again lay close to a plate margin, this time of constructive type, associated with the initiation and development of the present North Altantic. Structures of this age are thus essentially extensional in origin.

THE CALEDONIAN OROGENY

Structural investigation of the Lower Palaeozoic rocks of the Lake District over the last century or so has been reviewed by Hollingworth (1955), Mitchell (1956A, 1967) and Moseley (1972). Each of the major rock groups presents particular difficulties of interpretation. The Skiddaw and Eycott Groups show complex structures on all scales; a regional lithostratigraphic framework has been slow to emerge and in many areas palaeontological control is lacking. The Borrowdale Volcanics present all the problems of interpretive volcano-stratigraphy common to modern volcanic rocks, which must be investigated along with the structure; in addition, they are often strongly altered, rendering identification difficult and they lack interbedded fossiliferous sediments which elsewhere, for example in the Ordovician sequences of Snowdonia, have greatly assisted the definition and correlation of eruptive units. The Silurian greywacke sequences lack distinctive lithological markers, but their structure is simpler than that of the lower Ordovician sediments and its interpretation has proved less controversial. Each of these rock groups has a characteristic tectonic style. Investigations have usually concentrated on a limited area of one group, often with structural analysis not the primary objective. Attempts to correlate structures between the three rock groups have proved difficult, and, from the earliest days, controversial. Modern structural methods have been applied only in the last decade and the following regional synthesis is undoubtedly incomplete; it should be regarded as no more than a statement of the current position.

Three unconformities in the stratigraphical succession (pre-Borrowdale Volcanic, pre-Ashgillian and end-Silurian) reflect uplift and erosion associated with three important periods of Caledonian deformation.

The first two deformations are thought to relate to subduction of Iapetus oceanic crust along a plate boundary which lay to the north of the Lake District. The multiphase end-Silurian movements were much more important in terms of total strain and are believed to have resulted from plate collision.

PRE-BORROWDALE VOLCANIC STRUCTURES

Interpretive problems posed by the contrast in tectonic style between the Lower Ordovician sediments, with their multiphase folds and cleavages, and the Borrowdale Volcanic rocks with a single cleavage and open folds, have exercised geologists for many years. Major décollement at the base of the volcanics, pre-Borrowdale deformation of the slates and competency and layer thickness contrasts have all been invoked to explain the apparently greater deformation shown by the Skiddaw Slates.

A modern structural analysis was first attempted by Simpson (1967) who, extending his earlier work on the Manx Slates, described the structure in terms of three deformation episodes: F_1 — NE trending folds with an associated steeply inclined penetrative cleavage; F_2 — recumbent folds with a subhorizontal crenulation cleavage; F_3 — minor folds, kink-bands and spaced cleavages. This sequence has been recognized by other workers throughout the slate tract and has provided the basis for all subsequent structural studies, while the more extended variants of it proposed by Helm (1970) and Helm and Roberts (1971) have not been verified regionally. Simpson and his co-workers regarded F_1 and F_2 as pre–Borrowdale in age and interpreted the sub-Borrowdale junction as a major orogenic unconformity. This interpretation was challenged by Soper (1970) and Soper and Roberts (1971) who showed that the first cleavage is present in both the Skiddaw and Borrowdale rocks and that the second post-dates the Skiddaw Granite; F_1 and F_2 were, therefore, regarded as end-Silurian in age and the slate–volcanic junction interpreted as essentially conformable. The stratigraphical situation was soon shown to be more complex; Downie and Soper (1972) demonstrated that the 'northern' (Eycott Group) volcanics are older than the Borrowdales and are in sequence with the Skiddaw Slates, whilst Wadge (1972) and Jeans (1972) demonstrated an angular unconformity at the base of the Borrowdale Volcanic Group.

The base of the latter is now regarded as unconformable upon the Skiddaw and Eycott Groups (Chapter 6), whilst it now appears evident that the main Caledonoid cleavage in the slates and the volcanic rocks results from the same deformation episode (Moseley 1975), and that the whole F_1–F_3 sequence of Simpson is of end-Silurian age. What structures in the slates then relate to the pre-Borrowdale unconformity?

The question has been resolved by the discovery of pre-cleavage northerly trending folds in the Skiddaw Group by Roberts (1971, 1973, 1977), Jeans (1972, 1973) and Webb (1972, 1975). Roberts, describing the beautifully exposed structures in cordierite hornfelses in the River Caldew (220 326) recorded early, northerly-plunging folds deformed by later Caledonoid folds. He suggested (1973) a pre-Caradocian rather than a pre-Borrowdale age for these early folds, because at the time north-trending structures of pre-Caradocian age were thought to be important in the volcanic rocks and would be expected to continue into the Skiddaw Group, but he subsequently (1977) revised this interpretation in favour of a pre-Borrowdale age. The early folds have somewhat variable styles but are often straight-limbed with angular hinges (Fig. 21). They evidently pre-date the Caledonoid cleavage, which is associated with the later set of folds, although in the Caldew section the cleavage is largely obliterated by hornfelsing near the Skiddaw Granite.

Further south, Webb and Jeans working on adjacent parts of the main slate tract around Buttermere and Crummock Water recorded early fold structures which trend north to north–eastwards. The ENE cleavage is superimposed across them. Folds of this phase are well exposed on the Buttermere flank of Robinson [188 164] where Webb demonstrated complex interference patterns between them and the later Caledonoid folds (Fig. 22). The early folds are moderately tight with rounded hinges in the psammites, angular and almost isoclinal in the pelites and occur on all scales. Major and medium scale fold pairs verge westwards in this region and often have inverted common limbs, upon which the superimposed ENE folds face downwards on the cleavage. Jeans and Webb were able to relate the complicated outcrop patterns to this type of interference, within a single upward-fining sequence of Loweswater Flags and Mosser Slates, as envisaged by Rose (1954). Their interpretation is supported by detailed 'way up 'and geometrical evidence and seems preferable to Simpson's (1967) thicker and more complex succession with outcrops determined by one major F_1 structure, the Buttermere Anticline, whose existence Webb and Jeans were unable to verify.

Jeans (1972) described a locality in Newlands [230 158] where north-trending pre-cleavage folds in the Skiddaw Slates are truncated by the basal andesite of the Borrowdale sequence. This deformation episode is, therefore, pre-Borrowdale in age and provides one reason for the structural contrast between the slates and the Borrowdale Volcanics. Equally, contrasts in ductility and layer thickness must also be invoked, since massive psammites in the Skiddaw Group and thick lava flows in the Eycott Group show much smaller strains than the pelitic rocks.

Interpretive problems remain. In the Black Combe inlier the earliest structures described by Helm (1970) are tight N to NNE-trending folds with an associated cleavage which he correlated with F_1 of Simpson and regarded as pre-Borrowdale in age. Yet the slate–volcanic junction in this region appears to be conformable and the cleavage, if associated with F_1, is evidently end-Silurian. Are the pre-cleavage folds absent in this inlier? An alternative explanation, bearing in mind the manner in which the Caledonoid cleavage swings to NNE in the south-western part of the district, is that the pre-cleavage folds are present, but effectively coaxial with the 'cleavage folds' and thus difficult to identify in the monotonously pelitic Black Combe slates.

On a regional scale, no clear picture has yet emerged of the pre-Borrowdale structure of the Skiddaw Group. An interpretation presented by Helm and Roberts (1971) in which the regional outcrop pattern is dominated by the north-trending Buttermere Anticline is not supported by the available biostratigraphical evidence (compare Helm and Roberts op. cit., Fig. 3 with Wadge 1972, Fig. 1). From stratigraphical and structural evidence Downie and Soper (1972) suggested that the Skiddaw and Eycott Groups had been uplifted along an ENE–WSW axis and eroded in pre-Borrowdale time. This trend is compatible with the inferred existence of a similarly trending plate margin to the north, but is markedly oblique to the observed pre-Borrowdale structures in the slates described above.

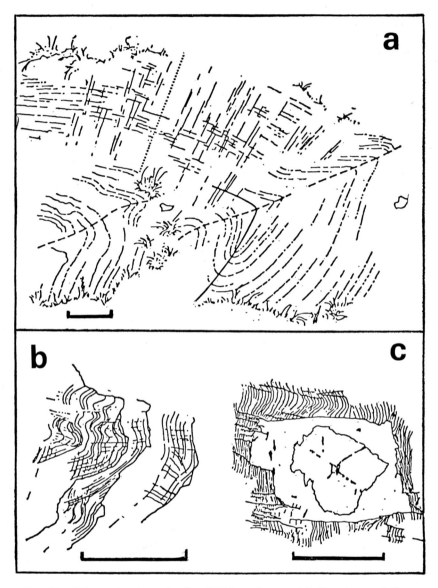

FIG. 21. Minor structures in the Skiddaw Group. (a) Structures of three superposed deformation episodes, Black Combe. Drawings (b) and (c) show relationships between the end-Silurian structures. In (b) the main cleavage (steep) is deformed by 'second phase' folds and cleavage. In (c) an andalusite porphyroblast in the Skiddaw Granite aureole has overgrown the main cleavage but is cut by 'second phase' spaced cleavage. The bars are 30 cm long in (a) and (b) and 1 mm in (c). (a) from Helm 1970, (b) and (c) from Soper and Roberts 1971.

PLATE 2. Oblique aerial view to the north-north-west across the Lake District fells. Little Langdale and Blea Tarn are in the foreground, with Great Langdale beyond. The Langdale Pikes and Stickle Tarn are north of Great Langdale, and Borrowdale, Derwent Water and Bassenthwaite Water can be seen in the distance. Mellbreak and Crummock Water are just visible in the west. *Photograph: C. H. Wood, Bradford*

A

B

1 metre

LATE ORDOVICIAN STRUCTURES

The pattern of crustal deformation which preceded the deposition of the Ash-gillian rocks (Ingham et al., Chapter 9) has been difficult to interpret in the past but has recently been clarified (Soper and Numan 1974). Green (1920) suggested that the unconformity which separates the Borrowdale Volcanics from the Coniston Lime-stone was related to NNE folding of the volcanic rocks in pre-Bala time. Mitchell (1929, 1934) adopted this interpretation in his mapping of the volcanic rocks between Troutbeck and Shap. Subsequent workers have also tended to relate northerly strikes in the Borrowdale and Skiddaw Groups to north-trending folds of this age. However, Mitchell's mapping of the volcanic structure between Coniston and Dunner-dale (1940, 1956B), while confirming the swing in strike shown on Aveline's map (1888), and its truncation by the Coniston Limestone unconformity, did not lend support to the hypothesis of NNE folding of the Borrowdale rocks. There is a major easterly trending syncline whose axial trace passes beneath the Coniston Limestone on Torver Moor. This fold was subsequently traced west across the Duddon by Firman (1957) and may be called the Ulpha Syncline. The geometry of this important intra-Ordovician fold was examined by Numan (1974). His analysis (Fig. 22) shows that the north limb has a strike of about 225° with a moderate to steep SE dip. Between Torver Moor and the Duddon valley the strike of the south limb averages N–S, dipping east, with local variations due to medium-scale folding. The fold is quite open, with an interlimb angle of 115°; the amplitude is approximately 4 km and the half wavelength 8 km. The axis plunges gently to the ENE and the trace is offset by several important faults. West of the Duddon the end-Silurian Black Combe Anticline is superimposed upon the south limb of the Ulpha fold. The regional cleavage is roughly congruous with this anticline but is superimposed obliquely across the earlier syncline, trending 20°–30° anticlockwise from the synclinal trace. Removal of the end-Silurian deformation in order to determine the original geometry of the Ulpha fold is difficult because of incomplete knowledge of the strains involved, but reasonable estimates lead to the conclusion that the fold was developed in late Ordovician time as a very open E–W trending syncline, with limb dips of about 30°. The only area where older maps showed NNE trending folds in the volcanic rocks truncated by the sub-Caradocian unconformity is near Kentmere (Mitchell 1929) but re-investigation of this area (Soper and Numan 1974) has shown that these folds do not exist. Northerly and northeasterly folds in the Skiddaw Group previously interpreted as pre-Caradocian structures (Roberts 1971, Jeans 1971, Moseley 1972) are now considered to be pre-Borrowdale in age (Jeans 1972, Webb 1972, Roberts 1977). Green's hypothesis of NNE-trending pre-Bala folding in the Lake District has, therefore, been abandoned. On the other hand, the importance in the Borrowdale Volcanic Group of E–W folding of late Ordovician age is apparent.

facing page

PLATE 3. A. Pre-cleavage, pre-Borrowdale, minor north-north-east trending folds in semi-pelitic Skiddaw Slates, Low Bank, Buttermere [176 175]. The locality lies close to the core of the major pre-cleavage syncline shown in figure 22b.

   B. End-Silurian minor fold with congruous cleavage in the Skiddaw Group, Hassness, Buttermere [186 159], Webb 1972. The rock surface is horizontal, the fold plunge vertical, and the cleavage vertical with an east-north-east trend.   *Photographs: F. Moseley*

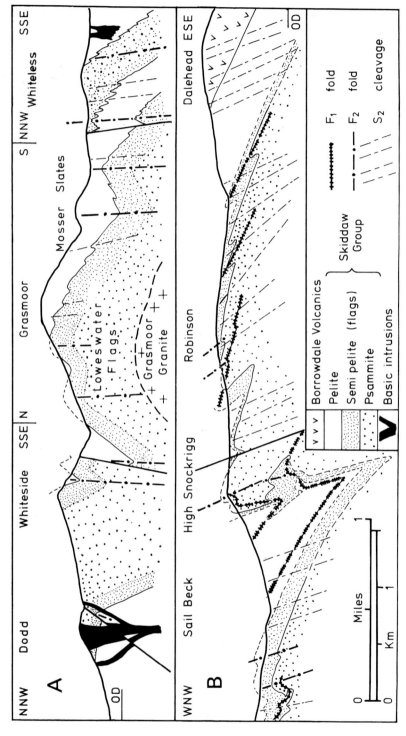

FIG. 22.   Sections across the Skiddaw Group in the Crummock Water–Buttermere area. (A – from P. J. F. Jeans 1974 and B – from B. C. Webb 1975.   Unpublished Ph.D. theses Birmingham and Sheffield Universities). F_1, F_2 and S_2 refers to the terminology of Webb and Jeans, in which F_2 and S_2 equals F_1 and S_1 of Simpson (1967).   The term Kirk Stile Slates is now used in preference to Mosser Slates, and has priority (Jackson, chapter 7).

The next major fold to the north of the Ulpha Syncline is the Wrynose–Nan Bield Anticline which trends ENE (Figs. 23 and 39). Since the two folds share a common limb, and the outcrop of the Coniston Limestone unconformity strikes obliquely to the anticlinal trace, a pre-Ashgillian age for the anticline is possible. The end-Silurian cleavage is also superimposed non-congruously across this fold (Soper and Numan 1974). The geometry of the anticline can be interpreted in terms of both late Ordovician and end Silurian strain episodes, as indeed can the next major fold to the north, the Scafell – Place Fell Syncline. Thus, although the Borrowdale Volcanics were considerably deformed by the end-Silurian earth movements, the three large folds in these rocks were probably initiated in late Ordovician times. Their original amplitude was approximately one half, and their wavelength three times, the maximum observed thickness of the Borrowdale succession, so they appear to represent primary compressional buckles which affected the whole thickness of the volcanic pile. Although there is no evidence of décollement at this time between the volcanics and the underlying Skiddaw rocks, no structures of comparable age have been detected in the latter. However, it is possible that medium and small-scale folds of this age exist in the slates and have not been differentiated from later structures of similar trend. It is also probable that late Ordovician uplift and erosion played a part in the development of the complex but essentially anticlinal zone now occupied by the main Skiddaw Slate tract in the Lake District.

END-SILURIAN STRUCTURES

The major tectonic evolution of the Lake District took place shortly after deposition of the Silurian turbidite sequences — the exposed succession is now believed to terminate in the Downtonian (Cocks et al. 1971) — and before deposition of the molasse-type Mell Fell Conglomerate probably in the Lower Devonian (Chapter 11). Polyphase deformation was accompanied by low greenschist-facies metamorphism, cleavage formation, faulting, granite emplacement and considerable uplift and erosion. A dominant element is the regional Caledonoid cleavage associated with folds on all scales. A superimposed set of open, sideways-closing folds with a sub-horizontal crenulation cleavage is developed sporadically in the Skiddaw Slates and very locally in the younger rocks. There are also several generations of kink-bands and associated structures, again largely confined to the Skiddaw Slates. These three sets of structures correspond to F_1–F_3 as recognized by Simpson (1967) in the Skiddaw Slates.

Caledonoid Cleavage

The main cleavage shows a pronounced swing in strike from NNE in the south-west regions through NE and ENE over much of the central fells to E–W in the southern Howgill Fells. The E–W trend is continued in the lower Palaeozoic rocks of the Cross Fell and Teesdale inliers and swings to ESE in the Ingleton and Ribblesdale inliers at the southern edge of the Askrigg Block. Moseley (1972) noted the parallelism of the cleavage trend in the Askrigg basement to the Craven Fault zone and a linear magnetic 'high' (Bott 1967) which trends ESE across the region.

The cleavage is deflected around certain intrusive bodies such as the Shap granite (Boulter and Soper 1973) and also near some faults, suggesting that part of

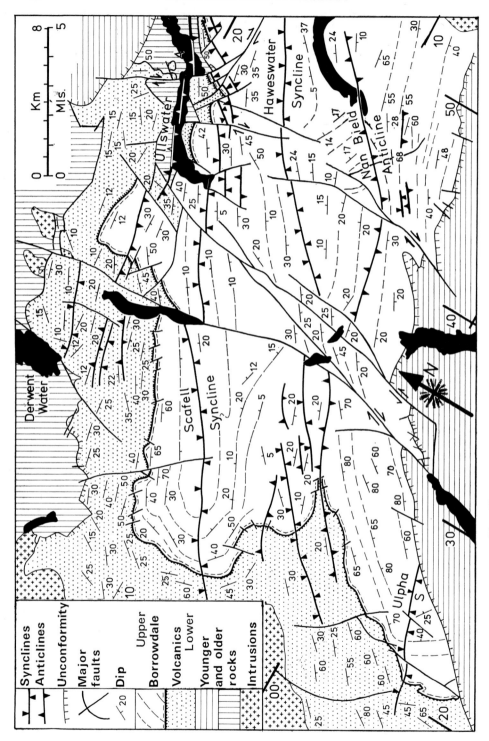

Fig. 23.   Structural map of the Borrowdale Volcanic outcrop.

the displacement on the faults occurred during the cleavage-forming strain episode. The cleavage is usually steeply inclined but in detail its orientation is much affected by contact strain adjacent to lithological boundaries with a strong competency contrast, and also locally by post-cleavage deformation particularly in the Skiddaw Slates.

The age of the cleavage is established by several lines of evidence. Stratigraphically, it post-dates the youngest Silurian strata exposed in the region and pre-dates the Devonian Mell Fell Conglomerate. It appears to have formed immediately before emplacement of the early Devonian Shap and Skiddaw Granites. At Shap, contact biotite grows across the cleavage but, as mentioned above, the cleavage is deflected around the granite stock. In the Skiddaw aureole (Soper and Roberts 1971 and Figure 21) andalusite has grown statically over the Caledonoid cleavage which is, however, slightly deflected around the porphyroblasts at some localities; cordierite entirely post-dates this cleavage. It appears that emplacement of these granites just overlapped the final stages of cleavage formation, suggesting that their rise was permitted by a dissipation of the main Caledonian compressive stress in the cleavage-forming strain episode. K–Ar ages of about 395 Ma on illite from Skiddaw Slates adjacent to the contact of the (Ordovician) Threlkeld Microgranite (P. R. Ineson and J. G. Mitchell personal communication and discussion to Wadge et al. 1974) support the end-Silurian or earliest Devonian age of the cleavage.

Both the nature and intensity of the cleavage show considerable variation, depending on lithology and strain state. It is convenient to describe these variations by reference to cleavage development in the three major rock groups.

In the Skiddaw and Eycott sediments, the intensity of the cleavage is clearly related to lithology, pelites showing the best development while massive psammites may be uncleaved. Depth of overburden (controlling hydrostatic stress and in part, P_{H_2O} and T) was also important, since the Llanvirn slates of the eastern Lake District are much less strongly cleaved than the Arenig slates further west. Although often described as a slaty or flow cleavage, the fabric is rarely truly penetrative; spaced types of cleavage predominate. In addition, the major plane of fissility is often parallel to the bedding, particularly in finely banded lithologies. Helm (1969) has described the textural development of these structures in the Black Combe inlier. He attributed the bedding fabric to mimetic crystallization of muscovite and chlorite under the influence of tectonic stress perhaps manifested as bedding slip, and described the transposition of this fabric into slaty cleavage through the intermediate stage of a crenulation cleavage. He also showed that in many tight minor folds the bedding fabric predominates in the limbs while an axial planar cleavage is largely restricted to the crests. This is a common feature throughout the Skiddaw Slates. It is perhaps unnecessary to invoke tectonic stress in the formation of the bedding cleavage. Water loss during diagenesis of sediments leads to volume reduction and hence uniaxial shortening normal to the bedding, thus intensifying any original bedding-parallel fabric, which would be further enhanced by mimetic crystallization of phyllosilicates during a weak thermal event.

Pressure solution has recently been recognized as an important cleavage-forming mechanism (McClay 1977). In many Skiddaw lithologies, particularly in semipelites, the spaced cleavage shows features consistent with an origin by pressure solution. Investigation is required of the interrelationship between the crenulation and other spaced cleavages and their transition to truly penetrative, slaty fabrics. Psammitic beds often show a spaced planar structure described as fracture cleavage, which may arise from pressure solution. However, in certain rock types in which pore water is unlikely to have played an important role, for example the Threlkeld Microgranite and hornfelses around other pre-cleavage intrusions, the Caledonoid cleavage also takes the form of closely spaced fractures. Their origin is not understood.

In the volcanic rocks the style and intensity of the cleavage is again closely related to lithology. Fine andesitic tuffs often show a strong penetrative cleavage, formed by the dimensional orientation of phyllosilicates and flattened lithic clasts. Strongly altered andesite flows are sometimes equally well cleaved. In coarse andesitic and dacitic tuffs, the larger lithic fragments may be flattened with a penetrative or spaced cleavage in the finer matrix around them, sometimes giving the appearance of two intersecting cleavages. Less altered flows often carry a spaced cleavage, but may also show evidence of ductile strain in the form of flattened vesicles and distorted polygonal joint patterns. Strongly-welded ash flows and fresh 'glassy' lavas are usually uncleaved but strongly jointed. The intensity of cleavage in the volcanic rocks seems to be related to the presence of phyllosilicate phases, principally chlorite, which could recrystallize during the cleavage-forming deformation episode.

Regionally, several zones can be recognized in the Borrowdale outcrop in which the cleavage is particularly intense. These extend for several kilometres along the strike and often coincide with important tuff units. The best known example lies in the Tilberthwaite Tuffs and crosses the Langdales; another is at Honister. They are evidently high strain zones. Several examples in the Honister Tuffs and another that coincides with the outcrop of lapilli (bird's eye) tuff (Pl. 4) between Kentmere and Wrengill (Soper and Numan 1974) appear to result from contact strain where tuff beds occur between massive lava flows. Others may be controlled by regional variations in strain state. One occupies the axial region of the Nan Bield Anticline in the Kentmere area and within it even massive lavas are strongly cleaved. The major working quarries are situated in these zones, although most of the slate is now sawn for building purposes or polished for decorative use, rather than split for roofing. When required for splitting the slate is worked when still 'green'; it is difficult to split after drying out. This suggests that ease of splitting is controlled by water films along the spaced cleavage (which may often be seen to accompany the slaty fabric in natural weathered outcrops) rather than the slaty cleavage itself. Outside the high strain zones cleavage is absent in the more resistant lithologies and even fine tuffs carry only a weak, spaced cleavage (Pl. 9).

In the Silurian greywackes cleavage development again greatly depends on lithology (Pl. 10). In argillites the cleavage is often penetrative and sufficiently strong to replace bedding as the main plane of fissility, as for example in the Burlington quarries near Grizebeck. In silty beds a weaker, spaced cleavage predominates and massive arenaceous units, such as the Coniston Grits, are often uncleaved. Moseley (1968) has recorded changes in cleavage types and orientation within graded beds which are seen as 'cleavage refraction', an expression of contact strain. He also recorded transitions from cleavage in mudstones to shear joints in interbedded arenites. Regionally the cleavage is less intense southwards in the higher members of the Silurian succession.

It has been suggested (Powell 1972) that the mechanism of tectonic dewatering (Maxwell 1962) was responsible for cleavage development in the Bannisdale Slates and Kirby Moor Flags, since the cleavage is parallel to siltstone dykes and other dewatering features. The mechanism involves the rotation of clay particles parallel to the direction of flow of upward-migrating pore water under tangential tectonic stress. Although the arguments presented by Powell are attractive, they have not found general acceptance since most of the observed features can be accounted for by superimposing cleavage on earlier dewatering structures. Tectonic dewatering cannot be a general mechanism for the production of slaty cleavage since many sedimentary sequences, for example the Skiddaw Slates, must have undergone considerable dewatering and lithification long before cleavage was imposed upon them. Nor can any type of cleavage in lavas be due to this mechanism. The role of pressure solution has not been explored in relation to cleavage formation in the Silurian greywackes but is likely to be important. Mention should also be made of the upper Ordovician sediments in this context: cleavage in the Coniston Limestone has every indication of a pressure solution origin (Pl. 10A) but again, no detailed investigations have been made.

While limited research has been directed towards the mechanisms of cleavage formation in Lake District rocks, in recent years important investigations of the strain state associated with slaty cleavage have been made utilizing accretionary lapilli tuffs in the Borrowdale Group. The lapilli are deformed into ellipsoidal shapes with their long and intermediate axes close to the cleavage plane. They are near ideal strain markers in that they are composed of similar material to the rock matrix and so may be expected to have deformed homogeneously with it, and they occur in large numbers so that statistically valid shape determinations can be made. An initial investigation was made by Green (1920) who assumed that the ellipsoid represented the strain responsible for the cleavage. He deduced a shortening of about 60 per cent normal to the cleavage. Modern investigations in the area were initiated by Oertel (1970). He assumed that the lapilli were initially spherical but, noting that the lapilli long axes do not lie exactly in the cleavage, invoked two deformations to account for their present shape and orientation. He was able to factorize the lapilli shape into a triaxial ellipsoid oriented in the cleavage frame (with the long axis parallel to a mineral lineation on the cleavage plane) representing the tectonic strain, and a uniaxial oblate ellipsoid in the bedding frame (short axis normal to bedding) representing compaction. Helm and Siddans (1971) criticised Oertel's methodology and presented a strain analysis utilizing the Rf/Ø technique of Dunnet (1969) and Dunnet and Siddans (1971) in which the lapilli shape was related to a tectonic strain associated with the cleavage and an initial shape factor, assuming no compaction strain. These two approaches reflect a fundamental difference in the interpretation of slaty cleavage as a strain-induced fabric: does slaty cleavage relate to the total (finite) strain to which a rock has been subjected throughout its history, as concluded by Ramsay (1967) and Siddans (1972), or does it reflect a particular, albeit important, component of the total strain? It would appear that in the case of sedimentary rocks which underwent volume reduction during compaction prior to tectonic deformation, the latter must be true.

Bell (1975) recognized four factors which contribute to the present shape of the lapilli: initial shape; 'compaction strain'; strains associated with buckling; and 'cleavage strain'. By selection of material Bell was able to partially eliminate the first factor. He developed Oertel's approach to provide an iterative computer technique to factorize the strains into the bedding and cleavage frames. His results suggest a maximum compaction strain of around 66 per cent normal to the bedding and a range of cleavage strains, all close to plane strain, with a maximum shortening normal to the cleavage of almost 70 per cent in samples from the high strain zones.

Main Caledonian Folding

Regional folding, on all scales, accompanied the main phase of cleavage formation in the Lower Palaeozoic rocks of the Lake District. Buckling presumably began

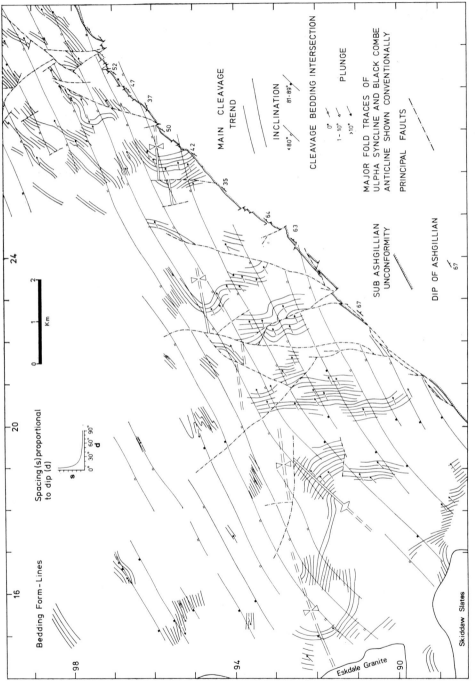

FIG. 24.   Structural map of the Borrowdale Volcanic Group and its junction with the Coniston Limestone Group in the Ulpha area (From N. M. S. Numan 1974. Unpublished Ph.D. thesis, Sheffield University).

before the onset of cleavage formation and the folds continued to tighten during the main compression. The fold traces broadly follow the arcuate trend of the cleavage across the region, but the cleavage is rarely exactly axial planar to the folds. The geometry, and to some degree the style, of folding depend on the earlier tectonic history of the rocks: the Skiddaw, Eycott and Borrowdale Groups had already been deformed whereas the younger Ordovician and Silurian strata had not. The major controls of fold style and scale appear, however, to have been associated with bedding anisotropy, particularly layer thickness and competence contrasts, whilst depth in the stratigraphic pile was less important. Strongly layered turbidite sequences in the Skiddaw Slates and Silurian display the best examples of fold structures in the Lake District and the following descriptions concentrate on them. Only brief mention is made of the pelitic rocks which show smaller scale folds of greater complexity, and of the massive arenaceous and volcanics sequences which are characterized by larger scale open folds.

Fold styles in psammite – pelite interbeds

Where thin psammites and pelites are interbedded, as for example in the transition beds between the Loweswater Flags and Mosser Slates, folds occur on all scales and their style is controlled by competency contrasts between psammite beds, commonly a few tens of centimetres thick, and the pelite or semipelite beds of similar thickness. Both show effects of buckling and 'flattening', the psammites with limited thinning of the limbs and sometimes a weak, outwardly-fanning spaced cleavage, the pelites approaching a true 'similar' fold style with strong, inwardly fanning, sometimes penetrative cleavage. Boudinage of favourably oriented psammite beds is common. Minor folds are upright, usually asymmetric, depending on their position in relation to major folds of the same generation, and roughly congruous with the cleavage.

It is this lithological facies which shows the clearest interference relationships between the main Caledonian and pre-Borrowdale folds. A fine example from Hassness, Buttermere (186 151) (Webb 1972) provided the key to understanding the geometry of interference between major folds in that area (Webb 1975). In areas of refolding, the later folds maintain a fairly constant axial plane orientation while their axes and the cleavage-bedding lineation change plunge by as much as 180° as they cross the earlier folds. The implications of this interference geometry in terms of outcrop pattern and detailed stratigraphy have been worked out in the Crummock–Buttermere–Newlands areas by Webb (op. cit.) and Jeans (1974), whilst investigations by Helm (1970) at Black Combe and Roberts (1971, 1973, 1977) at Mungrizedale indicate similar structural complexity throughout the slates.

Similar lithologies are found in the Silurian Bannisdale Slates and the transition beds into the underlying Coniston Grits. These have a comparable style of deformation, although the complexities due to superposed folding are absent. Excellent sections are provided by road cuts (55 05) along the A6 south of Shap Summit (Moseley 1968) and along the M6 and realigned A685 (61 02) in the Lune Gorge (Moseley 1972). In the Shap section (Fig. 29) the folds are upright, asymmetric, rather straight-limbed with interlimb angles of 60–90° and gentle plunges to the NE. Folds in the competent greywackes approach a 'parallel' style, but show limited thickening in the hinges. The more massive beds are uncleaved but strongly jointed. Mudstones thicken in the fold hinges, with inwardly fanning cleavage on the limbs. The combination of these two fold styles in the alternating sequence produces an overall 'similar' profile. The sections are ideal for the study of contact strain phenomena, the relationship of cleavage and joint patterns to fold geometry, as well as sedimentary features in the turbidites. The cleavage is not quite congruous with the folds, striking 5–10° clockwise from the fold traces. This is a common feature throughout the Silurian rocks of the southern Lake District and seems to indicate a component of rotational strain during the main Caledonian compressive phase.

Regionally, belts of minor and medium-scale folding alternate with steeply-dipping homoclinal zones. These are comparable in width to the zones of intense and weak cleavage in the volcanic rocks and there may be some connexion. They are generally coaxial with the main folds, though this is not true of the north–south belts of anomalous easterly dip near Underbarrow and Whitbarrow (Moseley 1972). The folds die out as they enter these belts but the cleavage continues through them (Fig. 25). They presumably pre-date the folding and may be related to some north-trending structure in the basement which could have been later reactivated to produce the steep monoclinal structures seen in the Carboniferous Limestone farther south.

The axial traces of the two major Caledonoid folds in the Silurian rocks, the Bannisdale Syncline and Selside Anticline, can be located by the change in asymmetry of the smaller folds. The latter do not plunge uniformly to the ENE across the whole region: reversals occur, for example across a plunge culmination near Tarn Hows (O'Connor and Stabler in unpublished theses), and there are much steeper NE plunges south of Coniston (Norman 1961 and Fig. 26). These variations in plunge have been attributed to the superimposition of gentle northerly folds, but may simply be due to inhomogeneous compression across the cleavage.

Fold styles in pelites

Where minor folds of the main Caledonian phase are superimposed upon pre-cleavage folds in Skiddaw pelites, complex interference patterns result. In many examples the term 'pattern' is a misnomer. Fold axial planes and axes cannot be traced for more than a few metres and disharmonic features are common. Only where the later folds have a roughly congruous cleavage can they be distinguished from earlier structures. Where the earlier folds are not strongly developed, for example in Warnscale Bottom [20 14], the main-phase minor folds are seen to be upright, tight to open, with gentle plunges to the NE and SW. Disharmonic effects are widespread and fold profiles change rapidly through the structures.

The least competent rocks in the higher part of the succession are the Stockdale Shales. Surprisingly, these rocks are largely devoid of folds and dip steadily south-east towards the Bannisdale Syncline. Northwesterly directed thrusts and reversed faults are however common at this level, which appears to have been a zone of partial décollement between the Silurian greywackes and the rigid, previously deformed and eroded volcanic rocks beneath.

Fold styles in competent rocks (psammites and volcanics)

Massive psammites occur towards the base of the Loweswater Flags in the Skiddaw Group and in the Silurian Coniston Grits. They tend to develop open folds, commonly with interlimb angles in excess of 90°, and often approach a true 'parallel' style. Cleavage is restricted to thin siltstone — shale interbeds and fractures of various types are common.

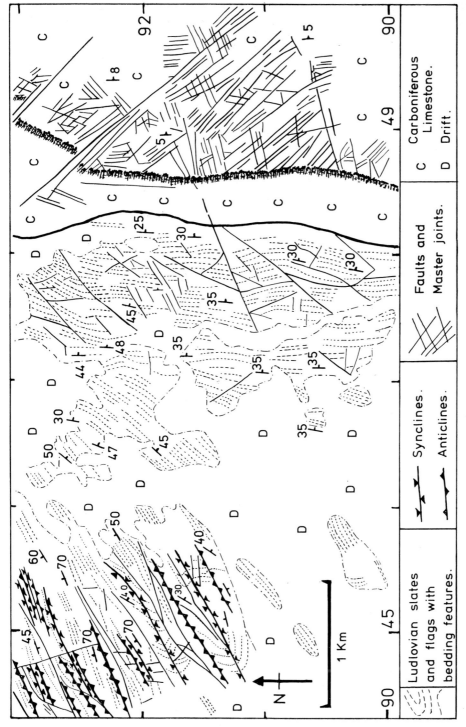

FIG. 25. Structural map of the Underbarrow area west of Kendal, showing one of the zones of northerly strike in the Silurian rocks.

| Ludlovian slates and flags with bedding features. | Synclines. | Faults and | Carboniferous Limestone. |
| | Anticlines. | Master joints. | Drift. |

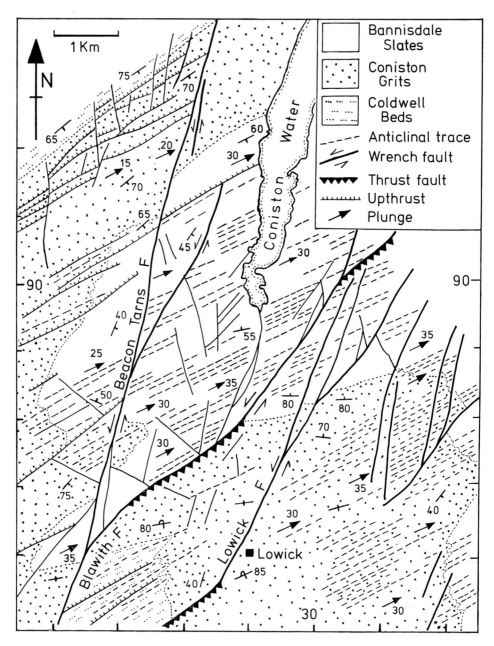

Fig. 26.   Structural map of the Silurian of the south Coniston area. Note the
composite wrench-thrust faults (From T. N. Norman 1961. Unpublished Ph.D.
thesis, Birmingham University). Plunge values are sub-area averages, and were
obtained from field measurements of (a) fold axes (8–15) and (b) bedding
orientations (80–230) for each of the areas centred on the plunge symbols.

The Borrowdale Volcanic sequence is braced by thick, competent lavas and ignimbrites and has large open folds associated with brittle fractures on all scales. As discussed above, some of these folds may be pre-Ashgillian in age but doubtless considerable modification of the earlier folds took place in end-Silurian time, due to renewed buckling, the initiation of new medium-scale folds, for example the Yoke Folds at Kentmere (Soper and Numan 1974) and the development of the steep monoclinal zones coincident with the high-strain cleavage belts described above. The more thinly bedded parts of the volcanic succession occasionally develop small-scale folds, often roughly congruous with the cleavage, but it is rarely possible to assign them unambiguously to the end-Silurian event.

The structure of the Eycott Group volcanics is quite different to that of the Borrowdales. The Eycotts strike just south of east on the north limb of the main Lake District Anticline and the dip increases from about 40° to the north at the base to vertical and steeply overturned higher in the sequence (Eastwood et al. 1968). This interpretation is confirmed by the palaeomagnetic evidence (Faller and Briden Chapter 2).

LATE CALEDONIAN DEFORMATION EPISODES

Simpson (1967) recognized that the main Caledonian cleavage and the bedding in the Skiddaw Slates are locally deformed by reclined folds with associated sub-horizontal crenulation or fracture cleavage. He designated these folds F_2 and regarded them as pre-Borrowdale in age. Subsequently Soper and Roberts (1971) showed that they postdate the growth of andalusite in the outer part of the Skiddaw aureole, which establishes their early Devonian age (Fig. 21).

Minor folds of this phase occur throughout the Skiddaw Slates wherever the principal plane of fissility, be it main cleavage, 'bedding cleavage' or bedding, is well developed and steeply inclined. The sub-horizontal 'second cleavage' is widely developed in pelitic and semipelitic lithologies, and in narrow zones may be so intense as to completely transpose the earlier planar fabrics. In places, where the second cleavage is of fracture type and intersects the main cleavage at a high angle, 'pencil slates' are formed. Even when the second cleavage is not strongly developed, a gently-plunging crenulation lineation is often present on the main cleavage planes in pelites. Structures of this generation are usually absent in massive rocks such as psammites, hornfelses and intrusions; nor have they been recorded from the Borrow-dale outcrops. They are perhaps found at a few localities in the Silurian greywackes, for example on the steep limbs of some of the main phase folds in the new road cuts in the Lune gorge.

A pair of major folds of this generation was mapped by Simpson near Crummock Water but subsequent remapping in greater detail by Jeans (1973) has not confirmed their presence. The episode has little overall effect on the outcrop pattern of the Skiddaw Slates; geometrically it produces local changes in the dip of the earlier planar structures but not in their strike. It represents a small vertical shortening and may be interpreted as gravity-induced, following tectonic thickening of the Lower Palaeo-zoic pile during the main Caledonian compression. It may have occurred early in the period of Devonian uplift before much erosional unloading had taken place.

Simpson recognized a further set of structures in the western part of the main Skiddaw outcrop. These are NW-trending minor monoclines and kink-bands. Late orogenic 'brittle' structures are in fact common throughout the Lower Palaeozoic rocks of the district. Helm (1970) described several sets in the Black Combe slates. They occur sporadically in appropriate lithologies in the Borrowdales and quite commonly in the Silurian rocks. Although sometimes striking in the field, particularly when associated with quartz veins, these structures represent only trivial deformation on the regional scale; they have a wide range of styles and orientations and they may not all be of Caledonian age.

CALEDONIAN FAULTING

It is probable that faulting accompanied all the Caledonian deformation periods with reactivation of older structures on each occasion, but these episodes are not readily differentiated. There is little evidence for pre-Borrowdale faulting providing one accepts the structural sequence presented here rather than that of Simpson (1967), but a pre-Ashgillian phase can be demonstrated for the Coniston Fault since there are different volcanic successions beneath the Coniston Limestone on either side of the fault (C. S. Stabler, unpublished report, Birmingham University). However, there is little doubt that the more important fault movements occurred during the end-Silurian episode.

Thrust Faults

Low- and high-angle thrusts are well developed close to the lower and upper junctions of the Borrowdale pile. The lower junction is one of great ductility contrast and it is understandable that orogeny would result in disharmony between the two groups. Décollement thrusts are a result and are well seen north-west of Ullswater where the junction is a moderately inclined plane which truncates steeply dipping volcanics and tight folds in the slates (Moseley 1964). Elsewhere, for example in the Newlands Valley and near Honister, there are numerous small low-angle thrusts and master joints in the Skiddaw Slates close to and sub-parallel to the unconformity at the base of the volcanics. Thrusts are also seen cutting Silurian strata, particularly within incompetent mudstones such as the Stockdale Shales. Most of these faults are composite wrench-thrust structures and are referred to below.

High-angle thrusts with ENE trends (upthrusts or reverse faults) are important in the Upper Ordovician and Silurian outcrops near to the Borrowdales, where there

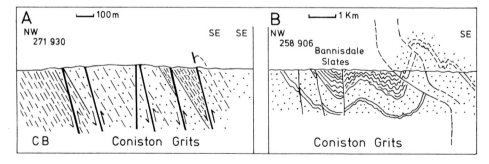

FIG. 27. Sections across parts of figure 26. A – in the north where upthrusts (or reverse faults) replace north dipping fold limbs. B – further south across the thrust component of a composite fault (From T. N. Norman 1961. Unpublished Ph.D. thesis, Birmingham University).

are relatively few folds (Figs 26, 27 and 28). They partly replace steep NW-dipping fold limbs but, like the cleavage, they cut obliquely across the folds having a more easterly trend. They are also parallel to cleavage, and likely to be related to it (Norman 1961).

Wrench Faults

These important faults have displacements ranging from a few metres to 2 km, with dextral and sinistral forms equally developed. It is probable that movements occurred during and after folding since they commonly separate different fold patterns. Oblique slip seems to have been more usual than strike slip, but can only be determined when circumstances are favourable; for example where there are composite wrench/thrust faults as described below or where it is possible to measure displacements of fold axes and marker horizons (Moseley 1960).

Composite Faults

These are structures with both wrench and thrust components. Their existence can be inferred from many of the old maps but the only detailed account was given by Norman (1961) who analysed fault movements south of Coniston. The Coniston and Brathay Faults (Fig. 28) are the two most obvious examples (they each displace the Coniston Limestone by nearly 2 km) yet the adjacent Upper Coldwell and Coniston Grits are unaffected. Mapping shows that the wrench structures rotate into thrusts about 5° oblique to bedding, the latter showing up clearly on aerial photographs. Further south Norman was able to determine the inclinations of the thrust components of composite faults (about 45°, Figs 26 and 27), and he concluded that the Blawith Fault, for example, was originally an oblique-slip fault with dip-slip and

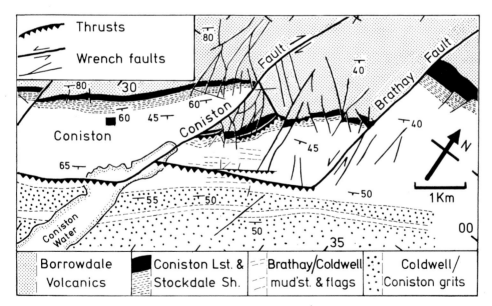

FIG. 28. Major and minor composite wrench-thrust faults NE. of Coniston. The thrust components tend to follow the strike of less competent strata.

strike-slip components each of about 1 km. There are also many small wrench faults which cross the Coniston Limestone outcrop and rotate into 'bedding thrusts' in the incompetent Stockdale Shales, and at the base of the Borrowdales a number of large faults terminate downwards at décollement thrusts separating the volcanics from the Skiddaw Slates (Moseley 1960, Fig. 9).

HERCYNIAN DEFORMATION

The Hercynian Orogeny was associated with closure of the mid-European Ocean and continental collision, but the Lake District was far to the north and deformation was mild. In northern England several periods of deformation have been recognized between Lower Carboniferous and Permian times, but in the Lake District the only important phase is the Saalian (end-Carboniferous).

The orientations and intensities of many Hercynian structures were posthumously derived from Caledonian structures reactivated by the renewed stresses. For example, NW wrench faults moved again as normal faults, and north-trending Caledonian (or older) structures (Fig. 25) were reactivated to form the pronounced zones of folding and faulting in the Carboniferous rocks of the southern Lake District. The Silverdale disturbance east of Morecambe Bay is a prime example (Moseley 1972). In addition the presence of the Caledonian batholith (Chapter 3) imparted rigidity to the Lake District and north Pennines resulting in the blocks and basins which influenced Carboniferous and later sedimentation. Thinner Carboniferous sequences were deposited on the blocks and thicker successions in the adjacent down-warping basins of the Central Pennines and Solway, but more important structurally was the lack of folding in the sedimentary cover of the rigid north Pennine and Lake District blocks (or massifs) compared with complex folds in the basinal sediments.

In the Lake District, Carboniferous rocks form a rim around the central Lower Palaeozoic core, and dip outwards at angles of less than 10 degrees from an original dome over central Lakeland. This is true everywhere except in the south near to Morecambe Bay where flat-lying beds are cut by north-trending monoclinal faulted folds, several with vertical or overturned strata (Moseley 1973) which largely die out as they are traced northwards towards the margin of the Lake District Granite (Chapter 3, Fig. 7). An example is shown on figure 25 which is a northward termination of the Silverdale Disturbance, but the Carboniferous Limestones now dip gently to the east (Moseley 1972, Fig. 12).

Numerically by far the most important Hercynian faults have NW trends, but there are also complementary NE faults (Moseley 1973, Fig. 3). Together they suggest

facing page

PLATE 4.   Specimens of accretionary lapilli ('birds-eye') tuff, Borrowdale Volcanic Group, Kentmere. The lapilli are composed of andesitic fragments and broken crystals identical to the tuffaceous matrix, and probably accreted in a volcanic ash cloud, falling either directly into shallow water or on to land with subsequent re-working.

A. The specimen is cut parallel to the slaty cleavage and shows the bedded and graded nature of the deposit. The elliptical shape of the lapilli reflects the strain parallel to the cleavage plane. The extension direction is marked by both the long axis of the lapilli and a mineral lineation in the matrix.

B. The specimen is cut across the cleavage and indicates a large compression normal to the cleavage, assuming the lapilli to have been originally statistically spherical.

Photographs: A. *F. Moseley* B. *A. M. Bell*

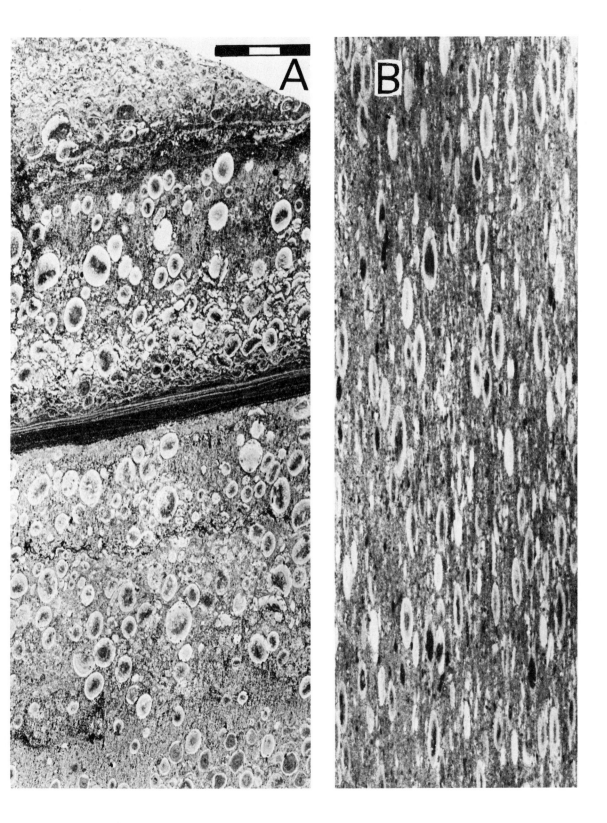

PLATE 5. Oblique aerial view northwards from near the northern end of Windermere, across Loughrigg Fell, to Grasmere (left) and Rydal Water (right). Thirlmere and Skiddaw are beyond and Helvellyn is to the right. Apart from Skiddaw the country is underlain by Borrowdale Volcanic rocks. The Coniston Fault, the largest dislocation in the Lake District (Fig. 23), traverses Grasmere, Dunmail Raise and Thirlmere.

Photograph: C. H. Wood, Bradford

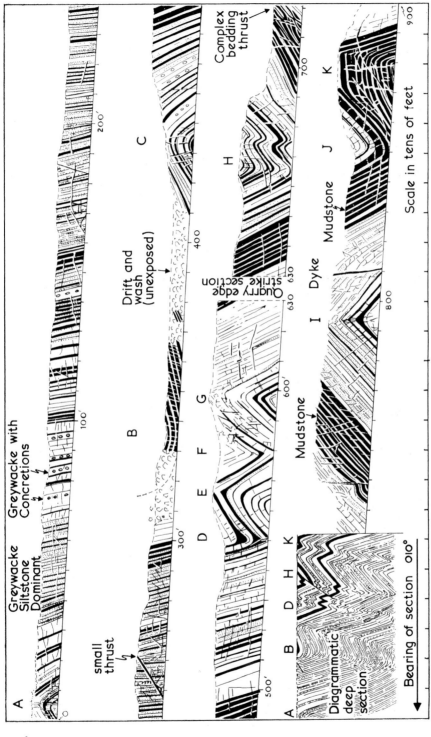

FIG. 29. Section in Coniston Grit – Bannisdale Slate transition beds on the A6, Shap. Black — mudstone, stippled — silt dominant, blank — greywacke. (From Moseley 1968.)

6

an E–W stress régime and this is supported by the presence of northerly trending folds and faults. Joints, impressive in the Carboniferous Limestones where they result in the grykes of limestone pavements, are parallel to the faults, and although their origins are different (Moseley and Ahmed 1967), they also indicate an E–W stress system.

POST TRIASSIC STRUCTURES

During later Mesozoic and Tertiary times the whole of northern Britain was under mild tension as the Atlantic Ocean widened and in the Lake District this resulted in normal faulting and gentle doming. The former was a rejuvenation of Hercynian and earlier faults, the trends being NW and NE and the down-throws predominantly to the SW and NW reflecting the uplift of the Pennine and Lake District massifs and the downwarp of the Irish Sea Basin (Moseley 1972). Some of the faults have large displacements although these are mostly outside the Lake District, the most important being the Outer Pennine and Craven faults. Those faults within and marginal to the Lake District have small displacements, of generally less than 30 m where they affect the Permo-Trias, although the same faults may displace Carboniferous and older rocks by much greater amounts, revealing a history of repeated movement. The doming of the Lake District during these earth movements is supposed to have initiated the radial drainage, which was subsequently imposed upon the Lower Palaeozoic rocks (Marr 1916). The dome itself is inferred from the gentle radial dips of the surrounding Permo-Triassic rocks.

REFERENCES

AVELINE, W. T., HUGHES, T. McK and STRAHAN, A. 1888. The geology of the country around Kendal, Sedbergh, Bowness and Tebay. *Mem. geol. Surv. U.K.*

BELL, A. M. 1975. A finite strain study of accretionary lapilli tuff in the Borrowdale Volcanic Group, English Lake District. Unpublished Ph.D. thesis, University of Sheffield.

BOTT, M. H. P. 1967. Geophysical investigations of the northern Pennine basement rocks. *Proc. Yorks. geol. Soc.* **36,** 139–68.

BOULTER C. A. and SOPER, N. J. 1973. Structural relationships of the Shap Granite. *Proc. Yorks. geol. Soc.* **39,** 365–69.

COCKS, L. R. M., HOLLAND, C. H., RICKARDS, R. B. and STRACHAN, I. 1971. A correlation of Silurian rocks in the British Isles, *Jl. geol. Soc. Lond.,* **127,** 103–36.

DOWNIE, C. and SOPER, N. J. 1972. Age of the Eycott Volcanic Group and its conformable relationship to the Skiddaw Slates in the English Lake District. *Geol. Mag.* **109,** 259–68.

DUNNET, D. 1969. A technique of finite strain analysis using elliptical particles. *Tectonophys.* **7,** 117–36.

—— and SIDDANS, A. W. B. 1971. Non-random sedimentary fabrics and their modification by strain. *Tectonophys.* **12,** 307–25.

EASTWOOD, T., HOLLINGWORTH, S. E., ROSE, W. C. C. and TROTTER, F. M. 1968. The Geology of the country around Cockermouth. *Mem. geol. Surv. U.K.*

FIRMAN, R. J. 1957. The Borrowdale Volcanic Series between Wastwater and the Duddon Valley, Cumberland. *Proc. Yorks. geol. Soc.* **31,** 39–64.

GREEN, J. F. N. 1920. The Geological structure of the Lake District. *Proc. Geol. Ass.* **31,** 109–26.

HELM, D. G. 1969. Microscopic and megascopic fabrics in the Skiddaw Group, Black Combe inlier, English Lake District. *Geol. Mag.* **106,** 587–94.

—— 1970. Stratigraphy and structure in the Black Combe inlier, English Lake District. *Proc. Yorks. geol. Soc.* **38,** 105–48.

—— and ROBERTS, B. 1971. The relationship between the Skiddaw and Borrowdale Volcanic Groups in the English Lake District. *Nature, Lond.* **232,** 181–83.

—— and SIDDANS, A. W. B. 1971. Deformation of a slaty lapillar in the English Lake District. Discussion. *Bull. Geol. Soc. Am.* **82,** 523–31.

HOLLINGWORTH, S. E. 1955. The geology of the Lake District — A review. *Proc. Geol. Ass.* **65,** 385–402.

JEANS, P. J. F. 1971. The relationship between the Skiddaw Slates and the Borrowdale Volcanics. *Nature Phys. Sci.* **234,** 59.

—— 1972. The junction between the Skiddaw Slates and Borrowdale Volcanics in Newlands Beck, Cumberland. *Geol. Mag.* **109,** 25–28.

—— 1974. The structure, metamorphism and stratigraphy of the Skiddaw Slates east of Crummock Water, Cumberland. Unpublished Ph.D. thesis, University of Birmingham.

MARR, J. E. 1916. The geology of the Lake District. Cambridge.

MAXWELL, J. C. 1962. Origin of slaty and fracture cleavage in the Delaware Water Gap area, New Jersey and Pennsylvania. *In* Engel, A. E. J., H. L. James and B. F. Leonard (Eds). Petrologic Studies: A volume in honour of A. F. Buddington. Geol. Soc. Am.

McCLAY, K. R. 1978. Pressure solution and coble creep in rocks and minerals. A review. *J. geol. Soc. Lond.* **134**, 57–70.

MITCHELL, G. H. 1929. The succession and structure of the Borrowdale Volcanic Series of Troutbeck, Kentmere and the western part of the Longsleddale. *Q. Jl geol. Soc. Lond.* **85**, 9–44.

—— 1934. The Borrowdale Volcanic Series of the country between Longsleddale and Shap. *Q. Jl geol. Soc. Lond.* **90**, 418–44.

—— 1940. The Borrowdale Volcanic Series of Coniston, Lancashire. *Q. Jl geol. Soc. Lond.* **96**, 301–19.

—— 1956A. The geological history of the Lake District. *Proc. Yorks. geol. Soc.* **30**, 407–63.

—— 1956B. The Borrowdale Volcanic Series of the Dunnerdale Fells, Lancashire. *L'pool Manchr. geol. J.* **1**, 428–49.

—— 1967. The Caledonian orogeny in Northern England. *Proc. Yorks. geol. Soc.* **36**, 135–38.

MOSELEY, F. 1960. The succession and structure of the Borrowdale Volcanic rocks south-east of Ullswater. *Q. Jl geol. Soc. Lond.* **116**, 55–84.

—— 1964. The succession and structure of the Borrowdale Volcanic rocks north-west of Ullswater. *Geol. J.* **4**, 127–42.

—— 1968. Joints and other structures on the Silurian rocks of the southern Shap Fells, Westmorland. *Geol. J.* **6**, 79–96.

—— 1972. A tectonic history of north-west England. *Jl geol. Soc. Lond.* **128**, 561–98.

—— 1973. Orientations and origins of joints, faults and folds in the Carboniferous Limestones of N.W. England. *Trans. Cave Res. Gp. Gt. Brit.* **15**, 99–106.

—— and S. M. AHMED, 1967. Carboniferous joints in the north of England and their relation to earlier and later structures. *Proc. Yorks. geol. Soc.* **36**, 61–90.

NORMAN, T. N. 1961. The geology of the Silurian strata in the Blawith area, Furness. Unpublished Ph.D. thesis, University of Birmingham.

NUMAN, N. M. S. 1974. Structure and stratigraphy of the southern part of the Borrowdale Volcanic Group, English Lake District. Unpublished Ph.D. thesis, University of Sheffield.

OERTEL, G. 1970. Deformation of a slaty lapillar tuff in the Lake District, England. *Bull. geol. Soc. Am.* **81**, 1173–87.

POWELL, C. McA. 1972. Tectonically dewatered slates in the Ludlovian of the Lake District, England. *Geol. J.* **8**, 95–110.

RAMSAY, J. G. 1967. Folding and fracturing of rocks. *McGraw-Hill, New York, N.Y.* 568 pp.

ROBERTS, D. E. 1971. Structures of the Skiddaw Slates in the Caldew Valley, Cumberland. *Geol. J.* **7**, 225–38

—— 1973. The structure of the Skiddaw Slates in the Blencathra–Mungrisdale area, Cumberland. Unpublished Ph.D. thesis, University of Birmingham.

—— 1976. Cleavage formation in the Skiddaw Slates of the Northern Lake District, England. *Geol. Mag.* **113**, 377–82.

—— 1977. The structure of the Skiddaw Slates in the Blencathra–Mungrisdale area, Cumbria. *Geol. J.* **12**, 33–58.

ROSE, W. C. C. 1954. The sequence and structure of the Skiddaw Slates in the Keswick–Buttermere area. *Proc. Geol. Ass.* **65**, 403–06.

SIDDANS, A. W. B. 1972. Slaty cleavage — A review of research since 1815. *Earth Sci. Rev.* **8**, 205–32.

SIMPSON, A. 1967. The stratigraphy and tectonics of the Skiddaw Slates and the relationship of the overlying Volcanic Series in part of the Lake District. *Geol. J.* **5**, 391–418.

SOPER, N. J. 1970. Three critical localities on the junction of the Borrowdale Volcanic rocks with the Skiddaw Slates in the Lake District. *Proc. Yorks geol. Soc.* **37**, 461–93.

—— and ROBERTS, D. E. 1971. Age of cleavage in the Skiddaw Slates in relation to the Skiddaw aureole. *Geol. Mag.* **108**, 293–302.

—— and NUMAN, N. M. S. 1974. Structure and stratigraphy of the Borrowdale Volcanic Rocks of the Kentmere area, English Lake District. *Geol. J.* **9**, 147–66.

WADGE, A. J. 1972. Sections through the Skiddaw–Borrowdale unconformity in eastern Lakeland. *Proc. Yorks. geol. Soc.* **39**, 179–98.

—— HARDING, R. R. and DARBYSHIRE, D. P. F. 1974. The rubidium strontium age and field relations of the Threlkeld Microgranite. *Proc. Yorks. geol. Soc.* **40**, 211–22.

WEBB, B. C. 1972. N–S trending pre-cleavage folds in the Skiddaw Slate Group of the English Lake District. *Nature Phys. Sci.* **235**, 138–40.

—— 1975. The structure and stratigraphy of the Skiddaw Slates between Buttermere and Newlands, Cumbria, and their relationship to the overlying volcanic rocks. Unpublished Ph.D. thesis, University of Sheffield.

N. J. SOPER, PH.D. Department of Geology, University, Sheffield

F. MOSELEY, D.SC., PH.D. Department of Geological Sciences, University, Birmingham

6

Classification and Stratigraphical Relationships of the Lower Ordovician Rocks

A. J. WADGE

The stratigraphical relationship between the Skiddaw and Borrowdale Volcanic groups was vigorously debated more than a century ago (Dakyns 1869, Nicholson 1869, Aveline 1869, 1872) and it remains controversial to this day. The rocks are undoubtedly difficult to classify, but the issue has been additionally confused over the years by the lack of a precisely defined Lower Ordovician lithostratigraphy. For example, the Borrowdale Volcanic Group is not rigorously defined and therefore has often been taken to include all the older volcanic rocks of the district. Similarly, the Skiddaw Group is commonly taken to include all the older sediments. This gives a simple bipartite stratigraphy in which volcanic rocks overlie sediments, but as detailed mapping of the Lake District has progressed, this succession has become increasingly difficult to apply. On the Caldbeck Fells, for example, complex thrusting has been invoked (Eastwood et al. 1968) to explain the widespread occurrence of sediments overlying lavas and tuffs.

The use of a bipartite stratigraphy has confused discussion of the relationship between the groups, since every upward transition from a sedimentary to a volcanic sequence in the Lake District is taken to be an example of the Skiddaw–Borrowdale junction. But there are exposed within the district many types of stratigraphical relationship between sedimentary and volcanic rocks, from widely different parts of the sequence, and much of the controversy has arisen when, wrongly, these have been compared. The re-classification of the Lower Ordovician sequence began with the recognition of two distinct volcanic sequences in the Lake District (Wadge 1972, Fig. 5), the Borrowdale and the Eycott Volcanic groups (Downie and Soper 1972). The modified classification used here is shown in figure 30.

CLASSIFICATION

Skiddaw Group

The terms 'Skiddaw Slates' and 'Skiddaw Group' were first used by Sedgwick (1832), but even earlier Otley (1820) had referred to the slates of 'Skiddaw, Saddleback, Grisdale Pike and Grasmoor with most of the Newlands mountains'. The constituent formations of the Group are described elsewhere (Chapter 8). The sequence totals several thousand metres of mudstones, siltstones and turbiditic greywacke-sandstones and its base is not exposed. Graptolites from the Group are restricted to the *Didymograptus extensus* and *D. hirundo* zones and the sequence is wholly Arenig in age.

The main outcrop stretches across the northern Lake District and smaller outcrops are present in the Black Combe and Cross Fell inliers. On Black Combe, the

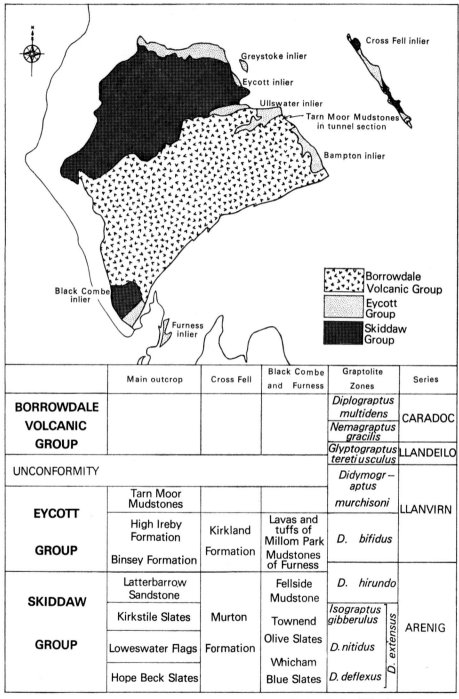

	Main outcrop	Cross Fell	Black Combe and Furness	Graptolite Zones	Series
BORROWDALE VOLCANIC GROUP				*Diplograptus multidens*	CARADOC
				Nemagraptus gracilis	
				Glyptograptus teretiusculus	LLANDEILO
UNCONFORMITY				*Didymogr—aptus murchisoni*	LLANVIRN
EYCOTT	Tarn Moor Mudstones				
	High Ireby Formation	Kirkland Formation	Lavas and tuffs of Millom Park	*D. bifidus*	
GROUP	Binsey Formation		Mudstones of Furness		
SKIDDAW	Latterbarrow Sandstone		Fellside Mudstone	*D. hirundo*	
	Kirkstile Slates	Murton	Townend	*Isograptus gibberulus*	ARENIG
GROUP	Loweswater Flags	Formation	Olive Slates	*D. nitidus*	
	Hope Beck Slates		Whicham Blue Slates	*D. deflexus*	

(the ARENIG cells also note *D. extensus* spanning the lower graptolite zones)

FIG. 30. The lower Ordovician stratigraphy of the Lake District. The classification differs from that proposed by Jackson (Chapter 7, figure 32). Jackson also prefers Kirk Stile to Kirkstile. The Llandeilo now includes most of the *N. gracilis* Zone.

rocks resemble the finer grained formations of the main outcrop, but complex folding makes it impossible to establish a stratigraphical order with certainty (Helm 1970). The record of *Tetragraptus bigsbyi* from these beds indicates a Lower Ordovician age, but the rocks cannot be dated more precisely. In the Cross Fell inlier, the Murton Formation (Burgess and Wadge 1974) includes both the striped siltstones of Murton Pike and Brownber and the flaggy greywacke sandstones around Cuns Fell. The siltstones are correlated tentatively with the Kirkstile Slates of the main outcrop and the sandstones with the Loweswater Flags.

Eycott Group

The term 'Eycott Volcanic Group' was first used when microfossil evidence showed the volcanic sequence of the Caldbeck Fells to be older than the Borrowdale Volcanic Group (Downie and Soper 1972). In this account, the name is shortened to Eycott Group. It includes all interbedded volcanic and sedimentary sequences correlating with the volcanic succession of Eycott Hill and the Caldbeck Fells. For some of these sequences the faunal and lithological evidence for correlation is good but in others, where fossils are few and the rocks are poorly exposed, correlation is less certain. Each succession is discussed below.

On the Caldbeck Fells and Eycott Hill, the Group is at least 2,600 m thick and consists of the Binsey and High Ireby formations (Chapter 8). The top of the Group is hidden beneath Carboniferous rocks, but the base is exposed in several sections. On Binsey it is probably best taken below the Lower Eycott lavas (Eastwood et al. 1968, p. 64) and in the Whitefield Cottage section [2377 3486], below the lowest mappable volcanic bed (op. cit. p. 66). On Eycott Hill (op. cit. p. 72), it lies below a 0.9 m tuffaceous sandstone in the stream section south of Fairy Knott [3841 3026]. Each of these sections shows an upward transition by intercalation from mudstones into tuffs and lavas, and it was mudstone partings near the base of the Group at Whitefield Cottage and the Overwater spillway [2590 3551] which yielded microfossils of high *D. hirundo* Zone age. It is probable that around Binsey, the lowest beds of the Group are uppermost Arenig in age and that the main part of the sequence dates from the lower Llanvirn (*D. bifidus* Zone). Supporting evidence comes from Eycott Hill where the volcanic sequence rests on mudstones yielding *D. bifidus* Zone graptolites.

The lack of thick sedimentary partings in the main part of the Binsey, Eycott Hill and Greystoke Park sections suggests that they accumulated close to a contemporary volcano, but sections of the Group elsewhere in the district suggest deposition farther from the eruptive centre. For example, the Kirkland Formation of the Cross Fell inlier contains thick mudstone sequences interbedded with massive beds of re-worked tuff and a few spilitic lavas, so this area appears to have lain off-shore from the volcanic islands. Neither the base nor the top of the formation is exposed, but at least 1,000 m of beds, yielding *D. bifidus* Zone graptolites and microfossils, are present around Wythwaite Top (Burgess and Wadge 1974).

A similar succession of interbedded graptolitic mudstones and massive, re-worked tuffs is present in the Bampton inlier. *D. bifidus* Zone graptolites have long been known from Thornship Beck and more detailed graptolite faunas and microfossils of the same age were collected from the Lanshaw-Tailbert tunnel (Skevington 1970). In the Ullswater inlier, the assignment of the 'Skiddaw Slates' underlying the low ground around the lake to the Group is more speculative. Again, the mudstones are interbedded with tuffs (Moseley 1964), but there are few sections beneath the extensive drift cover. *D. bifidus* Zone graptolites have been collected from Aik Beck and from mudstones interbedded with lavas and tuffs in Matterdale Beck (Wadge 1972), and W. C. C. Rose found graptolites 'typical of the Upper Skiddaw Slates' near Glencoyne Park (Moseley 1964, p. 128). The Tarn Moor Mudstones which were intersected by the Tarn Moor tunnel near Ullswater (Wadge et al. 1972), are younger than other beds in the Group since they yield *D. murchisoni* Zone graptolites. In the tunnel, these upper Llanvirn beds are faulted against lower Llanvirn rocks but the mudstones are not exposed at the surface and their stratigraphical relations with the rest of the Group are unknown. The Group may extend westwards from the outcrops in Matterdale Beck but there is no modern account of this largely drift-covered area. It is likely that the interbedded mudstones and tuffs cropping out below Clough Head [329 229] (Ward 1876, p. 69) also lie within the Group.

In the south-western Lake District, the rocks of the Furness inlier are poorly exposed, but graptolites of the *D. hirundo* Zone have been collected (Rose and Dunham in press) from the Askham Shale Quarry [2175 7540] and *D. bifidus* Zone microfaunas have been identified from mudstones on Hare Slack Hill [2319 7775]. Tuffs interbedded with mudstones are exposed in Bank House Gill [236 811], so there are both lithological and faunal grounds for including all the rocks of the inlier within the Group.

On the other side of the Duddon estuary, the Skiddaw Group of Black Combe is unconformably overlain to the north and east by the Borrowdale Volcanic Group, and it has been customary to regard the volcanic rocks of Millom Park as the south-westerly continuation of the Borrowdale outcrop (Green 1913). It is suggested here however that the rocks south of The Green [180 845] are better classified within the Eycott Group. They contain a high proportion of reworked tuff and appear to have little in common with the Borrowdale sequence farther north against which they strike almost at right angles. The gradational passage from mudstones into tuffs seen on Baystone Bank [1720 8577] and near Po House [1512 8271] compares with the sections through the base of the Group at Eycott Hill and Whitefield Cottage. Although no useful graptolites have been found in this area, Arenig acritarchs and chitinozoa have recently been recorded from several localities in the Whicham valley and on the lower slopes of Millom Park (Turner and Wadge in prep.).

Borrowdale Volcanic Group

The volcanic rocks forming the high fells of the central Lake District were recognized as an un-named division by Otley (1820) and termed the Green Slates and Porphyries by Sedgwick (1832). This name was replaced by Borrowdale Volcanic Series (Green 1920), which under the present rules of lithostratigraphical nomenclature is now the Borrowdale Volcanic Group. Apart from small fault-blocks in the Cross Fell inlier, it is restricted to the central Lake District, its base defined by an unconformity which overlies the Eycott Group in the east and the Skiddaw Group in the west, and the top delimited by the sub-Coniston Limestone unconformity.

The age of the Group is unknown, for the sequence is unfossiliferous. It must lie, however, somewhere between the upper Llanvirn of the Tarn Moor Mudstones and the middle Caradocian (Longvillian) of the Drygill Shales. It is customary to assign it to the Llandeilo and lower Caradoc, but volcanic deposits are generally built up comparatively quickly, and the sequence probably represents only a small fraction of that time.

STRATIGRAPHICAL RELATIONSHIPS BETWEEN THE GROUPS

The Skiddaw Group passes up into the Eycott Group without a break in sedimentation. Over most of the district the marine conditions of Skiddaw Group deposition persisted into the overlying Eycott Group although the proportion of volcanic detritus gradually increased. The boundary between the two groups is therefore arbitrary and is taken where the volcanic content of the sequence becomes significant; the uppermost part of the Skiddaw Group may also contain volcanic debris but only in minor amounts. In most places, the volcanic beds are re-worked tuffs but in the Caldbeck Fells, and in Millom Park, they also include lavas. In the Caldbeck Fells sections, the marine Skiddaw Group is succeeded by an Eycott volcanic sequence, the transition occurring by intercalation of mudstones and lavas over tens of metres of beds, but even here there is no sign of a major break in deposition.

The junction between the Borrowdales and the other two groups is more problematical and is therefore treated more fully below. A conformable passage from sedimentary into volcanic rocks was urged by Sedgwick (1832), Aveline (1869), Ward (1876) and Green (1920), and this opinion gained general acceptance despite the contrary views of Dakyns (1869), who suggested an intervening unconformity, and the preference of Goodchild (1885) for an overthrust and Marr (1916) for a lag-fault. The publication by Simpson (1967) of a revised stratigraphy and tectonics for the western Lake District revived interest in the problem and stimulated much new work. His tectonic interpretation, which envisaged pre-volcanic orogenesis, has not been generally accepted, nor has the extended lithostratigraphy which he proposed for the Skiddaw Group. His conclusion that the Borrowdales rest unconformably upon previously folded Skiddaw rocks is however correct, not only for the Buttermere area but for the whole region. The structural relationships are considered elsewhere (Chapter 5) but the stratigraphical evidence is outlined below.

The principal argument for an unconformity is provided by the regional distribution of the groups (Fig. 30). In the west, the Borrowdales rest upon Skiddaw Group rocks of Arenig age and farther east, they overlie Eycott Group rocks dating from the Llanvirn. The break is largest in the west where the Eycott rocks are completely absent. As alternatives to the regional unconformity, there appear to be two other explanations. Firstly, it has been suggested (Aveline 1869, Ward 1876) that the junction is conformable but everywhere faulted or thrust. If this was the case, the dislocation would have to be continuous across the district, either as a decollement zone or as a line of continuous faulting as shown on Ward's published 1-inch geological map (Keswick sheet, Old Series 101 SE, 1875). Faults and thrusts certainly mark the junction in many places, but there are sufficient unfaulted sections to show that the dislocations are not regionally continuous. The second explanation is that the junction is strongly diachronous, so that Eycott rocks in the east pass laterally into the lower Borrowdale succession farther west. In this view, many hundreds of metres of beds would change laterally in facies along the crop. A passage on this scale of mudstones into massive lavas or tuffs should be clear in the field but none has so far been detected. On the contrary, most unfaulted sections through the junction show a sharp transition from one group to another within a few metres of beds, even where the contact is difficult to fix precisely. It is concluded that neither of these alternatives to an unconformity is sustained by the field evidence.

Borrowdale Volcanics are in contact with older rocks across about 80 km of crop but most of it is obscured by boulder clay. Even on the high fells, the junction is commonly covered by scree from the crags of volcanic rocks above. Where exposed, the junction is commonly faulted with sufficient movement to obscure the stratigraphical relationship. Right across the district therefore, fewer than a dozen sections clearly show an undisturbed contact. These sections and the course of the junction through the district are described below.

Stratigraphical details of the sub-Borrowdale unconformity

The junction emerges from beneath the Carboniferous Limestone at the southern end of the Bampton inlier. The published 1-inch geological map (Appleby sheet Old Series 102 SW) shows Borrowdale rocks along the eastern side of the inlier, but these are now known to be Llanvirn tuffs (Skevington 1970) and are assigned to the Eycott Group. Along the south-west side of the inlier, the Ralfland Forest andesites overlie Eycott mudstones and tuffs. The junction is inclined westwards at 10 to 20 degrees, but is not exposed. Farther north, Nutt (1966, 1970) recorded exposures near Hungerhill [5145 1744], where andesites rest unconformably upon mudstones. In the north of the inlier, further evidence of unconformity is provided by the Bampton Conglomerate, which lies at, or close to, the base of the local Borrowdale succession and proved to be at least 130 m thick in the Tarn Moor tunnel (Wadge et al. 1972). It consists largely of rounded clasts of andesite and felsite set in an argillaceous matrix, but pebbles of mudstone are common in its lower part in the tunnel section and at Keldhead [4874 1970] (Nutt 1966).

In the Ullswater inlier there are no unfaulted sections through the junction (Moseley 1960, 1964). Green (1915) claimed that there was a conformable passage in Aik Beck [473 220] where mudstones with Llanvirn graptolites are succeeded by volcaniclastic sediments. He regarded the latter, the 'Mottled Tuffs', as passage beds up into the Borrowdale Group sequence, but they are now assigned to the Bampton Conglomerate and are separated from the Eycott mudstones in Aik Beck by a fault proved close by the Tarn Moor tunnel (Wadge et al. 1972). The confusion caused by the 'Mottled Tuffs' is not confined to this section, for Green applied the term to many outcrops of both Bampton Conglomerate and Eycott Group tuffs throughout the eastern Lake District. South of Great Mell Fell, the published 1-inch geological maps show Borrowdale rocks passing northwards beneath the Devonian conglomerates. It seems likely however that the Borrowdale Group rocks do not extend farther north than Matterdale End [39 23] and Priests Crag [42 23]. The scattered volcanic outcrops beyond may lie within the Eycott Group, since the base of the Borrowdale succession unconformably overlies Eycott Group rocks in Matterdale Beck [3888 2345]. This excellent section shows coarse Bampton Conglomerate resting with marked angular unconformity upon the eroded edges of steeply dipping Eycott mudstones (Wadge 1972). Sections through the unconformity between Matterdale and Keswick have recently been described (Wadge 1972, Wadge et al. 1974) and are not repeated here.

Towards Keswick the junction passes on to the outcrop of the Skiddaw Group. Along the eastern shore of Derwentwater the unconformity is thrown about 4 km to the south by a series of north-trending down-east faults. A section in the eastern bank of the Falls of Lodore [2650 1864] shows 1.1 m of conglomerate overlain by about 5 m of green–grey, fine-grained tuff and then by massive andesites. The conglomerate consists mainly of rounded mudstone clasts in a mudstone matrix with a few pebbles of green tuff. It rests upon grey Skiddaw mudstones, which are well cleaved but show no

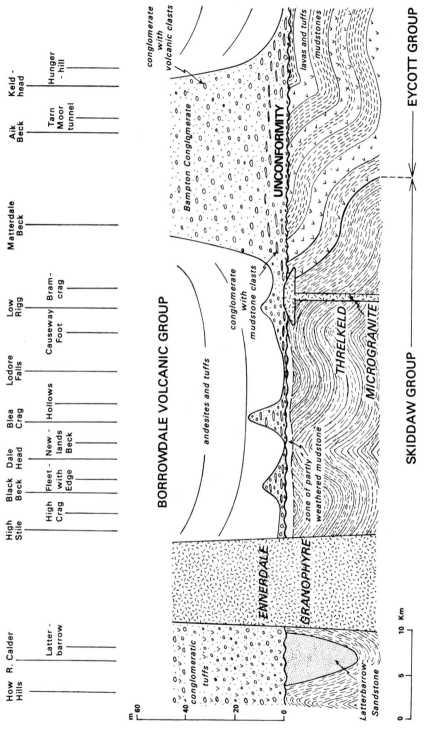

Fig. 31.   Diagrammatic section along the sub-Borrowdale unconformity.

recognizable bedding for several metres below the conglomerate. This seems to be due to pre-Borrowdale weathering of the mudstones. Similar weathering is present in several sections farther to the west but is not generally found in eastern Lakeland. Where a weathered zone is present the nature and position of the unconformity are difficult to establish, since the lithologies just above and below the junction are very similar, differing only in the amount of transport of the mudstone detritus. It is these sections that have been cited as demonstrating a conformable passage from sediments into volcanics.

The junction is well exposed on the west side of Borrowdale near Hollows Farm (Marr 1916), and on the steep hillside above towards Narrow Moor (Soper 1970, Moseley 1974). The unconformity traverses Scarbrow Wood, near Grange. Near the eastern edge of the wood [2493 1702] a low crag (Soper's locality B) shows the lowest Borrowdale beds to be massive, grey, andesitic bedded tuffs resting upon 2.5 m of conglomerate consisting of mudstone fragments in a mudstone matrix with a few tuff pebbles in the upper part. The clastics have been locally channelled and the finer debris re-worked. Again, the conglomerate is underlain by several metres of mudstone which are well-cleaved but lack discernible bedding, although crags lower in the wood show clear bedding in the Skiddaw rocks. Helm (in Mitchell et al. 1972) records tuffaceous material from the poorly bedded deposits but it is not clear whether this is derived from the thin tuff bands present in the local Skiddaw sequence (Soper 1970) or whether it was introduced into the weathered detritus at the onset of Borrowdale volcanicity. Alternative interpretations of these exposures suggest either that there is a conformable passage from sediments upwards into volcanics (Soper 1970) or that all the rocks in the crag fall within the Borrowdale succession and that the unconformity lies in the unexposed ground below (Helm and Roberts 1971).

A similar section is seen about 80 m to the west (Soper's locality C), but just beyond the western edge of the wood there is a different type of section. An eastward-trending fault, separating the two groups and throwing down to the south, can be readily traced across the fellside [2468 1698], but at its lower end just above the wood [2471 1699] the junction is again unfaulted (Jeans in Mitchell et al. 1972). Massive, green, aphanitic andesites pass down into bedded tuffs including thin bands of tuffaceous mudstone. Below the lowest mudstone lie 1.5 m of finely bedded tuff which thin rapidly westwards within a few metres. This seems to be the lowest bed in the Borrowdale succession, the conglomerate being absent. The underlying blue-grey mudstone is well cleaved, but again the bedding is obscure due to pre-Borrowdale weathering (Jeans op. cit.). Across the crag a few dips can be made out where thin paler silty bands cross fresh surfaces, and these are generally inclined at a high angle to the gently dipping tuffs above, but there is no obvious line for the junction in this exposure. It seems best to take it below the bedded tuffs.

The stratigraphical relationships are clearer higher up the fell. On the corner of the crag below High White Rake [2427 1703] at least 10 m of coarse conglomerate with prominent rounded mudstone clasts and andesite boulders crop out. Skiddaw mudstones are exposed just below the conglomerate, but the unconformity itself is obscured by scree. The conglomerate gradually thickens westwards and becomes even coarser but also includes several thin bands of sandstone and dark blue mudstone. Just above the path over High White Rake [2413 1705] the unconformity (Soper's locality h) can be continuously traced for nearly 50 m. The Skiddaw mudstones are unweathered below the break and their bedding is clear; they dip at 10 to 15 degrees more steeply than the conglomerate, which is about 8.5 m thick and is succeeded by massive andesites and tuffs. Farther up the crag the conglomerate passes into bedded tuff [2408 1709].

Along the foot of Blea Crag the unconformity is almost continuously exposed for about 250 m in a magnificent section. At the eastern end [2401 1714] the mudstones beneath the break are again weathered, although this zone is less than 1 m thick in the middle of the crag and is absent farther west [2383 1721] where the unconformity is best displayed by an angular difference of about 30 degrees. The conglomerate is 12 to 15 m thick, with mostly mudstone clasts and some volcanic boulders, up to 1 m across, set in grey, tuffaceous mudstone.

Above Castle Nook in the Vale of Newlands, the conglomerate is as thick as on Blea Crag and rests upon a 1-m zone of weathered mudstone. The clastics thin rapidly southwards along Eel Crags to about 4 m below Minum Crag [2325 1667], and are absent along the Long Work mineral vein [2317 1627], where fine-grained tuffs rest directly upon a 1-m weathered zone. In Newlands Beck [2296 1575] analysis of the different structures in the sediments and volcanics showed that the two groups are unconformable although the junction itself is obscured (Jeans 1972). The unconformity is next seen on Honister Crag where the field relations were long ago sketched by Dakyns (1869, Fig. 1). About 100 m north of the ridge above Fleetwith Edge, nearly 2 m of conglomerate rest unconformably upon unweathered siltstones [2047 1435], and on the ridge itself [2038 1429] less than 1 m of clastics is present as mudstone flakes in a tuffaceous mudstone matrix. Minor strike faulting here may cut out some of this section, but about 200 m to the south [2040 1408] 5 m of conglomerate rest unconformably upon a thin zone of weathered siltstones. The section in Warnscale Beck [2008 1353] is faulted (Moseley 1974), but in the next stream, Black Beck [1993 1344], grey silty Skiddaw Group mudstones, with the bedding picked out on fresh surfaces by paler sandy laminae, are unconformably overlain by 2.8 m of conglomerate and then by massive andesite with mudstone clasts near the base.

The unconformity skirts the foot of High Crag [1818 1456], where 3 m of conglomerate containing pebbles of bedded mudstone and andesite rest on weathered Skiddaw rocks. Unusually, the overlying andesites at this locality are inclined northeastwards into the junction and may represent local depositional dips. Traversing Burtness Combe the junction passes north of High Stile summit, where a thin conglomerate of volcanic pebbles rests on unweathered siltstones [1731 1510].

West of the outcrop of the Ennerdale Granophyre the junction crosses Kinniside Common but its position is uncertain. Between undoubted Skiddaw mudstones to the north and the Borrowdale volcanic sequence to the south lie the outcrops of the Latterbarrow Sandstone and the overlying conglomerates and tuffs (Eastwood et al. 1931). The Latterbarrow Sandstone is here assigned to the Skiddaw Group for there is some interbedding of mudstone and sandstone in the River Calder [0687 1180]. The conglomerates resemble the basal Borrowdale clastics elsewhere, and are tentatively correlated with them. The unconformity therefore probably lies below the conglomerates, but exposures are insufficient to disprove the suggestion (Simpson 1967) that it lies beneath the Latterbarrow Sandstone.

Farther south in the Black Combe inlier, sections through the junction are limited to Crookley Beck east of Bootle [115 883]. In recent years this section has been interpreted as a conformable passage (Soper 1970) and as an unconformity (Helm 1970). In the latter view, which is preferred here, a thick mudstone-conglomerate is the basal Borrowdale bed in places and elsewhere a vitric tuff lies directly upon the Skiddaw rocks.

Acknowledgement

This chapter is published by permission of the Director, Institute of Geological Sciences.

REFERENCES

ARTHURTON, R. S. and WADGE, A. J. *in press*. Geology of the country around Penrith. *Mem. geol. Surv.*

AVELINE, W. T. 1869. On the relation of the Porphyry Series to the Skiddaw Slates in the Lake District. *Geol. Mag.*, **6**, 382.

—— 1872. On the continuity and breaks between the various divisions of the Silurian strata in the Lake District. *Geol. Mag.*, **9**, 441–42.

BURGESS, I. C. and WADGE, A. J. 1974. The Geology of the Cross Fell area. H.M.S.O. London.

DAKYNS, J. R. 1869. Notes on the Geology of the Lake District. *Geol. Mag.*, **6**, 56–58; 116–17.

DOWNIE, C. and SOPER, N. J. 1972. Age of the Eycott Volcanic Group and its conformable relationship to the Skiddaw Slates in the English Lake District. *Geol. Mag.*, **109**, 259–68.

EASTWOOD, T., DIXON, E. E. L., HOLLINGWORTH, S. E., and SMITH, B. 1931. The geology of the Whitehaven and Workington District. *Mem. geol. Surv.*

—— HOLLINGWORTH, S. E., ROSE, W. C. C. and TROTTER, F. M. 1968. Geology of the country around Cockermouth and Caldbeck. *Mem. geol. Surv.*

GOODCHILD, J. G. 1885. Observations upon the stratigraphical relations of the Skiddaw Slates. *Proc. Geol. Ass.*, **9**, 469–81.

GREEN, J. F. N. 1913. *The older Palaeozoic succession of the Duddon estuary*. London.

—— 1915. The structure of the eastern part of the Lake District. *Proc. Geol. Ass.*, **26**, 195–223.

—— 1920. The geological structure of the Lake District. *Proc. Geol. Ass.*, **31**, 109–26.

HELM, D. G. 1970. Stratigraphy and structure in the Black Combe Inlier, English Lake District. *Proc. Yorks. geol. Soc.*, **38**, 105–48.

—— and ROBERTS, B. 1971. The relationship between the Skiddaw and Borrowdale Volcanic Groups in the English Lake District. *Nature Phys. Sci.*, **232**, 181–83.

JACKSON, D. E. 1916. Stratigraphy of the Skiddaw Group between Buttermere and Mungrisdale, Cumberland. *Geol. Mag.*, **98**, 515–28.

—— 1962. Graptolite zones in the Skiddaw Group in Cumberland, England. *J. Palaeont.*, **36**, 300–13.

JEANS, P. J. F. 1972. The junction between the Skiddaw Slates and Borrowdale Volcanics in Newlands Beck, Cumberland. *Geol. Mag.*, **109**, 25–28.

MARR, J. E. 1916. *The geology of the Lake District*. Cambridge.

MITCHELL, G. H., MOSELEY, F., FIRMAN, R. J., SOPER, N. J., ROBERTS, D. E., NUTT, M. J. C., and WADGE, A. J. 1972. Excursion to the northern Lake District. *Proc. Geol. Ass.*, **83**, 443–70.

MOSELEY, F. 1960. The succession and structure of the Borrowdale Volcanic rocks south-east of Ullswater. *Q. Jl geol. Soc. Lond.*, **116**, 55–84.

—— 1964. The succession and structure of the Borrowdale Volcanic rocks north-west of Ullswater. *Geol. J.*, **4**, 127–42.

—— 1974. Structural relations between the Skiddaw Slates and the Borrowdale Volcanics. *Proc. Cumb. geol. Soc.*, **3**, 127–45.

NICHOLSON, H. A. 1869. On the relations between the Skiddaw Slates and the Green Slates and Porphyries of the Lake District. *Geol. Mag.*, **6**, 105–08, 167–73.

NUTT, M. J. C. 1966. Field meeting in the Haweswater area. *Proc. Yorks. geol. Soc.*, **35**, 429–33.

—— 1970. The Borrowdale Volcanic Series and Associated Rocks around Haweswater, Westmorland. *Unpublished Ph.D. thesis, University of London.*

OTLEY, J. 1920. Remarks on the succession of rocks in the district of the Lakes. *Lonsdale Mag.*, **1**, 433.

ROSE, W. C. C. and DUNHAM, K. C. *in press*. Geology of South Cumbria and Furness. *Mem. Geol. Surv.*

SEDGWICK, A. 1832. On the geological relations of the stratified and unstratified groups of rocks composing the Cumbrian Mountains. *Proc. geol. Soc. Lond.*, **1**, 399.

SIMPSON, A. 1967. The stratigraphy and tectonics of the Skiddaw Slates and the relationship of the overlying Borrowdale Volcanic Series in part of the Lake District. *Geol. J.*, **5**, 391–418.

SKEVINGTON, D. 1970. A Lower Llanvirn graptolite fauna from the Skiddaw Slates, Westmorland. *Proc. Yorks. geol. Soc.*, **37**, 395–444.

SOPER, N. J. 1970. Three critical localities on the junction of the Borrowdale Volcanic rocks with the Skiddaw Slates in the Lake District. *Proc. Yorks. geol. Soc.*, **37**, 461–93.

TROTTER, F. M., HOLLINGWORTH, S. E., EASTWOOD, T., and ROSE, W. C. C. 1937. The Geology of Gosforth. *Mem. Geol. Surv.*

WADGE, A. J. 1972. Sections through the Skiddaw-Borrowdale unconformity in eastern Lakeland. *Proc. Yorks. geol. Soc.*, **39,** 179–98.

—— NUTT, M. J. C. and SKEVINGTON, D. 1972. Geology of the Tarn Moor tunnel in the Lake District. *Bull. geol. Surv. Gt. Br.*, **41,** 55–73.

—— HARDING, R. R. and DARBYSHIRE, D. P. F. 1974. The Rb/Sr age and field relations of the Threlkeld Microgranite. *Proc. Yorks. geol. Soc.*, **40,** 211–22.

WARD, J. C. 1876. The geology of the northern part of the English Lake District. *Mem. geol. Surv.*

A. J. WADGE, M.A.

Institute of Geological Sciences, Ring Road, Halton, Leeds LS15 8TQ.

7

The Skiddaw Group

DENNIS JACKSON

The Skiddaw Group are the oldest rocks exposed in the Cumbrian Mountains and crop out in four inliers. The largest is the Skiddaw Inlier, and extends from Cleator Moor in the west to Troutbeck in the east. Smaller inliers occur around Ullswater, Bampton and Black Combe. In recent years, interest has shifted from the main inlier to these smaller inliers and very important advances in knowledge have come from Helm (1970), Skevington (1970) and Wadge, Nutt and Skevington (1972) and some of their findings are described towards the end of this chapter. Notwithstanding modern accounts of these smaller inliers, it is the Skiddaw Inlier which forms the focal point of this chapter because in that area the stratigraphical succession is more complete and the palaeontological control is relatively good.

The subdivision and correlation of various formations within the Skiddaw Group has been controversial for more than a century. One hundred years ago, J. C. Ward (1876) worked out a stratigraphical succession whilst preparing a geological map (Sheet 101 SE) of the northern part of the Lake District. Subsequent to Ward's pioneering efforts, Dixon (1925), Eastwood et al. (1931), Eastwood (1933), and Rose (1955) advanced our understanding of the succession and distribution of rock units through the application of 'way-up' criteria and palaeontology, so that the number of 'formations' was reduced from five to two (Table 2) as it became clear that some of Ward's units were facies equivalents of others.

Since 1955, the trend towards simplification of the stratigraphical nomenclature has been reversed. Jackson (1961, 1962) recognized a new formation, below the Loweswater Flags which he called the Hope Beck Slates, whilst Simpson (1967) reverted to Dixon's (1925) five subdivisions plus three new formations (Table 2). Unfortunately, this latter work almost totally ignored a vast amount of graptolite data accumulated by earlier workers as well as the metamorphic effects of igneous intrusions believed to extend at depth between the Eskdale and Skiddaw Granites.

To this day, there remain tracts of Skiddaw rocks whose ages are debatable and whose stratigraphical assignation is problematical. For example, the 'grits' which crop out east of Cockermouth on Watch Hill, Elva Hill, Great Cockup, Burn Todd and Great Sca Fell have been assigned ages ranging from Precambrian (Rastall 1910) to late Ordovician (Green 1917). Similar arenaceous strata are exposed on Robinson, east of Buttermere and in Newlands. Graptolites collected by Survey geologists and by Jackson (1956) suggest that all these arenaceous strata correlate with the Loweswater Flags and are of early Arenig age. Other issues that arise out of Simpson's (1967) paper concern the validity of subdividing the Mosser–Kirkstile Slates (*sensu* Rose 1955) into Blakefell 'Mudstone', Kirkstile Slates, Mosser Slates and Sunderland Slates. The Blakefell Mudstone was established by Dixon (1925) as the oldest mappable

TABLE 2

Changes in stratigraphical nomenclature and ordering of formations in the Skiddaw Group in the Skiddaw Inlier during the last hundred years.

Ward (1876) Whitehaven to Mungrisdale	Dixon (1925) Ennerdale to Loweswater	Eastwood et al. (1931) North and south of Ennerdale	Eastwood et al. (1933) Cockermouth area
Black slates of Skiddaw	Mosser stiped Slates	Latterbarrow Sandstone	Kirk Stile Slates
Gritty bed of Gatesgarth, Latterbarrow, Tongue Beck, Watch Hill and Gt Cockup	Loweswater Flags Kirkstile Slates Blake Fell Mudstones	Mosser Slates with Kirkstile Slates and Watch Hill Grits Loweswater Flags Blakefell Mudstones	Mosser Slates Skiddaw Grits
Dark slates			
Sandstones of Grasmoor and Whiteside			
Dark slates of Kirk Stile			

Rose (1955) Keswick–Buttermere	Jackson (1961) Lorton–Mungrisdale	Simpson (1967) Egremont–Bassenthwaite	Jackson, this paper Skiddaw Inlier
Mosser–Kirkstile Slates	Mosser–Kirkstile Slates	Latterbarrow Sandstone	Latterbarrow Sandstone
Loweswater Flags	Loweswater Flags Hope Beck Slates	Sunderland Slates Watch Hill Grits Mosser Slates Loweswater Flags Kirkstile Slates Blakefell Mudstones Buttermere Flags Buttermere Slates	Kirk Stile Slates Loweswater Flags Hope Beck Slates

facing page

PLATE 6.   A. Fell country underlain by Borrowdale Volcanic rocks around Wasdale. Wast Water screes are seen in profile, with Kirk Fell, Great Gable, Lingmell and Scafell in the background.   *Photograph: Aerofilms Ltd*

B. View south-east from Great Gable. In the foreground (below Sprinkling Tarn) the strong bedding of the Seathwaite Fells Tuffs (Oliver 1961) is clearly visible. Immediately above Sprinkling Tarn the more massive Lincomb Tarns Ignimbrite can be seen, with the low rise of Allen Crags occupied by the bedded tuffs of the Esk Pike Hornstone. The path to Sprinkling Tarn and the gully to the right follow faults which converge across the col and descend Rossett Gill. Esk Hause is on the right, the Langdale Pikes on the left and Windermere is in the distance.   *Photograph: Sanderson & Dixon, Ambleside*

A

B

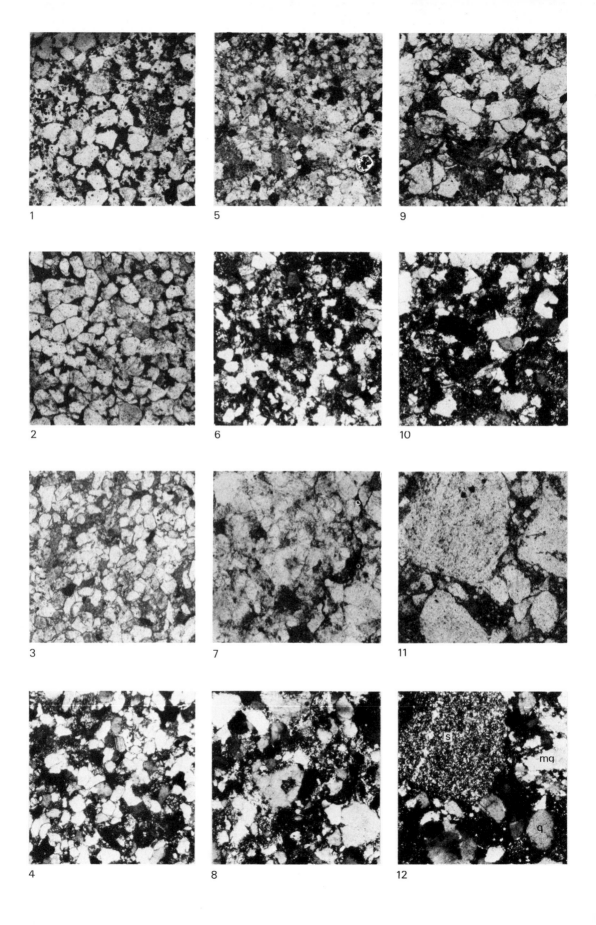

unit between lower Ennerdale and Loweswater. Although he recognized that it had been subsequently metamorphosed he believed that it represented a mappable formation which was laterally equivalent to the Loweswater Flags. Rose (in Hollingworth et al. 1955, p. 405) and Jackson (1961, p. 526) demonstrated that the unit represents low-grade thermally metamorphosed 'Mosser–Kirkstile Slates' and for this reason omitted it from the sequence (see also Eastwood et al. 1968, p. 23).

These different interpretations arise from the structural complexity, poor exposure, lack of distinctive marker horizons and the capricious distribution of graptolites. Many of the subdivisions proposed by Ward (1876) and Simpson (1967) are ill-defined and their stratigraphical relationships are obscured either by faults and/or by deeply glaciated strike valleys. Graptolites are the principal fossils of stratigraphical value. Unfortunately, they are rarely abundant and are not always zonally definitive, but they do allow most of the uncertainties about the rock succession and the number of valid subdivisions to be removed, and it is to this end that this contribution is largely directed.

LITHOLOGICAL SUBDIVISIONS

The Skiddaw Group embraces all those mudstones, siltstones, sandstones and conglomerates that lie beneath the Borrowdale Volcanics (Fig. 32). These rocks form a turbidite suite of sediments deposited in relatively deep water on the SE margin of proto-Atlantic Ocean. They are fossiliferous in places and have yielded more than 60

facing page

PLATE 7.

(1) Latterbarrow Sandstone from Tommy Gill [137 334], Redmain. Note well-sorted quartz grains, some quartz overgrowths and disseminated grains of iron oxide infilling pore spaces. Ordinary light.

(2) Latterbarrow Sandstone from Blakeley Raise [075 130] south to Ennerdale also showing iron oxide in the pore spaces. Ordinary light.

(3, 4) Turbidite in Loweswater Flags below felsite lava on Watch Hill [150 318]; under ordinary light and crossed-nicols respectively. Note well-sorted fine grained texture contains lithic clasts, twinned feldspars and abundant quartz grains set in a matrix of sericite mica.

(5, 6) Pale grey, parallel-laminated, well sorted, fine grained sandstone in Loweswater Flags from Jonah's Gill [191 341], near Sunderland; under ordinary light and crossed-nicols respectively.

(7, 8) Greywacke from Loweswater Flags in Scaw Gill Quarry [177 258], Lorton; under ordinary light and crossed-nicols respectively. Note poor sorting, ill-defined grain boundaries due to sericite-quartz intergrowths and stylolite.

(9, 10) Greywacke from Loweswater Flags on Robinson Crags [214 176]; under ordinary light and crossed-nicols respectively.

(11, 12) Basal portion of a massive graded greywacke within Loweswater Flags on Robinson Crags [214 176]; under ordinary light and crossed-nicols respectively. Note grains of siltstone (s), quartz (q) and quartzite (mg). *Photomicrographs: D. Jackson* species of graptolites and an abundance of *Cariocaris* as well as a few trilobites, *Lingula*, and trace fossils. In the main outcrop, the Group can be divided into the formations shown on Table 2 (Jackson).

Hope Beck Slates

The Hope Beck Slates were defined by Jackson (1961) as the dark bluish–black and grey striped, silty mudstones that crop out in Hope Beck 2 km SSE of High Lorton (Fig. 33). Here, on the north and west slopes of Dodd [166 233] and in Cold Gill [166 299] the formation is seen to be about 80 m thick although the base is never seen. They seem to have been laid down in quiet water. Upwards they grade into basal Loweswater Flags by a gradual increase in sandy intercalations. An arbitrary boundary was drawn on the east side of Dodd at about 1,100 ft OD but the boundary descends to about 900 ft OD on the west side. The formation is traceable NE from the type locality to High Swinside farm [169 266] and then to Blaze Beck [178 256] south of Scaw Gill Bridge. On the north side of Scaw Gill Bridge the unit is faulted against

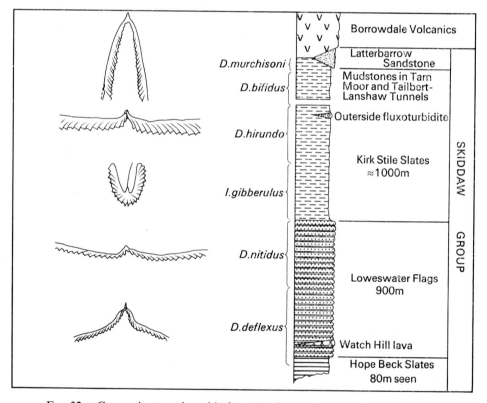

Fig. 32. Composite stratigraphical succession, of graptolite zones and relevant zonal designates in the Skiddaw Group. The upper limit of the Skiddaw Group shown here differs from that proposed by Wadge (Chapter 6), who inserts an Eycott Group (which includes the Tarn Moor Mudstones) between the Skiddaw and Borrowdale Groups.

Loweswater Flags but the transition beds are again seen in the north bank of Aiken Beck [179 259] between the bridge and Spout Force. They yield *D. deflexus* Zone faunas at Aiken Beck, Swinside and Dodd (Jackson 1962, p. 306).

The stratigraphical position of the Hope Beck Slates resembles that of the 'Dark Slates of Kirk Stile' in Ward's 1876 scheme (Table 2). Unfortunately, Ward did not designate a type locality for these rocks and we can only guess that he was referring to the dark striped mudstones in High Nook Beck [128 205] south of Loweswater. These mudstones were subsequently shown by Dixon (1931) to contain a *D. hirundo* fauna which places them above the Loweswater Flags, and meant that a new name was required for this formation. At the time of its formal definition, the author pointed out that knowledge of its distribution was hampered by the problem of distinguishing it from unfossiliferous 'Mosser–Kirkstile Slates'. This seemed the only reasonable explanation for the very restricted occurrence of the formation.

In Simpson's (1967) paper the Hope Beck Slates were subsumed in four other formations but this proposal has not been generally accepted. Paradoxically, however, some fossil data contained in Simpson's paper validates the Hope Beck Slates and will be discussed in a later section.

Loweswater Flags

Loweswater Flags is the name Dixon (1925) gave to a thick sequence of alternating grey flaggy sandstones and grey or greenish mudstones with occasional grits or fine conglomerates which crop out between Keltonfell Top north of Ennerdale and Loweswater. Both Rose (in Hollingworth et al. 1955, p. 390) and Jackson (1961, p. 518) equated this formation with the grits and sandstones on Watch Hill and Great Sca Fell on the basis of lithological and positional similarity (Figs 33 and 34). Subsequent palaeontological evidence confirms this view and shows that all the upfolded and upfaulted outcrops of arenaceous beds (excluding the Latterbarrow Sandstone) are of *D. deflexus* and *D. nitidus* zone ages.

Both the lower and upper boundaries of the formation are gradational. The lower boundary is best seen on the east slope of Dodd at about 1,100 ft OD [171 233] and at 900 ft OD on the west side [166 231] and is drawn arbitrarily at the lowest occurrence of sandstone interbeds which are more than 7 cm thick. Similarly, the upper boundary is drawn where the sandstones are thinner than 7 cm, and can be traced from Whiteless Pike to Grisedale Pike and Thornthwaite. These flags form a flysch facies containing mudstone, siltstone, sandstone, grit and conglomerate with an estimated thickness of about 900 m between Scaw Gill and Grisedale Pike. The unit is characterized by rhythmic alternation of grey or greenish–grey sandstones and striped bluish-black mudstone, the former ranging in thickness from a few centimetres up to 12 m. The sandstone–mudstone ratio varies vetween 1:4 and 4:1 with a 1:3 ratio most common. No systematic geographic variation has been established in thickness, sandstone–mudstone ratio or in grain size. Sandstones as coarse as those found on Watch Hill and Great Sca Fell are also found in Gasgale Gill [164 209] and Blea Crag, Little Dale [213 177]. The Loweswater Flags south of Newlands Pass tend to be more massive than elsewhere and the frequency and intensity of current-bedding, convolute-lamination and flute-marks increases southwards.

Small scale sedimentary structures found in the sandstones are typical of turbidity-current deposits. They include flute-marks and load casts on the undersides of beds and parallel lamination, cross-lamination, convolute-lamination and graded-bedding

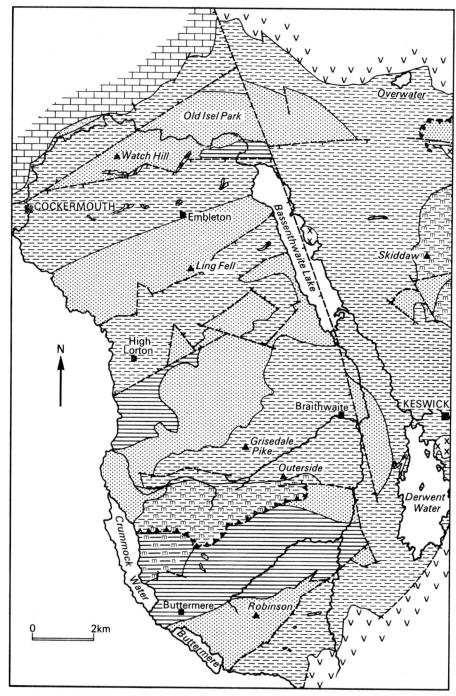

FIG. 33.   Distribution of formations in the Skiddaw Inlier between the Vale of
Lorton and Keswick. See legend in Text figure 3.

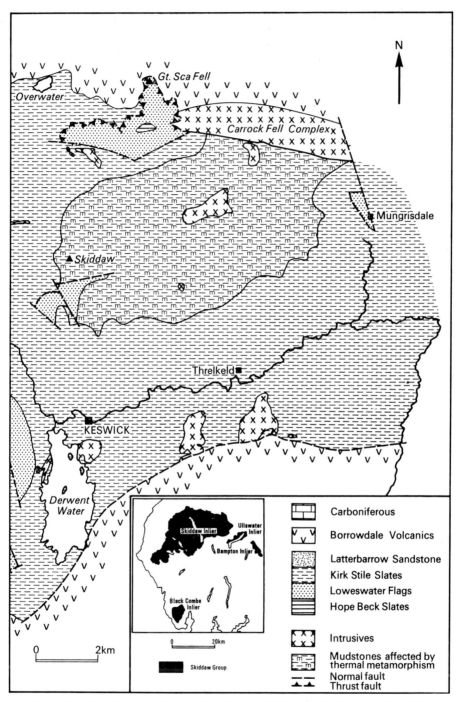

FIG. 34.   Distribution of formations in the Skiddaw Inlier between Keswick and Mungrisdale.

within beds, and have proved valuable in determining the 'way-up' of beds. In arenites 10 cm or more thick, these sedimentary structures commonly occur in a regular ascending sequence as follows (1 at the bottom):

7. Mudstone lying conformably upon 6.

6. Undisturbed cross-laminated silt laid down in the lower flow regime.

5. Convolute-bedding which may grade into 6 or may be separated from it by a surface of erosion.

4. Parallel horizontal laminae (with or without current-bedding grading upwards into 5) deposited under shooting flow.

3. Graded unlaminated coarse sand in basal part of arenite, perhaps deposited as antidunes under an upper flow regime.

2. Flute-marks, tool-marks or load-casts with intervening flame-structure on underside of arenite.

1. Surface or erosion cut in underlying mudstone.

Many variations on this motif have been noted and features 1, 2 and 6 are often absent. Similar sequences of associated sedimentary structures have been observed in turbidites elsewhere, e.g. Aberystwyth Grits (Wood and Smith 1959, pp. 170–72 and Martinsburg Slates of U.S.A. (van Houten 1954, p. 815).

Sole markings are not abundant in the Loweswater Flags and the list of types ranked by order of abundance comprise flute-casts, load-casts, groove-casts and bounce-casts. Jackson (1961, p. 521) has observed a total of 96 flute-casts from Scaw Gill Quarry [177 258], Gasgale Gill [164 209] and Mungrisdale [362 307] which collectively point to bottom currents flowing to the NNE at the time of their formation. Observations on the inclination of anticlinally-folded laminae in convoluted laminations seems to corroborate that these slope deposits were inclined northwards.

Thin-section studies of coarse grained beds show that the largest clasts are locally derived mudstones up to 5 cm long. Pebbles from an extraneous source tend to be well-rounded, poorly sorted and include sandstone, siltstone, chert, ortho-quartzite, granite, granophyre, quartz porphyry, rhyolite, brown andesitic glass, low-grade hornfels, gneissose and schistose rocks. Heavy mineral separations yielded purple, brown and colourless zircon, brown tourmaline, pale pink and colourless almandite garnet and orange–brown rutile. The arenites are generally well-cemented with secondary silica and the clays of the matrix have undergone diagenetic alteration to a mixture of chlorite and sericite (Pl. 7.7 and 7.8). These sediments fit the general characteristics of basin-floor turbidites in that individual beds extend the width of the largest outcrop, the current direction deduced from flute-marks appears to have been uniform and a suite of sedimentary structures arranged in Bouma sequences is present (Bouma 1962). The fact that such structures have been preserved suggests that the depth of water was such as to prevent reworking of the turbidites by traction currents. Finally, there is the lack of shelly fauna which may be due to either the depth of water or the rate of sedimentation being too rapid to permit proliferation of benthic organisms.

Although the coarse-grained varieties are poorly-sorted (Pl. 7.7 and 7.8) many finer-grained beds appear well-sorted (Pl. 7.3–7.6) and may represent deep-sea sands deposited by ocean bottom currents (Hubert 1964).

Contemporaneous volcanicity

The only record of contemporaneous volcanicity within the Skiddaw Group occurs in this formation 960 m west of the summit of Watch Hill [149 317] where a felsite lava flow is well exposed in an old quarry (see Eastwood et al. 1968, plate II). The lava has a maximum thickness of 10 m and outcrops for about 500 m. It consists of a pinkish–grey, microcrystalline felsite with well-developed columnar jointing. Small pockets of locally derived sandstone conglomerate lie upon the upper surface of the lava flow (Jackson 1961).

Kirk Stile Slates

The Kirk Stile Slates comprise black, blue–grey to blue–black striped silty mudstones which were first mapped in the area between Loweswater and Crummock by Ward (1876) and later recognized by Dixon (1925, p. 70) between Lower Ennerdale and Loweswater. Both workers originally considered the Kirk Stile Slates to be older than the Loweswater Flags but Dixon (in Hollingworth et al. 1931, pp. 14, 34) later believed that the Kirk Stile and Mosser Slates were facies equivalents. The reason for this correlation was Dixon's discovery that the Kirk Stile Slates in High Nook Beck, Loweswater contained *Isograptus gibberulus and Didymograptus hirundo*, typical of the 'Mosser Slates' of Mosser 6 km south of Cockermouth. This synonymy was subsequently referred to by Rose (1955, p. 405) and Jackson (1961, 1962). There is thus no justification for retaining the cumbersome term 'Mosser–Kirkstile Slates' and since the 'Dark slates of Kirk Stile' of Ward (1876) has priority of publication over the 'Mosser Slates' of Dixon (1925) that name is used throughout this Chapter.

The lower boundary is not well exposed in the type area; it is, however, easily followed between Thornthwaite and Whiteside. The writer believes that over most of the region the upper boundary of the Kirk Stile Slates corresponds with the base of the Borrowdale Volcanics except at Redmain NE of Cockermouth, where the Latterbarrow Sandstone crops out (Fig. 33). Rose (1955) has estimated a thickness of 750 m for these pelitic rocks, a figure which agrees with the writers calculations on Saddleback and on Southerfell. Survey geologists have attempted to distinguish a Kirk Stile facies (black silt-free mudstones) and a Mosser facies (bluish–grey silty striped mudstones) but experience has shown that such a distinction does little to advance mapping of the slates. Apart from small-scale cross laminations in the siltstone inter-beds, sedimentary structures are rare. Intra-formational mudstone breccias near the top of the unit are exposed 270 m east of Outerside [213 214] and on Narrow Moor at 1,800 ft OD NE of Castle Nook. They are very localized deposits and probably lie at different stratigraphical levels. According to Stanley and Unrug (1972, p. 303), till-like pebbly mudstones (fluxoturbidites) similar to these are known in flysch deposits of many mountain chains and represent arrested flows which just surpassed their liquid limits, but regained cohesion thixotropically before a true tubidity current developed. Stanley and Unrug regard such deposits as diagnostic of slope deposits.

Latterbarrow Sandstone

This formation was first defined by Eastwood et al. (1931) on Latterbarrow [073 114] SW of Ennerdale. In this area there are 200–300 m of clastics which on Boat How [087 104] were described by Trotter et al. (1937) as quartzitic sandstones which vary in colour from bottle-green to reddish brown or grey (Pl. 1, Fig. 1). According to

Shackleton (1975, pp. 35–38) the Latterbarrow Sandstone is interdigitated with dark shales of the Skiddaw Group beneath. Similarly, Hollingworth (in Eastwood et al. 1931, p. 44) and Shackleton (op. cit.) described a conformable gradation upwards into the overlying Borrowdale Volcanic Group near Caple Crag Beck [072 102]. Eastwood (1927, p. 53), working in the Cockermouth area, recorded a sandstone in Redmain Gill which, on the basis of stratigraphical position, he thought may be equivalent to the Latterbarrow Sandstone. In their survey of the Whitehaven and Workington area, Eastwood et al. (1931) retained the Latterbarrow Sandstone as the uppermost formation in the Skiddaw Group and this scheme was followed by Jackson (1961, p. 526). A different interpretation was advocated by Simpson (1967) namely that the Latterbarrow Sandstone rests with unconformity upon the Skiddaw Group and should be classed as the basal formation of the Borrowdale Volcanic Group.

The sandstone in Redmain Gill (Pl. 2.1) comprises 50 m of uniformly fine grained, parallel laminated, grey, flaggy quartzite devoid of shale interbeds except in Sunderland Gill where they probably grade downwards into Kirk Stile Slates. Microscopically, this rock is a fairly well sorted sandstone having subangular quartz grains with overgrowths of silica and a matrix with an abundance of iron oxide and small amounts of chloritic material and sericite (Pl. 2.1). It seems indistinguishable from some specimens of the formation in the type area (Pl. 2.2).

Conditions of deposition of the Skiddaw Group can be viewed as predominantly a period of pelagic sedimentation in which the variable factor was the input of coarse clastics from the south by way of turbidity currents. More than half the sedimentary pile accumulated through slow settling of clay and silt on to the floor of the proto-Atlantic (Iapetus). This was overprinted by an influx of arenaceous sediment during most of *D. extensus* time which formed the Loweswater Flags. These represent flysch sediments laid down as slope and/or basin floor tubidites that were derived from a land mass along the southern margin of the proto-Atlantic where Cambrian and/or Precambrian rocks were being eroded. Further work may show that Loweswater Flags in the northern part of the inlier represent distal turbidites in contrast to those of the Buttermere area which may be proximal turbidites. A southerly source for the Latterbarrow Sandstone also seems probable from a comparison of grain size distribution between Kinniside Common and Redmain.

PALAEONTOLOGY

Although mudstones of the Skiddaw Group yield a few genera of trilobites and brachiopods as well as the ubiquitous *Cariocaris* spp., our detailed knowledge of the stratigraphical succession and ages of these rocks comes exclusively from the graptolites.

It was largely through Marr (1894) and Nicholson and Marr's (1895) works on Skiddaw graptolites that the principle of progressive stipe reduction was formulated upon Marr's (1894, p. 122) recognition that *Bryograptus* beds were overlain by *Dichograptus* beds. This principle dominated biostratigraphical interpretation for four decades and received added support when Elles (1933) inserted a lower subzone of *Tetragraptus* (horizontal spp.) and an upper subzone of *Tetragraptus* (reclined spp.) at the top of the *Dichograptus* Zone and the base of the *D. extensus* Zone respectively. Thus the phylogeny for horizontal dichograptids was expressed by the series *Clonograptus-Loganograptus-Dichograptus-Tetragraptus-Didymograptus*. Both the zonal scheme and the phylogeny it was founded upon were subsequently assailed by Dixon (1931), Bulman (1941), and Thomas (1960). During a detailed study of the geology of the region between the Lorton–Buttermere Valley and Mungrisdale, Jackson (1962) revisited all of the classic localities and recorded 196 new ones. From this study there emerged an unambiguous picture of the distribution of graptolite zones and subzones of Arenig age and showed conclusively that the pre-*D. deflexus* zones were bogus. Jackson (ibid., p. 310) concluded that the lowest stratigraphic occurrence of graptolites between the Lorton Fells and Outerside, and probably the entire inlier, is in basal beds of the Loweswater Flags and lie in the *deflexus* Subzone. He went on to say: 'below this horizon the Hope Beck Slates have so far proved unfossiliferous, and the writer believes that there is no evidence for the existence of the Upper Subzone of *Tetragraptus* (reclined), the Zone of *Dichograptus* or the Zone of Bryograptus at the localities cited by Elles ————. It was found that in every case where Elles's locality of occurrence could be visited, careful collecting brought to light faunal assemblages (that) indicated a zonal position not older than the *Didymograptus extensus* Zone'.

A further anomaly that required explanation concerned old records of species of *Bryograptus* and *Clonograptus tenellus* Linnarsson by Elles and Wood (1901–18) from screes at Barf, Bassenthwaite Lake. Elsewhere in England and Norway, these particular species are regarded as truly Tremadoc graptolites and it was their occurrence on screes below outcrops in the *extensus* Zone that led Bulman to investigate these elements of the fauna. In 1941, Bulman re-allocated two of the bryograptid species (*B. callavei* and *B. divergens*) to his new genus *Adelograptus* and postulated that there were no beds of Tremadoc age in the Skiddaw inlier. And in 1971, Bulman reassigned the one remaining bryograptid — *B. kjerulfi cumbrensis* (Elles) — to the Arenig graptoloid genus *Pseudobryograptus*. He also discredited Elles and Woods (1901–18, Pl. 12, Fig. 2e) reference to *Clonograptus tenellus* Linnarsson from these same beds. Thus a revised list of graptolites from Barf looks indisputably early Arenig.

Arenig graptolite zones between the Lorton Fells and Outerside

The best place to view the zonal sequence is where the stratigraphic succession is most nearly complete and the structure is relatively well understood. Such an area exists between the Lorton Fells and Outerside, west of Keswick and can be considered the type area for the Skiddaw zones of Arenig age. Here, in the heart of the inlier, the structure comprises a number of small flexures superimposed upon the northern limb of a major syncline, the Braithwaite Syncline whose axis trends NE–SW about Braithwaite and plunges gently NE. One consequence of the southerly regional dip is that the oldest beds crop out in the Lorton Fells in the north and the younger Arenig strata crop out in the south. From north to south, the stratigraphical succession reads: Hope Beck Slates, Loweswater Flags and Kirk Stile Slates (Fig. 33). The reason why these older beds are widely exposed here and not so widespread to the west or to the east, except in a thrust sheet on Gt Sca Fell, is because erosion has proceeded to a lower structural level in this fault-bounded block than elsewhere.

The sources of much of the graptolite material collected over the past hundred years or so have been from quarries in the Loweswater Flags and from screes below outcrops of Kirk Stile Slates. Some of the screes such as Barf, Randel Crag and Outerside were known to the early collectors like Wright, Dover, Postlethwaite and Harkness. In defence of the use of scree material it is argued that it is firstly necessary, because attempts to hammer graptolites from outcrops seldom reward the effort, and secondly, permissible because of the great thickness of the individual zones and sub-zones. From this, however, it will be obvious that it is not possible to accurately map zonal boundaries or to determine the vertical ranges of species with anything like the degree of exactness demonstrated in the Silurian of the Howgill Fells by Rickards (1970).

A development of great importance during the 1950s which led to a new surge of collecting was the afforestation programme which resulted in the construction of many kilometres of new access roads. Many of these roads were cut in hillsides to the east of Lorton and contributed greatly to our knowledge of the local geology. Unfortunately, most of these new exposures were short-lived as banks and cuttings quickly became overgrown with grass.

Within the area under discussion it is possible to demonstrate the true metamorphic nature of the beds that have been mapped by Simpson (1967) as Blake Fell Mudstone. In a traverse from the summit of Outerside [211 215] southwards to the head of Stoneycroft Gill [211 211] one can observe a gradual change from bluish–black Kirk Stile mudstones to paler, discoloured spotted hornfels signifying the northern boundary of a thermal metamorphic aureole which trends eastwards from Crummock Water (Fig. 33). This aureole, first recognized by Dixon (in Eastwood et al. 1931, p. 39) in an area north of Ennerdale and subsequently referred to by Rose (1955, p. 405) and Jackson (1961, p. 526), was overlooked by Simpson (1976). The postulated existence of a large granitic body at depth beneath the aureole receives support from a recent gravity survey by Bott (1974, p. 319). Bott discovered a low Bouguer anomaly across much of the central and northern parts of the Lake District and interpreted this as a granite batholith. One of the ridges of granite is believed to lie below this aureole. Not surprisingly, graptolites are never seen where the mudstones are discoloured and consequently these beds are difficult to zone.

The description of the zones in this area will form a basis for the discussion of the zonation (Fig. 32) of the Skiddaw strata throughout the inlier in general. In this account, the *Didymograptus extensus* Zone with its three subzones is retained because there is little evidence that the subzones of *D. deflexus*, *D. nitidus* and *I. gibberulus* are of general application throughout the rest of the Atlantic Province.

Zone of *Didymograptus extensus*

The faunas of this zone are characterized by the presence of slender extensiform didymograptids of which the eponymic species *Didymograptus extensus* is but one. The *extensus* Zone is of the order of 1,200 m thick and can be subdivided into the following three subzones:

Subzone of *Isograptus gibberulus*
Subzone of *Didymograptus nitidus*
Subzone of *Didymograptus deflexus*

Subzone of *Didymograptus deflexus*

The fauna of the subzone is characterized by deflexed and declined didymograptids like *D. deflexus* (Fig. 1), *D.* aff. *D.* v-*fractus* and forms related to *D. balticus* (see Jackson 1962, pp. 306–07).

The *deflexus* Subzone embraces the lower half of the Loweswater Flags and probably exceeds a thickness of 300 m. The base of the subzone in the area around Lorton coincides approximately with the base of the Flags for in this area the underlying Hope Beck Slates have so far yielded no graptolites. However, collecting by Temple (in Simpson 1967) shows that the subzone extends down into the underlying formation in the Buttermere area.

The subzone is best exposed along the north bank of Aiken Beck in a number of small quarries (Jackson 1962, p. 306) once worked for dry-wall stone a few metres north and NE of Scaw Gill Bridge, Lorton. Scaw Gill quarry, north of this bridge, contains a bench of sandstone which is covered with a tangled mat of *D. deflexus*. About 500 m NE of the Bridge the subzone is faulted out by a NW–SE trending fault. Southward from Scaw Gill, the *deflexus* Subzone crops out in two small quarries 1 km NE of High Swinside Farm [179 250] and again on the eastern slopes of Dodd [171 231]. Elsewhere in the area, the *deflexus* Subzone is recognized 300 m north of the summit of Barf [215 270] where it appears to directly underlie the *nitidus* Subzone which crops out on the south side of Barf.

East of the Bassenthwaite Valley the *deflexus* Subzone has long been known (Elles 1933) to exist in a fault-bounded horst of Loweswater Flags that crop out in Raven Crags, Mungrisdale [363 306]. Here a roadside quarry 60 m NNE of Mungrisdale School yielded an *in situ* assemblage containing *Didymograptus* cf. *D. deflexus* (see Elles 1933, Fig. 13) and *D.* aff. *D.* v-*fractus*.

Subzone of *Didymograptus nitidus*

The first appearance of the subzonal designate *D. nitidus* (Fig. 32) in the upper part of the Loweswater Flags is taken as the base of the *nitidus* Subzone. However, the species is long ranging and occurs as high as the upper part of the *hirundo* Zone. The subzone can be defined as that part of the range of *D. nitidus* which is not overlapped by *Isograptus gibberulus*.

The *nitidus* Subzone spans the upper part of the Loweswater Flags and is about 400 m thick. It is best exposed on the southern slopes of Barf below Slape Crag (Jackson 1962, p. 306) and it was from these screes that Marr and Elles collected assemblages which led them to formulate the pre-*extensus* zonal scheme that has since been discredited. The style of folding seen below Slape Crag is quite chaotic and appears to have been misinterpreted by Elles (1933) as indicative of thrust faulting. It now seems reasonable to regard these folds, many of which have large voids along their axial crests, as being the result of post-glacial gravity sliding. The subzone can be traced via numerous small screes across Broom Fell, Todd Fell and Aiken and is probably present on Hobcarton End (see Jackson 1962, Text fig. 3). Other isolated occurrences on the margins of this area are a small quarry in Stoneycroft Gill, 1 km west of Stair and Hodgson How Quarry, Portinscale (Jackson 1962, p. 306).

Subzone of *Isograptus gibberulus*

The subzone of *I. gibberulus* (Fig. 32) is defined as that part of the *gibberulus* range-zone which is not overlapped by *Didymograptus hirundo*. The subzone is characterized by a preponderance of *Phyllograptus* spp., mutants between *D. nitidus* and *D. hirundo* and *Glyptograptus dentatus*.

The *gibberulus* Subzone occurs in the lower part of the Kirk Stile Slates and may be of the order of 500 m thick. It is widely distributed between Whiteside and the village of Thornthwaite and has been identified on Hobcarton, Grisedale Pike, Heavy Sides, Comb Gill and Knott Head. In the Lorton Fells, the Subzone has been recorded from Lords Seat and Aiken (see Jackson 1962, pp. 304–05).

Zone of *Didymograptus hirundo*

The assemblage of the *hirundo* Zone is characterized by the presence of *Didymograptus hirundo* (Fig. 32) which is restricted to the zone and also by the abundance of *Glyptograptus* spp., *Cryptograptus antennarius*, and *Tristichograptus ensiformis*. Bulman (1968) also showed that the zone contained the earliest records of *Climacograptus* cf. *C. biformis* and *Pseudoclimacograptus cumbrensis*.

The *hirundo* Zone embraces the upper part of the Kirk Stile Slates (Fig. 32) in the main inlier and may have a thickness of 500 m. The Zone is known to exist only around Outerside on four small screes along the 1,500 foot contour north and east of the summit (see Jackson 1962, pp. 303–05). The distribution of rocks of this age may be quite extensive along strike from Outerside but thermal metamorphism has obliterated all trace of graptolites.

Llanvirn Graptolite Zones

Although Elles (1933) recorded the *Didymograptus bifidus* Zone from Outerside the occurrence has been questioned by Jackson (1962). This means that apart from a locality south of Mungrisdale, reliable records of the *bifidus* Zone are confined to Thornship Beck, the Tailbert–Lanshaw tunnel and possibly Aik Beck. The *Didymograptus murchisoni* Zone is known only from Tarn Moor tunnel SE of Ullswater. All of these occurrences are in mudstones belonging to the Skiddaw Group and in fact could legitimately be referred to the Kirk Stile Slates (but see Wadge, chapter 6).

Didymograptus bifidus Zone

Earliest records of the *bifidus* Zone (Elles 1898) refer to faunas from Thornship Beck in the Bampton Inlier and from Aik Beck. But by far the most informative account emerged from Skevington's (1970) work on the Tailbert–Lanshaw tunnel in the Bampton Inlier west of Shap village. The tunnel is 1220 m long and cuts through an unnamed unit of grey–black shales of the Skiddaw Group that are seen to be interleaved with tuffaceous beds of the Borrowdale Group. The cutting yielded some 400 specimens of which the most notable are: *Cryptograptus tricornis schaeferi* 37%, *Glyptograptus dentatus* 12%, *Aulograptus cucullograptus* 7% and *Nicholsonograptus fasciculatus* 3%. Here it is worth noting that only six specimens of didymograptids could be said to belong to or have affinities with *D. bifidus*.

Didymograptus murchisoni Zone

Recent construction of the Tarn Moor tunnel SE of Ullswater has brought to light the existence of upper Llanvirn faunas in the uppermost beds of the Skiddaw Group. The geology has been described by Wadge et al. (1972) as black graptolitic mudstones as well as 'greyish brown to greyish green, rusty-brown-weathering silty mudstones' and similar to the lithology of exposures in the nearby Aik Beck section. The graptolites described by Skevington (in Wadge et al. 1972) occur in two discrete stratigraphic intervals. The lower of the two faunas yielded 20 specimens amongst which Skevington identified *Pseudoclimacograptus angulatus magnus, P. angulatus micidus*?, and *Pterograptus elegans*; species that are definitive of the *murchisoni* Zone.

The higher fauna is dominated by *Amplexograptus confertus* in association with *Cryptograptus tricornis schaeferi, Glyptograptus* sp., *Pseudoclimacograptus* cf. *P. scharenbergi, Didymograptus* cf. *D. artus, D.* aff. *D. bifidus, D. murchisoni geminus, D. murchisoni speciosus*, and *D.* spp. (extensiform). 'The known ranges of the grapto-lites in this fauna do not entirely preclude reference of the total assemblage to the late *D. bifidus* Zone' wrote Skevington (in Wadge et al. 1972, p. 68) who went on to note the absence of late *D. bifidus* Zone indicators in what can best be regarded as a fauna from the *D. murchisoni* Zone.

Distribution of graptolite zones across the rest of the main Skiddaw inlier

With the exception of two fault-bounded inliers of Loweswater Flags in the Great Sca Fell area and on Ravens Crag near Mungrisdale (where the *deflexus* Subzone outcrops) the outcrops east of Bassenthwaite Lake are predominantly of Kirk Stile Slates (Fig. 34). The *hirundo* Zone has an extensive distribution around the western and southern flanks of the Skiddaw Granite metamorphic aureole and has been identified by Jackson (1956) from Dead Crags, Dead Beck, Broad End, Randel Crag, Buzzard Knott, Watches, Ling How, Carlside, Dodd, Howgill, Lonscale Crags, Knowe Crags, Gategill Fell, Halls Fell, Scales Fell, White Horse Bent, Bannersdale Crags and Souther Fell. At a few scattered localities such as the Watches, White Horse Fell and Randel Crag the outcrops may expose the *gibberulus* Subzone. South of the Glendera-mackin River, the *hirundo* Zone is probably present in Mosedale Beck upstream from Lobb. According to Bulman (1968) a locality 1 km south of Mungrisdale yields a fauna indicative of the basal *Didymograptus bifidus* Zone. This is the youngest recorded fauna from the main inlier of the Skiddaw Group.

West of the Vale of Lorton, some important collections were acquired by A. Templeman for Geological Survey geologists in the 1920s. These collections were studied by G. L. Elles and it is apparent from her identification that the oldest fossili-ferous horizon of the area is in an old quarry [116 216] south of Hudson Place Farm, Loweswater where the *nitidus* Subzone crops out.

Apart from this collection, all others made by Templeman have been referred by Elles to either the *gibberulus* Subzone or to the *hirundo* Zone. Published records of the *hirundo* Zone include the following localities in the Kirk Stile Slates: in High Nook Beck [128 205] near High Nook Farm, Loweswater (Dixon in Eastwood et al. 1931, p. 25); an old quarry on the NW shoulder of Knockmurton Fell, 225 m SE of South Felldyke; and at Abbey Gate about 2 km east of Eaglesfield (Eastwood et al. 1931, p. 27). Three other questionable records include: left bank of tributary of Kirk Beck, 113 m NNW of road junction at Wilton, East of Egremont (IGS records); an old quarry [107 241] 145 m NW of Whinnah Farm, east of Sosgill (Eastwood et al. 1931, p. 27), and the southern bank of a gully 245 m south of Wath Mill, Cleator Moor (Eastwood et al. 1931, p. 30).

THREE CRITICAL AREAS FOR INTERPRETATION OF THE STRATIGRAPHY

Different interpretations of the rock succession between Bassenthwaite Lake and Lorton Vale (e.g. Rose 1955, Simpson 1967) stem in part from the way different workers have interpreted the stratigraphy and structure of three critical areas. These areas comprise: (1) the area SE of Sail Beck, Buttermere, (2) Wythop Moss — Ling Fell and Sale Fell, and (3) those outcrops between Watch Hill near Cockermouth and Great Sca Fell. All three areas are structurally very complex and have yielded few fossils. However, sufficient graptolites have recently come to light to permit a new assessment to be made.

The area east of Buttermere represents one of the most structurally complex areas in the entire inlier and the interpretation of the stratigraphy and structure turns on two graptolite collections on Robinson and near Buttermere village. The Robinson collection (Jackson 1956) is from mudstones intercalated in flagstones 100 m WNW of the summit of Robinson [199 169] and contain *Didymograptus filiformis* Tullberg, *?Pseudobryograptus* sp. and fragments of dichograptids. In Sweden and Norway (Monsen 1937) *Didymograptus filiformis* is confined to the *D. validus* Zone and *D. balticus* Zone which are generally agreed to be equivalent to our *deflexus* Zone. In Gueizhou Province of China, Lee and Chen (1962) actually map a *Didymograptus filiformis* Zone immediately above the *Tetragraptus approximatus* Zone of the Tonggao Shale. The implication of the Robinson fauna is that these transitional beds are roughly correlative with the basal Loweswater Flags at Scaw Gill, Lorton and therefore the 'grits of Robinson' are indeed Loweswater Flags and that the transition beds on the north side of Robinson grade downwards and northwards into older Hope Beck Slates. Confirmation that this is so comes from the Buttermere fauna. The Buttermere collection was made by J. Temple (in Simpson 1967) from a roadside quarry [173 172] 333 m NW of Buttermere village. His *in situ* collection from what Simpson designated 'Buttermere Slates' contained: *Didymograptus balticus vicinatus* Monsen or *D.* v-*fractus scandinavus* Monsen, *D. decens* Törnquist or *D. suecicus* Tullberg, *D. nichosoni* Lapworth, *D. serpens* Monsen, *D.* cf. *uniformis* Elles and Wood, *Tetragraptus fruticosus distans* Monsen, T.? *gracilis* Monsen and *T. quadribrachiatus*.

Temple correctly assigned the fauna to late 3bA of Oslo (Erdtmann 1965) which is generally regarded as of *D. deflexus* age. Such an age makes these striped mudstones and siltstones at Buttermere about the same age as those beds on Dodd that are transitional between Hope Beck Slates and Loweswater Flags. From this conclusion two facts must follow. Firstly, the *deflexus* Zone embraces the lower part of the Loweswater Flags and also the upper part of the underlying mudstones: Secondly, the 'Buttermere Slates' of Simpson (1967, p. 394) is another name for the Hope Beck Slates that occur 7 km to the north and therefore Simpson's term is a junior synonym which should be suppressed.

Such an interpretation of the stratigraphy has considerable bearing upon the structural interpretation of the area between Buttermere and Keswick. It seems most probable that the unfossiliferous mudstones and siltstones that crop out along strike on Knott Rigg, Ard Crags and Aikin Knott, north of Newlands Pass belong to this, the oldest formation in the Skiddaw Group (cf. Jackson 1956). Such rocks must mark an anticlinal fold having an east–west trend flanked by outcrops of Loweswater Flags on Rowling End and in Stoneycroft Gill west of Stair [237 213] on the northern limb and across Robinson Crags and in Scope Beck [214 175] on the southern limb (Fig. 33). It will be noted however that both Webb and Jeans (unpub. Ph.D. theses, Sheffield and Birmingham Universities, and Chapter 5) decided on structural grounds that this area is synclinal.

The second area of interest centres on Wythop Moss SE of Ling Fell (Fig. 33) where Simpson (1967 Pl. 19) mapped anticlinally folded Kirk Stile Slates with flanking younger Loweswater Flags to the north (through Harrot — Ling Fell — Sale Fell) and to the south across the Lorton Fells. But the palaeontological findings of Jackson (1956) totally contradict this interpretation of the stratigraphy and structure. It has long been known that the Loweswater Flags of the Lorton Fells yield *deflexus* and *nitidus* Subzone faunas (Jackson 1962). However, the bluish striped mudstones of Lords Seat [203 266], Beck Wythop [212 283], Lowthaite Side [209 292], Chapel Wood [198 292], Broom Fell [189 275] and southern slope of Ling Fell [178 282] represent Kirk Stile Slates which all yield faunas assigned to either the *gibberulus* Subzone or *hirundo* Zone (Jackson 1956). Some key fossils found in the Kirk Stile Slates on the southern slopes of Ling Fell include *Isograptus gibberulus* and *Glyptograptus dentatus* and it is fairly certain that an E–W striking reverse fault separates the *synclinally folded* Kirk Stile Slates of Wythop Moss from the anticlinally folded Loweswater Flags.

The third area is that structurally enigmatic tract north of the Embleton Valley and Bassenthwaite village (Fig. 34) which contains the much discussed 'Watch Hill Grits' NE of Cockermouth, the 'Great Sca Fell Grits' as well as the 'Sunderland Slates' of Simpson (1967). Various ideas concerning the stratigraphical positions of the grits have been based on reasoning such as structural interpretation, geological setting and lithological similarity to clastics elsewhere. No one has so far marshalled any palaeontological evidence to support any of the various interpretations. This will now be attempted. First of all we can consider the outcrops due east of Cockermouth from Slate Fell [146 303] to Seathwaite How near Embleton [176 309]. Generally, these rocks are dark blue–black striped mudstones except adjacent to igneous intrusions where they are pale buff coloured hornfels very much like the so-called Blakefell Mudstones. Jackson (1956) reported graptolites from Embleton Quarries on Seathwaite How which he assigned to the *D. hirundo* Zone implying that these strata represent Kirk Stile Slates. To the north lie the northward dipping flaggy turbidites of Watch Hill and Elva Hill. Similar strata crop our along strike in Howgill Wood and Messengermire Wood in Old Isel Park (Fig. 33) and again reappear in the Uldale Fells particularly on Great Cockup, Burn Todd and Great Sca Fell (Fig. 34). Two questions that require answers are whether these similar looking beds represent the same formation and whether they lie conformably above the Kirk Stile Slates of Seathwaite How or low down in the stratigraphic pile. In 1955, Rose tentatively considered this turbidite facies to represent Loweswater Flags. Jackson (1956) also arrived at this conclusion primarily because graptolites collected from an old quarry west of Setmurthy Church [182 324] contained two species of horizontal *Tetragraptus* which were tentatively thought to indicate an early Arenig age. Hope that these beds could be more accurately dated was recently rekindled when the writer examined some old collections in the Institute of Geological Sciences and the Sedgwick Museum, Cambridge.

One survey collection (Ht 1250 to 1256, and Ht 1260) from flaggy beds in Jonah's Gill, Howgill Wood [approximately 190 341] yields well preserved specimens of *Didymograptus deflexus*. Because these strata dip and young northwards they can be regarded as a faulted continuation of the grits on Watch Hill and, even more important, they are correlative with Loweswater Flags seen in the Scaw Gill area in the Lorton Fells. Eastwards from Isel Old Park, the strike of this formation changes direction to a NW–SE trend near Burthwaite Bridge [217 332]. No outcrops of Loweswater Flags are encountered between Burthwaite Bridge and Little Cockup in the Uldale Fells. In the latter area, flags and grits of the Skiddaw Group have been thrust northwards over the Kirk Stile Slates and the Volcanics (see Eastwood et al.

1968, p. 39) and the thrust sheet is bounded on the south by a major steep-angled fault running along the southern slopes of Great Sca Fell (Fig. 34). Within this thrust sheet the rock types are practically identical to those seen on Watch Hill. Graptolites have been collected from two localities in Frozenfell Gill [280 334 and 284 335] as well as in Burn Tod Gill [279 335] and the southern slopes of Great Sca Fell [288 337]. Collectively these assemblages include *Didymograptus balticus, D. nicholsoni, Tetragraptus* spp., *Dichograptus* sp. and *Temnograptus* sp. In the Oslo region, Erdtmann (1965, p. 465) established that *Didymograptus balticus* has a range that is confined to the uppermost 3bB (*sensu* Erdtmann) which he correlates to the lower part of the *D. deflexus* Subzone of England. From this discussion it is clear that Rose (1955) was correct in referring the 'Watch Hill Grits' and the 'Great Sca Fell Grits' to the Loweswater Flags.

Lastly, mention must be made of bluish mudstones that crop out between Loweswater Flags in Isel Old Park and the volcanics of the Binsey Formation (Fig. 34). A collection made by S. W. Hester in 1929 and now housed in the Institute of Geological Sciences and labelled 'Scalegill, 1½ ml east of Sunderland' (HS 1331 to 1338) contains *Didymograptus hirundo* and *Phyllograptus angustifolius* together with *Cyclopyge*. This suggests that these beds are Kirk Stile Slates thus invalidating the term 'Sunderland Slates' of Simpson (1967).

Within the main inlier the youngest datable beds in the Skiddaw Group are of early *bifidus* age and the general absence of Llanvirn zones has been regarded by some as support for the presence of a widespread unconformity at the top of the Group (Wadge *in* Helm 1970, p. 137). A more probable explanation for the 'missing' zones might be, firstly, that nowhere in the main inlier have we been able to date the Kirk Stile Slates that lie immediately below the Borrowdale Group (cf. Wadge, op. cit.) and secondly, that the base of the Borrowdale Group is probably diachronous.

Acknowledgements

I am grateful to Mr W. C. C. Rose of the Institute of Geological Sciences for unpublished data concerning the geographical locations and identifications of graptolites (by G. L. Elles) collected by Hester and Templeman during the 1920s and 1930s. I also wish to thank Dr A. Rushton and Mr R. Clark of the Institute for helping to retrieve some of these very important collections, and Mr T. Shipp of Workington for collecting specimens of Latterbarrow Sandstone and Redmain Sandstone shown in Plate 7.

REFERENCES

Bott, M. H. P. 1974. The geological interpretation of a gravity survey of the English Lake District and the Vale of Eden. *Jl geol. Soc. Lond.* **130,** 309–31.
Bouma, A. H. 1962. *Sedimentology of Some Flysch Deposits.* Elsevier, Amsterdam. 168 pp.
Bulman, O. M. B.
—— 1941. Some Dichograptids of Tremadocian and Lower ordovician. *Ann. Mag. nat. Hist.* **7,** 100–21.

facing page

PLATE 8. A. Flow brecciated andesitic lava, Norman Crag, Gowbarrow [400 222].
B. Coarse andesitic tuff, Hayes Water [43 12]. The identification of these rocks is assisted by weathering. In the lava the fragments are softer than the matrix and become recessed, whereas in the tuff they are harder and project. *Photographs: F. Moseley*

A

B

—— 1958. The sequence of graptolite faunas. *Palaeontology,* **1,** 159–73.

—— 1971. Some species of *Bryograptus* and *Pseudobryograptus* from Northwest Europe. *Geol. Mag.* **108,** 5, 361–71.

DIXON, E. E. L. 1925. In *Summ. Prog. Geol. Surv. Gt Br.* for 1924, 70–71.

EASTWOOD, T. 1927. In *Summ. Prog. Geol. Surv. Gt Br.* for 1926, p. 53.

—— 1933. In *Summ. Prog. Geol. Surv. Gt Br.* for 1932. Pt. I., p. 59.

—— DIXON, E. E. L., HOLLINGWORTH, S. E. and SMITH, B. 1931. The geology of the Whitehaven and Workington district. *Mem. Geol. Surv. Gt Br.* 304 pp.

—— HOLLINGWORTH, S. E., ROSE, W. C. C. and TROTTER, F. M. 1968. Geology of the Country around Cockermouth and Caldbeck. *Mem. Geol. Surv.,* x + 298 pp.

ELLES, G. L. 1898. The graptolite-fauna of the Skiddaw Slates. *Q. Jl geol. Soc. Lond.* **54,** 463–539.

—— 1933. The Lower Ordovician graptolite faunas with special reference to the Skiddaw Slates. *Summ. Prog. Geol. Surv. Gt Br.* for 1932, Pt. II, 94–111.

—— and WOOD, ETHEL M. R. 1901–18. A monograph of British graptolites. *Palaeotogr. Soc.* (*Monogr.*), 526 pp.

ERDTMANN, B. D. 1965. Outline stratigraphy of graptolite-bearing 3b (Lower Ordovician) Strata in the Oslo region, Norway. *Norsk Geol. Tidsskr.* **45,** 481–547.

GREEN, J. F. N. 1917. The age of the chief intrusions of the Lake District. *Proc. Geol. Ass.* **28,** 1–30.

HARRIS, D. E. 1960. The zonal distribution of Australian graptolites. *Jl Proc. Roy Soc. New South Wales.* **94,** 58 pp.

HELM, G. G. 1970. Stratigraphy and structure in the Black Coombe Inlier, English Lake District. *Proc. Yorks. geol. Soc.* **38,** 105–48.

HOLLINGWORTH, S. E. 1955. The geology of the Lake District — a review. *Proc. Geol. Ass.* **65,** 385–402.

HUBERT, J. F. 1964. Textural evidence for deposition for many western North Atlantic deep-sea sands by ocean bottom currents rather than turbidity currents. *J. Geol.* **72,** 757–85.

JACKSON, D. E. 1956. Geology of the Skiddaw Slates between the Lorton–Buttermere Valley and Troutbeck, Cumberland. *Unpubl. Ph.D. thesis*, Newcastle University.

—— 1961. Stratigraphy of the Skiddaw Group between Buttermere and Mungrisdale, Cumberland. *Geol. Mag.* **98,** 515–28.

—— 1962. Graptolite zones in the Skiddaw Group in Cumberland, England. *J. Palaeont.* **36,** 300–13.

—— 1964. Observations on the sequence and correlation of Lower and Middle Ordovician graptolite faunas of North America. *Bull. geol. Soc. Am.* **75,** 523–34.

LEE, C. K. and CHEN, X. 1962. Cambrian and Ordovician graptolites from Sandu, S. Guezhou (Kueichou). *Acta Palaeont Sinica.* **10,** 12–27 (in Chinese), 27–30 (in English).

MARR, J. E. 1894. Notes on the Skiddaw Slates. *Geol. Mag.* **31,** 122–30.

MONSEN, A. 1937. Die Graptolithenfauna im unteren Didymograptus-Schiefer Norwegens. *Norsk geol. Tidsskr.* **16,** 57–226.

NICHOLSON, H. A. and MARR, J. E. 1895. Notes on the phylogeny of the graptolites. *Geol. mag.* **42,** 529–39.

RASTALL, R. H. 1910. The Skiddaw Granite and its metamorphism. *Q. Jl geol. Soc. Lond.* **66,** 116–41.

RICKARDS, R. B. 1970. The Llandovery (Silurian) graptolites of the Howgill Fells, Northern England. *Palaeontogr. Soc.* (*Monogr.*) 108 pp.

ROSE, W. C. C. 1955. The sequence and structure of the Skiddaw Slates in the Keswick–Buttermere area. *Proc. Geol. Assoc.* **65,** 403–06.

—— 1969. Stratigraphical sub-divisions of the Skiddaw Slates of the Lake District (Correspondence) *Geol. Mag.* **106,** 293–94.

SHACKLETON, E. H. 1975. Geological excursions in Lakeland. Dalesman Publishing Co. 125 pp.

SIMPSON, A. 1967. The stratigraphy and tectonics of the Skiddaw Slates and the relationship of the overlying Borrowdale Volcanic Series in part of the Lake District. *Geol. J.* **5,** 391–418.

SKEVINGTON, D. 1970. A Lower Llanvirn graptolite fauna from the Skiddaw Slates, Westmorland. *Proc. Yorks. geol. Soc.* **37,** 395–444.

—— 1974. Controls influencing the composition and distribution of Ordovician graptolite faunal provinces.

SMITH, B. 1930. Cumbrian District (Cockermouth Sheet 23). *Summ. Prog. Geol. Surv.* (for 1929), Pt. III, 37–41.

facing page

PLATE 9. Aqueously-deposited andesitic bedded tuff, Borrowdale Volcanic Group. St Sunday Crag [365 134]. Note the convolute bedding in several units. The area lies in a low strain zone and the 'main phase' cleavage consists of widely spaced fractures.

Photograph: F. Moseley

STANLEY, D. J. and UNRUG, R. 1971. Submarine channel deposits, fluxoturbidites and other indicators of slope and base-of-slope environments in modern and ancient marine basins. *In* Rigby, J. K. and Hamblin, W. K. (editors), *Recognition of ancient sedimentary environments. Soc. Econ. Paleont. and Mineral.* Spec. Pub. **16**, 287–340.

TROTTER, F. M., HOLLINGWORTH, S. E., EASTWOOD, T., and ROSE, W. C. C. 1937. The geology of the Gosforth District. *Mem. Geol. Surv.*

WARD, J. C. 1876. The geology of the northern part of the English Lake District. *Mem. Geol. Surv. Gt Br.* 132 pp.

WADGE, A. J., NUTT, M. J. C., and SKEVINGTON, D. 1972. Geology of the Tarn Moor Tunnel in the Lake District. *Bull. Geol. Surv. Gt Br.* **41,** 55–73.

D. E. JACKSON, PH.D.
Department of Earth Sciences, The Open University,
Milton Keynes MK6 7AA

8

The Eycott and Borrowdale Volcanic Rocks

D. MILLWARD, F. MOSELEY and N. J. SOPER

The Skiddaw Slates are succeeded by a considerable pile of volcanic rocks which outcrop over some 900 km² in the Lake District, whilst their correlatives are seen in the Cross Fell Inlier. Along with volcanic rocks of similar age in Wales and Ireland, they form part of a major petrographic province whose relationship to a postulated Ordovician plate margin has been the subject of interest and speculation in recent years.

The map of the Lake District shows that the volcanic rocks occupy two principal areas of outcrop on either flank of the main anticlinal slate tract. Until recently they were all referred to as the Borrowdale Volcanic Series, but it is now known that the rocks exposed in the smaller northern area differ petrographically and chemically from those of the main outcrop (Fitton and Hughes 1970), and are entirely older. A separate formal name, the Eycott Volcanic Group, was therefore proposed for them (Downie and Soper 1972), now modified to Eycott Group to include associated sediments (Wadge, Chapter 6), whilst the name Borrowdale Volcanic Group is used for the rocks of the main outcrop.

THE EYCOTT GROUP

Volcanic rocks of the Eycott Group occupy less than 50 km² in the northern part of the Lake District; indeed, the whole outcrop is covered by a single 1-inch geological map (Sheet 32, Cockermouth) and much detail is provided by the accompanying memoir (Eastwood et al. 1968).

The largest continuous part of the outcrop is a narrow, E–W trending belt between Bothel [18 39] and Linewath [36 34] about 20 km to the east. The rocks dip steeply northward and in places are vertical or even overturned. Their southern margin against the underlying Skiddaw Slates is largely faulted, but the original junction is preserved to the south of Binsey between 220 352 and 237 348, and is described below.

To the north the volcanic rocks are overlain unconformably by Lower Carboniferous beds, although the junction is usually faulted. Nothing is known of the subsurface extent of the volcanic rocks to the north and west beneath the Solway Plain, but their eastward extent is better known. Although the main outcrop is terminated just east of Linewath by the southward swing of the Carboniferous unconformity, the volcanic rocks reappear at Eycott Hill [39 29], where they strike N–S with a moderate easterly dip, at Greystroke Park [41 33]. Rocks correlated with the Eycott Group are known from the Cross Fell Inlier, so the minumum east–west extent of the group was originally at least 50 km.

PETROGRAPHY

The Eycott Group is dominated by basalt and basaltic andesite flows of transitional calc-alkaline to tholeiitic type with a limited development of acid flows and tuffs (Fitton and Hughes op. cit.). A brief description of the principal rock types follows, largely summarized from Eastwood et al. (op. cit.).

Basalt, basaltic andesite, and andesite lavas

The bulk of the sequence is composed of microporphyritic and non-porphyritic basaltic and andesitic flows. The former (Berrier type) contain altered labradorite phenocrysts a few millimetres in length, small glomeroporphyritic aggregates of chloritized orthopyroxene and sometimes fresh augite. The proportion of phenocrysts to groundmass is variable and there is a transition to non-porphyritic types, many of which closely resemble the groundmass of the Eycott-type lavas. In the field all these basic lavas are dense, flinty and dark grey when fresh, but often purple or dark green when altered. Some flows have vesicular, autobrecciated tops and are flow-banded.

Of limited development are microporphyritic flows in which the feldspar is oligoclase and ferromagnesian phenocrysts are absent. Other types contain a little primary quartz or cryptofelsitic intergrowths of quartz and alkali feldspar in the groundmass.

The Eycott-type basaltic andesite

This is a distinctive variety of highly feldsparphyric rock which was first described by Teall (1888) from Eycott Hill and appears to be unique to the Eycott Group. Typical samples contain idiomorphic labradorite phenocrysts up to 3 or even 4 cm in length which are presumably of cumulate origin. Smaller phenocrysts of chloritized orthopyroxene and sometimes fresh augite are also present. The groundmass is basaltic in composition (Eastwood et al. and Fitton and Hughes op. cit.), glassy to microcrystalline, and usually contains small laths of labradorite and skeletal augites.

Flows sometimes exceed 30 m in thickness. They are usually dark green and flinty, with pale green feldspars, and often have a rubbly-weathering vesicular upper portion. Parts of certain flows are remarkably fresh in comparison with typical Lake District lavas but elsewhere hydration and carbonate metasomatism have been widespread. Three flow groups of Eycott-type lavas occur in the succession at Eycott Hill and are valuable for the purposes of subdivision and correlation.

Acid lavas

The succession also contains a few flows classed as rhyolites one of which attains a thickness of almost 100 m in the Greystoke Park inlier. These rocks are conspicuously pale-weathered, but dark and flinty when fresh. They range from glassy to microfelsitic; some carry small phenocrysts of alkali feldspar but ferromagnesian minerals are absent. Although no clear vitroclastic or eutaxitic texture has been recorded, some units are streaky in appearance and may be ignimbrites.

Pyroclastic rocks

Medium to coarse tuffs and agglomerates constitute almost a quarter of the exposed succession. They are composed dominantly of lithic fragments with a preponderance of felsitic over basic clasts, which far exceeds the proportion of acid flows in the lava sequence. Some units are crystal-lithic tuffs in which sodic feldspars occur to the exclusion of the calcic types which characterize the basic flows. Fragments of Skiddaw Slate are present in some of the lower tuff horizons and one bed contains rounded quartzitic pebbles identical petrographically to the Latterbarrow Sandstone.

Some of the tuffs, particularly those towards the base of the succession show crude bedding, a clastic matrix, rounding of the larger fragments and sometimes interbedding with Skiddaw-type mudstones, which suggests reworking and deposition in water. A few thin beds show fine, parallel lamination which may indicate airfall directly into still water. None show the delicate sedimentation structures which characterize many of the aqueously-deposited andesitic tuffs of the Borrowdale Group. Most of the pyroclastic units are unbedded. Some have glassy matrices which contain feldspar microlites, but more commonly they consist of angular lithic and crystal fragments set in an obscure, altered vitreous matrix, and it seems probable that the majority represent unwelded and poorly-welded ignimbrites, perhaps accompanied by airfall deposits. It is evident that effusion of the pile of basic lavas was punctuated by violent eruptions of more acid material.

GEOCHEMISTRY

Major element data for lavas of the Eycott Group are limited to a few analyses presented in the Geological Survey memoir (Eastwood et al., op. cit., pp. 58–59), analyses in Fitton and Hughes (1970), and unpublished analyses (Fitton 1971). Most of the Eycott Group lavas are silica-oversaturated, containing quartz, hypersthene and occasionally corundum in the norm, although some contain olivine as pseudomorphs. Compared to the calc-alkaline Borrowdale Group the Eycott Group lavas are richer in Fe and Ti but poorer in K, Rb and Ba. Figure 35 shows that there is a pronounced trend towards iron enrichment. Fitton (in Fitton and Hughes op. cit.) concluded that the Eycott Group should be regarded as transitional between calc-alkaline and tholeiitic in chemical affinity. In addition to aphyric lavas the AFM plot includes separated basaltic matrices of two Eycott-type feldsparphyric flows. Total rock data from the feldsparphyric lavas (not shown) scatters off the differentiation trend, suggesting that these rocks are of cumulate origin. It is interesting to note in this context that in certain of the Eycott-type flows the plagioclase phenocrysts are concentrated towards the top (Fitton and Hughes 1970).

Unpublished trace element data from Fitton (op. cit.) shows that the Eycott Group lavas have higher Sr, Zr and K/Rb values (240–460 in basaltic andesite and basalt) but lower La and Y values than the Borrowdale lavas. Ni and Cr levels are typically low. Plots of some elements (Fig. 36) suggest that different melting and crystallization histories have produced the chemical differences. This is particularly noticeable with the plot of La_n/Y_n against Y_n in which two distinct trends are displayed (Fig. 36). The flat trend shown by the Eycott Group lavas is indicative of a major influence of plagioclase fractionation, a feature noted earlier with the plagioclase rich Eycott-type basalts.

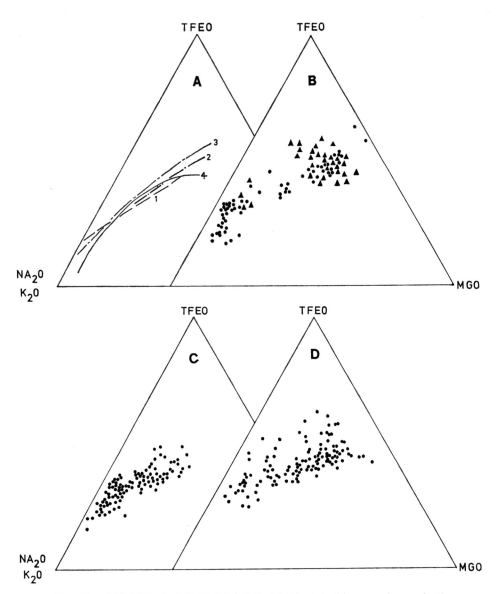

FIG. 35.   AFM (Na₂O + K₂O, total FeO, MgO) plot of lavas and pyroclastic rocks from the Eycott and Borrowdale volcanics.

A. Comparison with modern calc-alkaline provinces. Trend 1–Alaska; 2–Andes; 3–Central America; 4–south-west Aegean.

B. Lavas. Filled circles — Borrowdale volcanics, triangles — Eycott volcanics.

C. Ignimbrites — Borrowdale volcanics

D. Pyroclastic fall deposits — Borrowdale volcanics

TFEO — total Fe calculated as FeO.

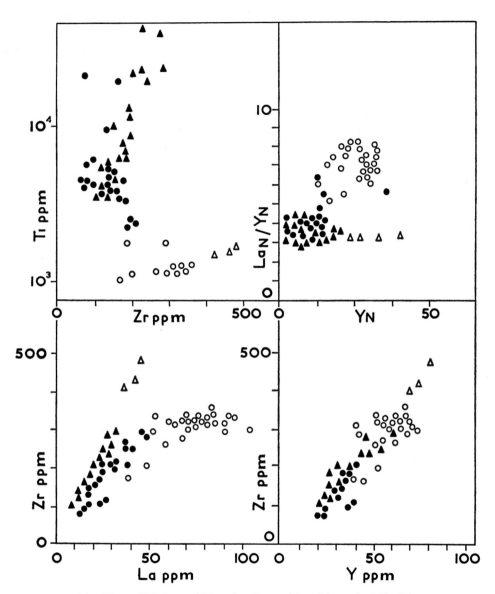

FIG. 36.   Plots of Ti, La and Y against Zr, and La_N/Y_N against Y_N. Elements in ppm; La_N and Y_N — ratios of La and Y abundances in the volcanics to that in chondrites. Circles — Borrowdale volcanics; triangles — Eycott volcanics.
Open symbols — lavas greater than 65 wt.% SiO_2
Filled symbols — lavas less than 65 wt.% SiO_2

STRATIGRAPHY

The most complete succession is exposed between Binsey and High Ireby [23 35 to 23 37] where it totals almost 2500 m and was divided into a lower Binsey Group and an upper High Ireby Group by Eastwood et al. (1968). As a first step towards rationalization of Lake District volcanic stratigraphy in line with the recommendations of the International Subcommission on Stratigraphic Classification (Hedberg 1972), these two useful subdivisions have been adopted, without modification, as constituent formations of the Eycott Group (Downie and Soper 1972).

Basal relationships and age

Mudstones interbedded with the lower Eycott lavas south of Binsey have yielded assemblages of marine microfossils (acritarchs and chitinozoa) which, by comparison with assemblages from graptolite-bearing sequences in the Cross Fell Inlier were referred by Downie and Soper (op. cit.), to a level low in the *D. bifidus* Zone (earliest Llanvirn). This evidence suggests that the lower part of the Binsey Formation was formed in submarine conditions and follows conformably above the Skiddaw Slates which are here of *D. hirundo* (late Arenig) age. Further, since the lower Eycott lavas fail eastwards and the much thinner volcanic succession at Eycott Hill stratigraphically overlies slates of *D. bifidus* age, the evidence indicates that the Binsey Formation dies out towards the south-east, to be replaced by time-equivalent marine mudstones of the Eycott Group. A re-examination of the dated samples in the light of further knowledge of Ordovician microfloras suggests that they can be assigned to the *D. hirundo* zone (G. Booth personal communication). It is thus probable that the Eycott Group spans the Arenig–Llanvirn boundary.

East of Ullswater graptolitic mudstones are known to extend up into the *D. murchisoni* Zone (Wadge et al. 1969). These are the youngest beds of the Eycott Group. Thin volcanic horizons interbedded with *D. bifidus* Zone slates below the Borrowdale unconformity at Matterdale Beck (Wadge 1972) must be regarded as time-equivalents of part of the Eycott Group. So too must the numerous tuff horizons within the Kirkland Formation of the Cross Fell Inlier, which are also of lower Llanvirn age. The occurrence of an Eycott-type flow in this inlier is noted below, but its stratigraphic relationship to the Kirkland Formation is not known.

Binsey Formation

This formation is about 1200 m thick on Binsey, and commences with three feldsparphyric flows of Eycott-type (the lower Eycott lavas) which are interbedded with tuffs and thin marine mudstones, the latter identical to the underlying Skiddaw mudstones except for a small admixture of volcanic material. A 30 m tuff follows, which locally contains pebbles of Latterbarrow Sandstone (Skiddaw Group). The upper part of the formation consists of flinty, non-porphyritic lavas interbedded with tuffs and agglomerates, followed by microporphyritic flows with plagioclase and pyroxene. The Binsey Formation thins markedly eastwards. On Eycott Hill, which provides the only other sections, the lower Eycott lavas are absent and the formation is reduced to about 120 m. It consists of andesitic tuffs which are interbedded with tuffaceous mudstones and sandstones towards the base and contains a 90 m sequence of micro-porphyritic andesites, similar to those in the upper part of the formation on Binsey

High Ireby Formation

The second and major group of Eycott-type flows (middle Eycott lavas) marks the base of the High Ireby Formation. This flow group is over 200 m thick north of Binsey, about 300 m at Linewath where five flows are interbedded with microporphyritic pyroxene andesites but only 75 m (two flows) on Eycott Hill. The middle Eycott lavas evidently did not extend much further south than this, but a small patch of Eycott-type andesite in the Cross Fell Inlier near Melmerby [626 374] may be correlated with them.

The High Ireby Formation is not well exposed in its type area but about 1300 m appear to be present. The succession above the middle Eycott lavas consists largely of aphyric and microporphyritic andesites with at least one rhyolite and a single Eycott-type flow (upper Eycott lava) which dies out to the east. Exposure is again poor in the eastern part of the main crop around Roughton Gill, but there appears to be an increase in the proportion of acid flows and tuffs. The Greystoke Park inlier exposes some 780 m of the High Ireby succession, mainly andesites and tuffs, but containing a 90 m nodular rhyolite flow. At Eycott Hill a 580 m succession is well exposed above the middle Eycott lavas. It consists largely of andesites of which the most prominent are a 100 m group of porphyritic flows which form the main scarp of the hill. Thin, well-bedded tuffs are also present and a 12 m rhyolite may represent the thick flow of Greystoke Park.

BORROWDALE VOLCANIC GROUP

The Borrowdale Volcanic rocks form most of the best known mountains and crags of the Lake District including the Sca Fell, Helvellyn, High Street and Coniston ranges, and whilst there are reliable geological maps of most of the outcrop, there are gaps which add to the general difficulties of correlation in a volcanic pile, where several volcanic centres may have been operating at any one time, and it is not made easier by the effects of the Caledonian folding and faulting (Fig. 39). The formations are characterized by a common association of lithologies rather than a particular rock type, and it must not be thought, for example, that the Wrengill Andesite is all andesite, nor was it erupted from only one vent.

LAVAS

Flow brecciation is perhaps the most common characteristic of all the lava flows of whatever composition either affecting the whole flow or the upper and frontal parts. Such features are characteristic of modern block lava flows, for example the andesite volcanoes of the Andes and Central America. In the early days of Lakeland geology flow brecciated lavas were confused with pyroclastic deposits, but it is now known that the difference can generally be resolved by the fact that the fragments in flow breccias tend to be softer than the matrix which stands proud on weathering. In many cases simple block lava flows form small crags up to 15 m high separated by thin bands of pyroclastic fall deposits which crop out on the intervening bench features. The more massive crags 60 or 70 m high may represent several simple flows, or one or more compound lava flows.

Other characteristics of the lavas include flow banding and flow jointing, sometimes with complex flow folds such as those displayed on Haystacks [19 13], and less commonly amygdaloidal textures, with chlorite and calcite vesicular infillings. The lavas are frequently porphyritic, the most common of the phenocrysts being feldspar which is often altered to sericite, calcite and epidote. Pyroxene and amphibole phenocrysts are to be found in the more basic rocks although they also are generally altered. The total phenocryst content is variable, from 5 to 30%, indicating some degree of crystal fractionation in the magma prior to eruption. The matrix is almost always fine grained and altered, but some fresh specimens of andesites and basalts show that the matrix is composed of feldspar, pyroxene and opaques, the feldspars frequently indicating a flow texture. Perlitic cracking reveals the original glassy nature of some flows.

Basic and intermediate lavas

Basalts and andesites were the most abundant of the lavas erupted during the Borrowdale volcanic episode varying chemically from high alumina basalt to high potassium andesite. The most basic of the lavas are basalts and basaltic andesites; distinctive rocks in the field with a fine grained dark blue, grey or green matrix and small phenocrysts of labradorite and pyroxene, the latter usually pseudomorphed by chlorite. Individual flows generally form prominent easily mapped features which show up well on aerial photographs. Closely spaced platy jointing parallel to the lower and upper surfaces is particularly common, and flow breccia is to be seen occasionally, as for example in the lowest of the Birker Fell basaltic andesites of Rosthwaite Fell [258 137], where several flows are completely brecciated. Grey andesites are the commonest lavas, and although mostly massive, a tendency to flow brecciation makes them more susceptible to cleavage which is often visible on the weathered tops of the crags. In thin section cleavage is seen to be a result of crystallization of mica films through the matrix.

Acid lavas

These rocks range from dacites to rhyolites, vary in colour from dark grey to pale grey, cream or pink, are generally fresh, and have an extremely fine-grained flinty matrix, with occasional small percentages of feldspar (often oligoclase) phenocrysts. The matrix is composed of quartz and alkali feldspar which in some cases shows perlitic cracking, whereas in the more massive examples the texture is of the snowflake variety (Torske 1975). This texture comprises an equidimensional mosaic of complex crystal aggregates of quartz and alkali feldspar with random optical orientation, formed as the result of devitrification. Quaternary obsidian-rhyolite flows are generally simple, with a small length to thickness ratio, and many of the acid flows of the Borrowdale Volcanics show features typical of these modern counterparts (Hartley 1932, Oliver 1961). Brecciation is common in most flows, and is especially well developed at the tops where there may be large blocks many metres in size. Flow banding is another characteristic feature seen as a continuous streaking on weathered surfaces or as a strong platy structure frequently contorted by complex flow folds.

Many of these features can be best explained in the context of ramp structuring, a common development in modern acid lavas. They can be identified, for example, within the Airy's Bridge lavas of Rosthwaite Fell [263 123]. The ramp structure is formed by shearing within the flow, with higher, later and hotter parts sliding over material beneath and in front of the lava flow along paraboloid fractures. In such a flow a rubble flow breccia is formed in front of the unit and at the base. The shearing motion also provides a mechanism for the formation of the flow banding. As material is pushed along the paraboloid fractures the cooler material becomes contorted into coaxial folds of varying amplitude and low plunge, contorted blocks of flow folded and banded material breaking off from the brittle top part of the flow to form an extremely massive surface breccia of jumbled material. Analysis of the orientations of flow folds in a massive part of the flow should reveal flow directions normal to the average axial trends. The Rosthwaite Fell example displays these features, with the flow folds indicating movement from the north.

Geochemistry of the lavas

The lavas vary in composition from high alumina basalt to rhyolite (Table 3). The distribution, however, is bimodal, with basalt and andesite distinct from dacite and rhyolite. The typical calc-alkaline trend of low iron enrichment of these lavas is shown in figure 35 and can be compared with lavas from some modern calc-alkaline provinces. One of the main problems in comparing Ordovician rocks with their modern counterparts is that the chemistry of the former may well have been modified considerably since emplacement, and this is especially so for the more basic material.

Table 3

	Basalt	Basaltic Andesite	Andesite	Dacite	Dacitic Ignimbrite	Rhyolite
No of Analyses	18	46	46	37	95	7
SiO_2	51.04	55.74	59.98	67.50	66.77	72.21
TiO_2	1.29	1.09	0.91	0.31	0.33	0.31
Al_2O_3	16.26	17.81	17.76	16.60	16.46	14.72
Fe_2O_3	11.11	8.84	7.54	4.22	4.37	3.17
MnO	0.24	0.24	0.20	0.09	0.10	0.06
MgO	7.73	4.63	2.83	0.62	0.55	0.75
CaO	8.68	6.53	4.75	1.12	1.35	0.58
Na_2O	1.88	2.32	2.67	3.00	2.83	2.36
K_2O	1.33	2.42	2.97	4.60	4.84	5.55
P_2O_5	0.21	0.19	0.19	0.30	0.29	0.21
Cr	410	123	62	5	6	3
Ni	145	38	22	2	2	2
Rb	49	97	124	218	232	237
Sr	346	310	270	101	122	85
Y	27	34	40	63	64	65
Zr	141	179	227	315	340	282
Nb	11	12	14	25	25	17
Ba	349	586	747	1057	1072	1096
La	19	22	35	76	77	46
Ce	45	54	62	118	124	
Pb	14	23	22	23	24	
Th	10	12	16	26	28	

Table of average compositions of the Borrowdale volcanic rocks. Basalt, basaltic andesite, andesite and rhyolite compositions from Fitton (1972). Dacite lava and ignimbrite averages and all La, Ce, Pb and Th data from unpublished analyses (Millward 1976) Oxides in wt. per cent, elements in ppm. Total Fe expressed as Fe_2O_3.

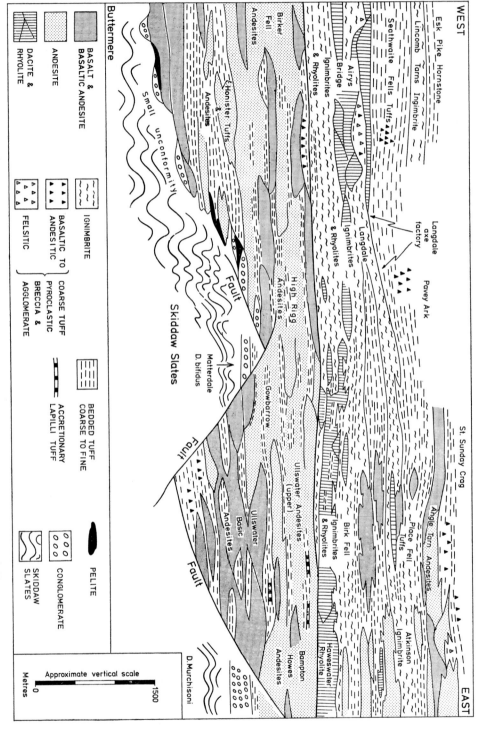

FIG. 37. The lower part of the Borrowdale volcanic succession between Buttermere and Haweswater.

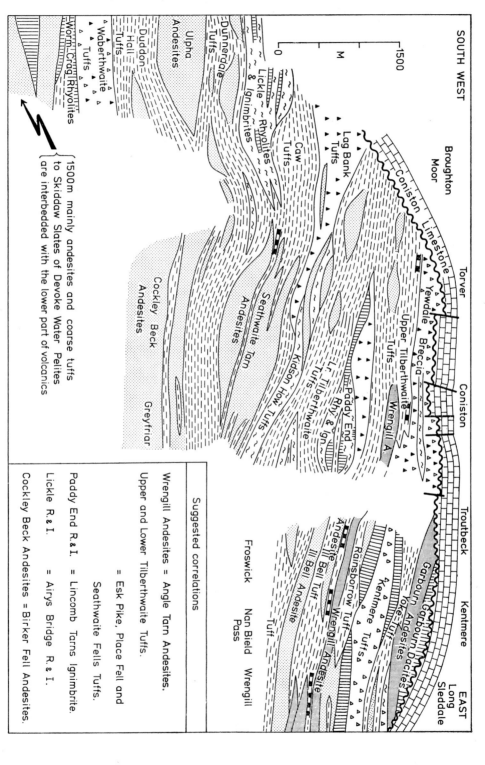

FIG. 38. The Borrowdale volcanic succession between the Duddon Valley and Long Sleddale. Suggested correlations between figures 37 and 38 are given. Legend is shown on figure 37.

The lavas are quartz and hypersthene normative and the more acid members also contain corundum in the norm. In comparison to more recent calc-alkaline lavas, for example the Andean lavas of Chile (Thorpe et al. 1976), the basic lavas have low soda. The K_2O/Na_2O ratios are invariably greater than unity and although both potassium and sodium are easily redistributed the high ratios are nevertheless consistent with those found in continental margin calc-alkaline provinces (Jakes and White 1972). The dacitic lavas show alkali modification features similar to those described in the section on ignimbrites. Fe/Mg ratios range from 1.5–3.0 for the basalts and andesites to 15 for the acid lavas. Calc-alkaline provinces in general are also characterized by low Ni and Cr and this is a feature of the Borrowdale Volcanics. Ni and Cr have good correlation between themselves and with Fe/Mg indicating an involvement of pyroxene and possibly spinel in the magmatic history. The Borrowdale lavas are high in the large ion lithophile elements, particularly Rb and Ba, the radiogenic elements Pb, Th and U and the rare earth elements (REE) (cf. Northern Chile, Siegers et al. 1969). They are very low in Sr in comparison to the Chilean examples and Sr varies inversely with SiO_2, a feature explained by feldspar fractionation. The basic rocks are low in Zr but the acid rocks high in Zr when compared to the Chilean examples. Considerable variation is present in some of the element ratios: Ba/Sr is always greater than unity, and K/Rb ratios vary from 170–300 and are comparable with average crustal values (Taylor and White 1965).

The acid and basic materials are comagmatic, and must be related at least in source, as indicated by the continuous plots shown in the variation diagrams of figure 36. The rare earth element patterns, although enriched are abnormal in that they have a marked depletion in Eu relative to the other elements (La–Lu). Such a feature is commonly associated with the fractionation of plagioclase, a mineral which preferentially includes Eu in its structure over the other rare earths. Since the garnets in the ignimbrites also possess such an anomaly the Borrowdale Volcanics must have been formed from a source previously depleted in Eu.

PYROCLASTIC DEPOSITS

Sparks and Walker (1973) recognized three distinct groups of pyroclastic deposits; (i) pyroclastic flow deposits (ignimbrites), (ii) pyroclastic fall deposits and (iii) ground surge deposits. To these can be added the mass flow deposits (including volcanic mudflows) typical of most recent volcanic provinces.

Ignimbrites

In the early days of geological survey in the Lake District these rocks were understandably referred to as rhyolites, and their true nature was not appreciated until Oliver (1954) recognized them as similar to ignimbrites from his native New Zealand (Marshall 1932). It is now accepted that these deposits were associated with a nuée ardente type of eruptive mechanism, similar to that of Mont Pelée (Martinique) in 1902. The nuée ardente consists of an ash cloud and a ground-hugging pyroclastic flow component and results in rocks which are extremely variable in field, petrographic and geochemical characteristics.

A large proportion of the ignimbrites in the Borrowdales are strongly welded, the rocks have a hard, fine grained, flinty appearance and form distinct crags.

These are the 'streaky' rocks mentioned by numerous authors since Ward (1876). The elongate streaks (fiamme) vary in their length to thickness ratios, the smaller ratios being referred to as a eutaxitic texture, whereas further elongation of the fiamme, perhaps by secondary flow produces a parataxitic structure. The fiamme in ignimbrites have been regarded as flattened pumice (Ross and Smith 1961) or as blobs of unvesiculated magma (Gibson and Tazieff 1967). Many of the Lake District examples contain coexisting fiamme and unflattened pumice fragments, supporting the latter view.

Typical ignimbrite units have been described by Smith (1960) from Nevada, U.S.A. They include both strongly welded (eutaxitic) and unconsolidated, non-welded material; some of the flows are in fact totally unwelded. In contrast most of the ignimbrites exposed in the Lake District are strongly welded, but there are several reasons why these may not be typical of the original deposits. In the first place non-welded deposits would be far more likely to be removed by contemporaneous erosion, secondly these deposits are less massive and more highly cleaved resulting in poor exposure, and thirdly hand specimens of non-welded ignimbrites are not easily distinguished from some ash fall tuffs and in some accounts they may have been misrepresented. It has also been noted in other parts of the world where there are younger ignimbrites (e.g. Smith 1960), that a thick sequence may contain a number of cooling units each divided into several flow units. This characteristic has been noticed in the Lake District, for example on Rosthwaite Fell [26 12] the sequence is 170 m thick and has been divided into several cooling units, each of which is made up of a number of flow units. The recognition of single flow units is more difficult than with lavas, and most of the mapping features and escarpments represent cooling unit boundaries or boundaries between different facies such as welded and non-welded tuffs. Nodular units are another, but rare feature of ignimbrites, occurring in the Lake District in the Atkinson Ignimbrites of Haweswater (Nutt 1968) and on Kidsty Pike. The nodules are commonly found at the base and top of their units and are considered to be late stage vapour-phase products. More common features seen in the field are columnar cooling joints well displayed in the Airy's Bridge Ignimbrites of Rosthwaite Fell and Base Brown [22 11] and by the Yewdale Ignimbrite near Coniston [300 982 and 275 963].

In thin section the fiamme are seen to be elongate areas of chlorite, quartz and feldspar, sometimes with flame-like ends. The outer rims are occasionally of layered quartz and feldspar only, whilst in some fiamme and in the matrix there are circular or elongate areas of coarse quartz and feldspar, probably the result of crystallization from a vapour phase. Fiamme vary from 0–45% of the rock by volume. Primary crystallization in the form of spherulites is found in some fiamme especially in the Yewdale Ignimbrite of Long Crag, Coniston. The crystal component, comprising feldspar, altered mafics, garnet and opaques, range from less than 1 to 30 per cent by volume. Large systematic increases of this type upwards in an ignimbrite unit may indicate the progressive evacuation of a zoned magma body. The feldspars are frequently highly corroded indicating disequilibrium conditions before deposition. Extraneous lithic fragments form 0 to 30 per cent by volume of the deposits and consist of small basic to acid fragments. Some deposits contain very few lithic clasts whereas others have large and variable amounts. The matrix is a fine grained devitrification quartz-alkali feldspar mosaic with perlitic cracking and distinct vitroclastic structure commonly preserved. The degree of welding is indicated by the shape and orientation of the now devitrified bicuspate, tricuspate or rod-shaped shards.

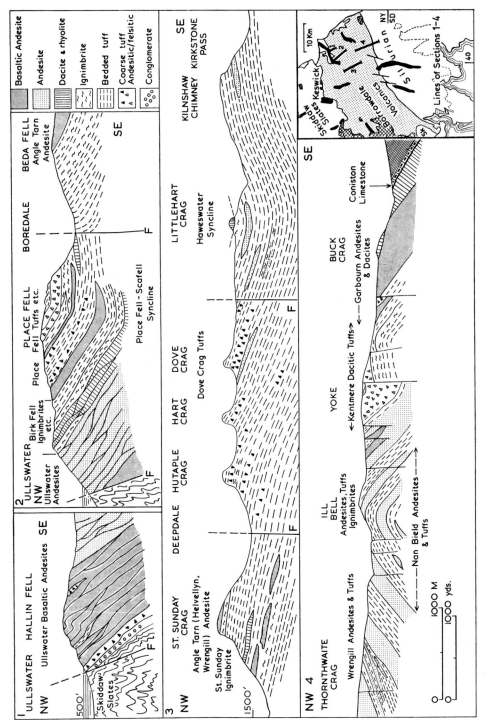

Fɪɢ. 39.   Sections across the Borrowdale volcanic outcrop from Ullswater to Kentmere.

Ignimbrite geochemistry

The bulk chemistry of the Lake District ignimbrites is extremely variable, from basaltic andesite to rhyolite, but more important than regional variations are those which occur within the individual units. They can be compared with similar variations recorded in other parts of the world, for example in Tertiary ignimbrites (Lipman 1965, 1967 and Scott 1966, 1971). The three main types of chemical variation recognized in the Lake District are as follows:

(i) Variation in the composition of units, such as those of the lower ignimbrites of Rosthwaite Fell, from dacite at the base to basaltic andesite at the top, coupled with systematic changes in certain element ratios and the crystal concentration towards the top, suggests that these units were formed by the evacuation of a magma body zoned by crystal settling, the deposits being in reverse order to their layering in the magma chamber. A similar origin is assigned to the ignimbrite near the summit of St Sunday Crag [368 137].

(ii) Large variations of crystals and lithic fragments within a sequence may be responsible for some chemical variation. For example the Yewdale ignimbrite of Torver Beck, Coniston contains up to 30 per cent lithic clasts of a basic or intermediate composition which must make some contribution to the chemistry.

(iii) Post-emplacement modifications are by far the most easily recognizable chemical changes. During cooling the ignimbrites may be subjected to chemical modification by ion-exchange processes. Orville (1963) found experimentally, and Scott (1966) found practically that as the temperature of the unit was reduced, there was an exchange between Na and K in the glass and aqueous pore fluids on an ion for ion basis. Plots of Na-K and Na-Rb and Ba show negative correlations with the slope Na/K approximating to -1 (Fig. 40). As with other post emplacement modifications, pore space is essential to the operation of such processes. Similar alkali relationships are to be found in the ignimbrites of Rosthwaite Fell, Langdale and Coniston. It has also been shown that these modifications are dependent on the degree of welding, which is inversely proportional to the original but not to the present day pore space. Changes are also associated with secondary hydration and devitrification of the glass but these processes are difficult to ascertain since the ignimbrites of the Borrowdale Volcanics are all devitrified. Groundwater leaching is also considered to be an important process in the chemical modification of ignimbrites, and in this respect the more porous tops of units show leaching of Na, K and Si and enrichment of Ca and Mg, relative to the lower parts.

Other pyroclastic rocks

Ashfall tuffs of various kinds form the largest part of the Borrowdale succession. Their compositions vary like those of the lavas from basic to intermediate, but chemical analyses have to be treated cautiously since the large pore-space makes them prone to more alteration than massive lavas, and the composition of an ash fall may change significantly away from the vent with decrease in grain size (Macdonald 1972).

Coarse tuffs, and breccias containing fragments up to one metre in diameter and often rudely bedded in units up to several metres in thickness are common. In some cases the breccias form thick lenses interbedded with other tuffs, and were probably deposited near to the vents; whereas other breccias which are completely transgressive

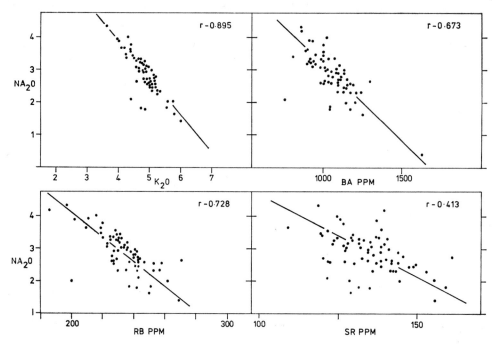

FIG. 40.   Plots of Na_2O against K_2O, Rb, Ba and Sr for the predominantly welded upper ignimbrites of Rosthwaite Fell, Borrowdale. Best fit lines through the points are shown; the slope Na/K = −1.06. Oxides in wt. per cent, elements in ppm. r – correlation coefficient.

to adjacent lavas and tuffs are best interpreted as vent agglomerates. Most of the coarse tuffs are monogenetic, highly fragmented and non-pumiceous, and in these respects resemble vulcanican type deposits. Excellent examples of the interbedded coarse tuffs are to be seen east of Dove Crag [37 10] whereas examples of possible vent agglomerates have been described by Mitchell (1963), Oliver (1961) and Clark (1964). There are similar transgressive breccias on Pavey Ark [28 08] for which a hyaloclastic origin is suspected.

Ground surge deposits have so far not been described in relation to the Borrowdale Volcanics. Such deposits are difficult to identify since they are usually thin and are rarely extensive laterally. They are typically cross bedded, mantle topography and are unlikely to survive unless rapidly covered by suceeding material (Sparks and Walker 1973). In some of the rudely bedded sequences there are thin units containing low angle cross-bedding which may have this origin.

In contrast, poorly sorted unstratified polygenetic pyroclastics are common within the Borrowdales. They contain basic and felsitic fragments, and probably represent volcanic mudflows (lahars). They have often been referred to (Hartley 1925, 1932, 1941, Mitchell 1940, Soper and Numan 1974) although there have been no detailed descriptions. The characteristics and causes of volcanic mudflows have been discussed in detail by Macdonald (1972).

Perhaps the most striking of the fragmental deposits however are the well-known green slates which have been widely quarried for ornamental stone near Coniston, Langdale, Kirkstone and Honister. They are mostly bedded lithic tuffs of varying composition, but also include crystal and 'vitric' tuffs. Since they are fine grained and often have a high chlorite content they tend to be the most strongly cleaved of the Lake District volcanic rocks, and it is the complex bedding patterns of finer pale green, and coarser dark grey alternations, so well seen on cleavage planes which give the rocks the highly desired ornamental value. Bedding features and internal structures displayed are of great variety, and in many cases of obvious subaqueous origin. Horizontal and convolute lamination, current ripples showing cross lamination, other types of small scale cross-bedding, graded bedding, flute marks, load casts with sand balls and flame structures are some of the features which show they were deposited in water. In some cases, for example in the quarry at Hodge Close [31 01], there are small faults and complex slump folds with amplitudes up to several metres easily confused with tectonic structures. Another striking variety of bedded tuff, is the accretionary lapilli tuff, composed of spherical to ellipsoidal lapilli up to one cm in diameter, the central part being fine to medium tuff, surrounded by a finer rim. Their origin is not absolutely certain but it is probable that they accreted in saturated ash clouds, and fell into shallow water either as hailstones or as water bound droplets during the heavy thunderstorms associated with volcanic eruption.

STRATIGRAPHY

Basal relations

The junction between the Borrowdale Volcanics and the older rocks has been described in Chapter 6, and little further comment is necessary. The age of the Skiddaw and Eycott rocks immediately below the volcanics is variable since these rocks had been moderately folded and eroded before the start of the Borrowdale volcanic episode. However it also appears that vulcanicity did not begin simultaneously over the whole region, and that the basal Borrowdale rocks differ in age from one area to another. Figure 37 indicates that the rather basic lavas and tuffs of the west and south-west were amongst the first to be extruded, and possibly formed a pile exceeding 1000 m in thickness before eruptions further east had started. Such variability is common in modern volcanic regions, and can be demonstrated in the Borrowdales by detailed mapping. To give one example, it is easy to follow the Birker Fell Andesites from near Buttermere 8 km across the fells to eastern Borrowdale, and to observe that the great thickness of the Honister Tuffs and Andesites disappears in this short distance.

Generalized sequence

The Borrowdales have a considerable diversity of composition which at first sight appears to be an almost random alternation of basaltic, andesitic and dacitic lavas and tuffs, including ignimbrites, totalling upwards of 6000 m. Closer scrutiny however reveals an order with several basic to acid magmatic trends (Fig. 41). Basic rocks tend to occur low in the sequence, for example in the Honister, Birker Fell and Ullswater andesites and basalts (Fig. 37), with these rocks followed by the thick accumulations of ignimbrite and acid lavas of the Airy's Bridge, Langdale, Lickle, Birk Fell and Haweswater sequences. Above these acid rocks and in places strongly unconformable upon them are andesitic to basaltic tuffs and lavas, but this is not a strong development and is quickly followed by renewed eruption of acidic material, mostly as ignimbrite, with the Lincomb Tarns, Paddy End and Atkinson acid lavas and pyroclastics as approximate correlatives. A more important return to basaltic and andesitic magmas is marked by the Wrengill, Angle Tarn and Helvellyn andesites and associated tuffs which may have formed from separate vents during one continuous eruptive phase and possibly indicates the start of a new magmatic cycle. The top of the Borrowdale Volcanic sequence contains both intermediate and acidic members which are found within such units as the Yewdale Breccia and the Kentmere and Garbourn formations and, together with the underlying Wrengill Andesites, could represent a basic to acid trend similar to that for the lower part of the sequence (Figs. 37, 38, and 41). There is therefore an indication of two, or possibly three major cycles within the Borrowdale pile, each one variable and complex, which is understandable if one considers the probability that different eruptive centres would be likely to

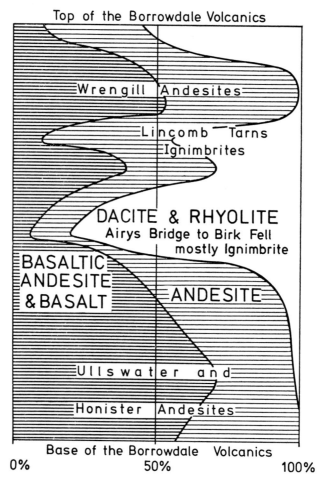

FIG. 41.   Relative variation of rock types within the Borrowdale volcanic sequence, illustrating the major phases of volcanism.

build up separate and possibly overlapping cones each with its own distinctive sequence.

It is also possible to interpret the sequences in terms of three or more major eruptive centres, each with its quota of subsidiary vents some of which may have given rise to large parasitic cones, the accumulations of one overlapping and inter-digitating with those of the next thus accounting for much of the complexity. In the east the Ullswater, Haweswater and Kentmere sequences may have originated largely from the major centre of the Haweswater Complex, although Nutt (1968) observed two quite different successions in the Haweswater district. Further west the predominantly pyroclastic deposits of vulcanian type between Helvellyn and Kirkstone (Fig. 39) are different again, perhaps indicating a large parasitic structure, or even another major vent. In the extreme west the sequence including the Honister and Birker Fell basalts, andesites and tuffs and the Airy's Bridge Ignimbrites (Oliver 1961) is quite distinct. Basic intrusions into the Skiddaw Slates, such as Squatt Knotts, Newlands may be old vents. Suthren (1977) suggested from an examination of pyro-clastic material in the Borrowdale–Thirlmere area that an eruptive centre was situated there producing much of the basaltic and andesitic material of Buller Fell, whilst Oliver (1961) and Clark (1964) show several areas of vent agglomerate.

In the SW Firman (1957) and Mitchell (1956) have described sequences so different from that of Oliver (op. cit.) that another major centre must be suspected, and the Greenscoe Vent (Soper 1970) may have been a part of this. It is also possible that the locations of some volcanic centres at the beginning of the episode differed from those operating towards the end, and it may be that the laccolithic intrusion of the St John's microgranite, which has now been dated at 445 ± 15 Ma (Wadge et al. 1974) had surface expression as a vent supplying acidic material during the late stages of the volcano.

Correlation problems

The complexities and variability of the volcanic sequences just referred to are important hazards when correlations such as those shown on figures 37 and 38 are attempted. The considerable lateral variation in the volcanic sequences, with most lava flows and pyroclastic deposits of limited extent, is expected from the nature of eruptions of this type. There are unconformities within the succession, for example between the Rainsborrow Tuffs and Kentmere Dacites (Soper and Numan 1974), indicating that extensive erosion frequently occurred between successive eruptions, and it is impossible to determine the proportion of material deposited during one episode and removed before the next. There are many localities where sequences deposited subaerially are followed abruptly by sub-aqueous deposits and vice versa. Sequences of bedded tuff with sub-aqueous structures underlain or followed by ignimbrite are common, and probably indicate either the sporadic rise and fall of volcanic islands, or alternatively the periodic formation of extensive but temporary lakes. To all these complexities must be added the later Caledonian tectonism with folding, cleavage and strong faulting. The latter can be particularly confusing since it may be uncertain whether different sequences on the opposite sides of a drift-filled valley are a result of abrupt lateral variation, or of faulting along the valley, and the most detailed mapping may fail to resolve the problem. Indeed in many cases both factors may operate since an abrupt change of facies may have provided the discon-tinuity along which a fault has most easily developed. For all these reasons figures

37 and 38 must be regarded as simplifications, and although some of the major correlations seem obvious, the possibilities of separate events giving comparable sequences at different times must not be discounted. For example, it has been indicated above and on figure 37 that the great development of the Airy's Bridge Ignimbrites are laterally equivalent to others across the width of the Lake District, mostly on the basis of similar successions. In detail, however, these successions are sufficiently different to suggest at least that they were extruded from several different eruptive centres, and once this possibility has been admitted their equivalence in time cannot be assumed.

Other features of the vulcanicity

One feature of particular interest which has attracted attention in recent years has been the nature and origin of the ignimbrites which form such an important part of the Borrowdale succession. There are two types of sequence. The first, typified by the Airy's Bridge Ignimbrites of Borrowdale and Langdale, consists of thick sheets, often preceded and followed by acidic lava. The association of acid lava with ignimbrite characterizes the formation of calderas and it is possible that the widespread Airy's Bridge lavas and ignimbrites were erupted from a separate 'acid' volcano, which collaped after evacuation of large amounts of pyroclastic flow material. An intraformational unconformity between the ignimbrites and later acid lavas on Rosthwaite Fell may mark such a caldera collapse whilst the impressive but restricted breccias (possibly hyaloclastic) of Pavey Ark have textures which suggest they could have been formed by subsequent eruption in a crater lake. The second type of ignimbrite consists of thin units of dacitic material associated with thick ash fall pyroclastics. In many recent volcanic areas, including northern Chile (Francis et al. 1974) similar occurrences have shown that thin ignimbrite units have been erupted from central vents during an eruptive history of mainly andesitic material. It is therefore suggested that the ignimbrites of St Sunday Crag and Kidsty Pike were erupted from the same vents as the underlying and overlying pyroclastic fall material.

Another interesting problem not yet completely resolved is the question of subaerial versus subaqueous deposition of many lavas and tuffs. There is no doubt that the beautifully bedded tuffs of Honister, Langdale and Coniston were subaqueous, because of sedimentary structures they display, but there are doubts about other bedded tuffs and in cases where they overlie ignimbrites this could be important in view of recent claims for subaqueous ignimbrite. Furthermore whether the environment of the subaqueous tuffs was lacustrine or marine is important from the geotectonic viewpoint. Marine microfossils have recently been found in mudstones interbedded with the Duddon Hall Tuffs in Holehouse Gill, [182 926] (Numan 1974). This evidence of marine deposition, plus the occurrence of other mudstone horizons within the lowest kilometre or so of the Borrowdale volcanic succession, adds weight to the view that the Ordovician volcanic rocks of the Lake District were formed in an ancient island arc, although the islands were undoubtedly underlain by continental crust. Thus the overall picture which emerges is that of a chain of volcanic islands arranged along the continental margin of ancient Europe. These islands were predominantly stratovolcanoes composed of interbedded lava and pyroclastic deposits, but with acid shields of ignimbrite and acid lava. The newly erupted deposits must have been continuously subjected to destructive erosion, and redistributed within the subaqueous basins around the islands.

Acknowledgements

The authors wish to thank J. G. Fitton for permission to use unpublished data on the Eycott lavas in figure 36, and to A. J. Wadge for critically reading early drafts of the manuscript. One of us (DM) gratefully acknowledges financial assistance in the form of a NERC research studentship.

REFERENCES

CLARKE, L. 1964. The Borrowdale Volcanic Series between Buttermere and Wasdale, Cumberland. *Proc. Yorks. geol. Soc.* **34**, 343–56.

DOWNIE, C. and SOPER, N. J. 1972. Age of the Eycott Volcanic Group and its conformable relationship to the Skiddaw Slates in the English Lake District. *Geol. Mag.* **109**, 259–68.

EASTWOOD, T., HOLLINGWORTH, S. E., ROSE, W. C. C. and TROTTER, R. M. 1968. Geology of the country around Cockermouth and Caldbeck. *Mem. Geol. Surv. Great Britain.*

FIRMAN, R. J. 1957. The Borrowdale Volcanic Series between Wastwater and the Duddon Valley, Cumberland. *Proc. Yorks. geol. Soc.* **31**, 39–64.

FITTON, J. G. 1971. The petrogenesis of the calc-alkaline Borrowdale Volcanic Group, Northern England. *Unpubl. Ph.D. Thesis Univ. Durham.*

—— 1972. The Genetic Significance of Almandine-Pyrope Phenocrysts in the Calc-alkaline Borrowdale Group. Northern England. *Contrib. Mineral. Petrol.* **36**, 231–48.

—— and HUGHES, D. J. 1970. Volcanism and Plate Tectonics. *Earth & Planet. Sci. Lett.* **8**, 223–28.

FRANCIS, P. W., ROOBOL, M. J., WALKER, G. P. L., COBBOLD, P. R. and COWARD, M. 1974. The San Pedro and San Pablo Volcanoes of Northern Chile and their hot avalanche deposits. *Geol. Rundschau.* **63**, 357–88.

GIBSON, I. L. and TAZIEFF, H. 1967. Additional theory of Origin of Fiamme in Ignimbrites. *Nature* **215**, 1473.

HARTLEY, J. J. 1925. The Borrowdale Volcanic Series of Grasmere, Windermere and Coniston. *Proc. Geol. Ass.* **36**, 203–26.

—— 1932. The Volcanic and other igneous rocks of Great and Little Langdale. *Proc. Geol. Ass.* **43**, 32–69.

—— 1941. Geology of Helvellyn and the southern part of Thirlmere. *Jl Geol. Soc. Lond.* **97**, 129–62.

HEDBERG, H. D. 1972. International Subcommission on Stratigraphic Classification (ISSC) of IUGS Commission on Stratigraphy: *Int. Union. geol. Sci., Int. Subcomm. Stratigr.* No. 43, 59 pp.

HELM, D. G. and SIDDANS, A. W. B. 1971. Deformation of a slaty lapillar tuff in the English Lake District. Discussion. *Bull. geol. Soc. Am.* **82**, 523.

JAKES, P. and WHITE, A. J. R. 1972. Major and trace element abundances in volcanic rocks from orogenic areas. *Geol. Soc. Amer. Bull.* **83**, 29–40.

LIPMAN, P. W. 1965. Chemical composition of glassy and crystalline volcanic rocks. *U.S. geol. Surv. Bull.* **1201–D**, D1–24.

—— 1967. Mineral and chemical variations within an ash-flow sheet from Aso Caldera, Southwestern Japan. *Contrib. Mineral. Petrol.* **16**, 300–27.

MACDONALD, G. A. 1972. Volcanoes. Prentice Hall 510 pp.

MARR, J. E. 1900. Notes on the Geology of the English Lake District. *Proc. Geol. Ass.* **16**, 449–83.

MARSHALL, P. 1932. Notes on some volcanic rocks of the North Island of New Zealand. *N.Z. Jl Sci. Tech.* **13**, 198–202.

MITCHELL, G. H. 1940. The Borrowdale Volcanic Series of Coniston, Lancs. *Q. Jl geol. Lond.* **96**, 301–19.

—— 1956. The Borrowdale Volcanic Series of the Dunnerdale Fells, Lancashire. *L'pool Manchr geol. Jl.* **1**, 428–49.

—— 1963. The Borrowdale volcanic rocks of the Seathwaite Fells, Lancashire. *Geol. J.* **70**, 289–300.

MILLWARD, D. 1976. Petrology, vulcanicity and geochemistry of selected areas of the Borrowdale Volcanics of the English Lake District. *Unpubl. Ph.D. Thesis Univ. Birmingham.*

MOSELEY, F. 1960. The succession and structure of the Borrowdale Volcanic Series South-East of Ullswater. *Q. Jl geol. Soc. Lond.* **116**, 55–84.

—— 1964. The succession and structure of the Borrowdale Volcanic Rocks North-West of Ullswater. *Geol. J.* **4**, 127–42.

—— 1972. A tectonic history of North-West England. *Jl geol. Soc. Lond.* **128**, 561–98.

NUMAN, N. M. S. 1974. Structure and stratigraphy of the southern part of the Borrowdale Volcanic Group, English Lake District. *Unpub. Ph.D. Thesis*, University of Sheffield.

NUTT, M. J. C. 1968. Borrowdale Volcanic Series and associated rocks around Haweswater, Westmorland. *Proc. geol. Soc.* **1649**, 112.

OLIVER, R. L. 1954. Welded Tuffs in the Borrowdale Volcanic Series, English Lake District, with a note on similar rocks in Wales. *Geol. Mag.* **91**. 473–83.

—— 1961. The Borrowdale Volcanic and associated rocks of the Scafell area, English Lake District. *Q. Jl geol. Soc. Lond.* **117**, 377–417.

ORVILLE, P. M. 1963. Alkali ion exchange between vapour and feldspar phases. *Am. Jl Sci.* **261**, 201–37.

ROSS, C. S. and SMITH, R. L. 1961. Ash flow tuffs; their origin, geologic relations and identification. *U.S. geol. Surv. Prof. Pap.* **366**.

SCOTT, R. B. 1966. Origin of Chemical Variations within Ignimbrite Cooling Units. *Am. Jl Sci.* **264**, 273–88.

—— 1971. Alkali exchange during devitrification and hydration of glasses in ignimbrite cooling units. *Jl Geol.* **79**, 100–10.

SIEGERS, A., PICHLER, H. and ZEIL, W. 1969. Trace element abundances in the 'Andesite' Formation of Northern Chile. *Geochim. Cosmochim Acta.* **33**, 882–87.

SMITH, R. L. 1960. Zones and Zonal Variations in welded ash flows. *U.S. Geol. Surv. Prof. Paper.* **354** F.

SOPER, N. J. 1970 Three critical localities on the junction of the Borrowdale Volcanic rocks with the Skiddaw Slates in the Lake District. *Proc. Yorks. geol. Soc.* **37**, 461–93.

—— and NUMAN, N. M. S. 1974. Structure and stratigraphy of the Borrowdale Volcanic rocks of the Kentmere area, English Lake District. *Geol. J.* **9**, 147–66.

SPARKS, R. S. J. and WALKER, G. P. L. 1973. The Ground Surge Deposit: a third type of pyroclastic Rock. *Nature*, **241**, 62–4.

SUTHREN, R. 1977. Sedimentary processes in the Borrowdale Volcanic Group. In: Conference Report — Palaeozoic volcanism in Great Britain and Ireland, C. J. Stillman. *Jl geol. Soc. Lond.* **133**, 410–11.

TAYLOR, S. R. and WHITE, A. J. R. 1965. Geochemistry of Andesites and the growth of Continents. *Nature.* **208**, 271–73.

TEALL, J. J. H. 1888. British Petrography; with special reference to the igneous rocks. London.

THORPE, R. S., POTTS, P. J. and FRANCIS, P. W. 1976. Rare Earth Data and Petrogenesis of Andesites from the North Chilean Andes. *Contr. Mineral. Petrol.* **54**, 65–78.

TORSKE, T. 1975. Snowflake texture in Ordovician rhyolite from Stord, Hordaland *Norges geol. Unders.* **319**, 17–27.

WADGE, A. J. 1972. Sections through the Skiddaw–Borrowdale unconformity in Eastern Lakeland. *Proc. Yorks. geol. Soc.* **39**, 179–98.

WADGE, A. J., NUTT, M. J. C., LISTER, T. R. and SKEVINGTON, D. 1969. A probable Didymograptus murchisoni zone fauna from the Lake District. *Geol. Mag.* **106**, 595–98.

WADGE, A. J., HARDING, R. R. and DARBYSHIRE, D. P. F. 1974. The rubidium-strontium age and field relations of the Threlkeld Microgranite. *Proc. Yorks. geol. Soc.* **40**, 211–22.

WARD, J. C. 1876. The Geology of the northern part of the Lake District. *Mem. Geol. Surv. England and Wales.*

D. MILLWARD, PH.D.
Institute of Geological Sciences, South Kensington, London SW7

F. MOSELEY, D.SC., PH.D.
Department of Geological Sciences, University, Birmingham

N. J. SOPER, PH.D.
Department of Geology, University, Sheffield

9

The Upper Ordovician and Silurian Rocks

J. K. INGHAM, K. J. McNAMARA and R. B. RICKARDS

THE CONISTON LIMESTONE GROUP

J. K. INGHAM AND K. J. McNAMARA

By early to mid Caradoc times the area of deposition and outpouring of the rocks of the Borrowdale Volcanic Group had become primarily subaerial and was undergoing active erosion, particularly in the south of the Lake District, for the overlying late Ordovician sediments of the Coniston Limestone Group are markedly transgressive across this mid-Ordovician landscape. They rest unconformably on Borrowdale rocks over much of the area but in the extreme south the group not only shows pronounced overlap but also oversteps on to pre-Borrowdale strata (Fig. 42a) (see also Ingham and Rickards 1974, p. 32). The area south of Coniston shows the greatest angular discordance between the two groups.

Caradoc strata of the northern Lake District and Cross Fell Inlier

The transgression was essentially from north to south for the earliest known post Borrowdale volcanic sediments, of mid Caradoc (Longvillian) age, are the Drygill Shales of Carrock Fell in the northern Lake District and the Corona Beds and Melmerby Beds of the Cross Fell Inlier, also of Longvillian age (Dean 1959, 1963). The Drygill Shales form a small, faulted outlier within the Carrock Fell igneous complex and consist of cleaved, calcareous ashy mudstones containing a diagnostic trilobite and brachiopod fauna, whereas the Corona Beds are purple and grey mudstones and the Melmerby Beds consist mainly of dark grey calcareous mudstone with occasional limestone nodules. *Broeggerolithus* is the most notable trilobite genus present. These equivalent units, together with the overlying Dufton Shale Formation, evidently comprise a thick succession but the Cross Fell Inlier consists of such faulted and variably exposed ground that accurate thickness assessments cannot be made. Nevertheless, all the stages of the upper part of the Caradoc Series are represented, together with the Pusgillian and part of the Cautleyan Stages of the Ashgill Series (Ingham and Wright, *in* Williams et al., 1972, p. 46). Reliable correlation with the stages of the type Caradoc Series in Salop has been affected by means of the well-documented shelly faunas of Anglo-Welsh-Baltic aspect (Dean 1959, 1962).

The Howgill Fells succession

In the Howgill Fells, the lithology of almost 600 m of late Ordovician strata is a monotonous grey calcareous mudstone with variably developed limestone nodules (Cautley Mudstone Formation). It is similar to that of the Dufton Shale Formation

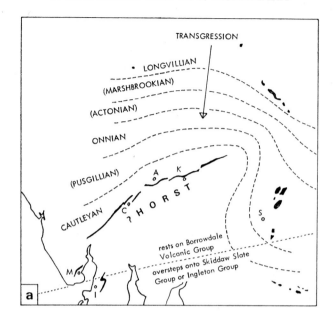

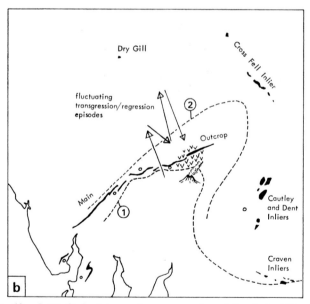

FIG. 42.   Late Ordovician palaeogeography of the Lake District.

a.   The southerly overlap and overstep at the base of the Coniston Limestone Group in terms of the ages of the oldest Coniston Limestone strata resting on earlier rocks; interpolated information in parentheses. (Modified from Ingham and Rickards 1974, fig. 9a, in the light of recent work on the main outcrop by McNamara).

b.   Two Ashgill shorelines (1, mid-Cautleyan; 2, mid-Rawtheyan) indicating areas affected by elevation and submergence of a probable horst region in the southern Lake District. The volcano suggested could be the source of the mid-Cautleyan Yarlside Rhyolite and, in late Rawtheyan times, the ash of the western Lake District and the Cautley Volcanics of the Cautley and Dent inliers.
I, Ireleth; M, Millom; C, Coniston; A, Ambleside; K, Kentmere; S, Sedbergh.

in the Cross Fell Inlier but in the Howgill Fells, where Ordovician rocks reach the surface in a number of inliers in the Cautley and Dent districts, adjacent to the Dent Fault belt, the basal, locally sandy beds are younger. In the north-easternmost inlier, the basal Cautley Mudstone Formation, of Onnian age appears to follow a purple andesite which is probably referable to the Borrowdale Volcanic Group (Ingham 1966, p. 463; 1974, p. 31). The bulk of the Cautley Mudstone Formation of this district comprises much of the type succession for the Ashgill Series (Ingham and Wright 1970, pp. 233–34): stages and biozones are recognized and defined in this most complete late Ordovician sequence in northern England (Fig. 43). The trilobite faunas which are largely documented (Ingham 1966, 1970, 1974, 1977) have already provided a basis for correlation both with other Coniston Limestone Group successions in the region and with sequences further afield in Britain and abroad.

The strata of Onnian age are locally very fossiliferous and some horizons are crowded with the trinucleid trilobite *Onnia superba pusgillensis* Dean and also *Flexicalymene onniensis lata* Ingham. The unweathered beds are fairly dark and are followed conformably by similarly dark Cautley Mudstones with diagnostic low Ashgill (Pusgillian) faunas. The beds of Cautleyan age are usually a little lighter in colour and sometimes silty: locally, near Dent, there is a thick sandstone member within the second biozone. Shelly faunas are more diverse in the zones of this stage than in any other part of the Ashgill Series. Typical trilobite genera include *Tretaspis*, '*Diacalymene*', *Gravicalymene*, *Prionocheilus*, *Acidaspis*, *Primaspis*, *Platylichas*, *Illaenus*, *Stenopareia*, *Stygina*, *Atractopyge*, *Cybeloides*, *Celtencrinurus*, *Chasmops* and *Calyptaulax*. Strophomenoid and orthoid brachiopods abound in numbers and variety and the distribution of species suggests an approximate correlation with the Lower Drummuck Group of Girvan in south-western Scotland.

The Cautley Mudstones of Rawtheyan age contain more restricted faunas with *Tretaspis*, *Kloucekia*, *Sphaerocoryphe*, *Dindymene* and *Trinodus* being the dominant trilobites. Graptolites, referable to the *Dicellograptus anceps* Zone occur in the higher beds.

Towards the top of the Cautley Mudstone Formation, the beds become increasingly ashy, particularly in Zones 6 and 7 and within this part of the succession there is a well developed, predominantly rhyolitic ash unit, the Cautley Volcanic Formation, which thickens towards the west. The thin Cystoid Limestone, of latest Rawtheyan age, follows the Cautley Mudstone Formation unconformably but in the northernmost inliers this break is thought to be small. Locally however, the Cystoid Limestone rests on strata of Rawtheyan Zone 5 age suggesting the existence of a small, positive, perhaps fault-bounded area, active during late middle Ashgill times.

The Cystoid Limestone, as its name implies, contains a varied assortment of diploporitan and rhombiferan cystoids, together with trilobites and brachiopods of essentially Rawtheyan aspect (Ingham and Wright, *in* Williams et al., p. 47). The dalmanitid trilobite *Mucronaspis* makes its first appearence in the Howgill Fells at this level (*M. olini* (Temple)), although the genus is known from Pusgillian beds in the Cross Fell Inlier. Above the Cystoid Limestone is a thick (over 90 m) succession of relatively non-calcareous strata, the Ashgill Shale Formation, which contains a restricted *Hirnantia* brachiopod assemblage and an even more restricted trilobite fauna, consisting of but one species, *Mucronaspis mucronata* (Brongniart). Near the top of this unit is a calcareous sandstone and conglomerate lens which develops towards the south or south-east. The Ashgill Shale Formation is correlated with the highest recognized Ashgill stage, the Hirnantian, whose type section is the Foel-y-Ddinas

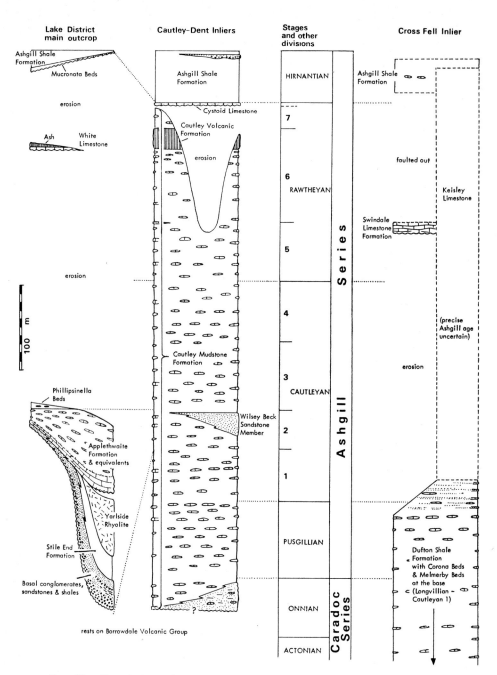

FIG. 43.   Correlation of the Lake District and nearby sequences with the
British late Ordovician standard.

Mudstone of the Bala district in North Wales (Ingham and Wright 1970). Above the Ashgill Shale Formation is a thin argillaceous limestone at the base of the Stockdale Shales which is currently regarded as representing the lowest Silurian in the area, although the precise position of the Ordovician/Silurian boundary is still a matter for international debate and a boundary stratotype has not yet been selected (see below).

The Ashgill Series in the main Lake District outcrop

The main outcrop of the Coniston Limestone Group is in the southern part of the Lake District and extends from near Shap, in the north-east, as a faulted, but linear, strip for some 50 km towards the Duddon Estuary in the south-west. The succession is not a thick one: this reflects both the relatively young age of the basal beds and also the two substantial stratigraphical breaks which have been recognized in the sequence (Fig. 43). The succession is far from complete when compared with that at Cautley in the Howgill Fells, but more varied lithologically, and this is almost certainly a reflection of deposition on, or along the north-west flanks of a pene-contemporaneous horst-like structure which seems to have extended in the direction of the Cross Fell Inlier and partly separated the Howgill Fells area from the main body of the Lake District. The effect of this structure is evident from details of the Lake District Ashgill succession and it seems to have affected early Silurian sedimentation also (see below.)

Recent work by one of us (K. J. McN.) on the Coniston Limestone Group of the main outcrop has necessitated a major revision, particularly of the lower part of the succession. Previously, the Stile End Formation, which has usually been interpreted as including the sandstone and conglomerate at the base of the sequence, was correlated with the late Caradoc Actonian Stage of Salop (Dean 1963) and this interpretation was followed by Ingham and Rickards (1974) when they attempted a palaeogeographical reconstruction. It is now clear that not only do the Applethwaite Formation and overlying Phillipsinella Beds correlate with the mid Cautleyan Stage (Zones 2 and 3) of the Howgill Fells but the Stile End Formation and the basal beds equate with Cautleyan Zone 2 also. Obviously, the intervening Yarlside Rhyolite is of this age too.

This affects the overlap reconstruction to a considerable extent (Fig. 42a). The horst-like structure, whose effect had been recognized previously (Ingham and Rickards 1974, figure 9b) owing to its influence on mid-Ashgill successional details, now becomes evident at the initial transgressive phase.

The basal beds of the Coniston Limestone Group consist of variably thick (up to 35 m) conglomerates and sandstones with some subordinate red and green shale and siltstone horizons. This unit, here regarded as a formation separate from the overlying Stile End Formation is probably diachronous and its varying thickness reflects a transgression over a fairly irregular Borrowdale Volcanic surface. It is mostly unfossiliferous and initially probably reflects a fluviatile regime, particularly in the north-east where conglomerate-sandstone-shale cycles are well developed. The higher sandstones, however, are certainly marine and at Pull Beck, the trilobites *Decoroproetus piriceps* (Ingham) and *Celtencrinurus cornutus* (Ingham) are known, whereas south-west of High Pike Haw, *Chasmops marri* (Reed), *Decoroproetus* and corals have been recovered (Fig. 44). South-west of High Pike Haw the development of the basal beds varies somewhat but near Millom the thickness again increases to about 40 m.

The Stile End Formation, used here in the restricted sense following Harkness and Nicholson (1877), is only present in the eastern part of the outcrop, mainly to the north-east of Kentmere, where up to 28 m of grey, calcareous siltstones, coarsening westwards, are developed. A marked diachronous relationship to the basal sandstones and conglomerates is suspected. The trilobite fauna is an essentially mid Cautleyan (Zone 2) assemblage dominated by *Chasmops marri*, *Decoroproetus piriceps*, *Celtencrinurus cornutus*, *Erratencrinurus kingi* (Dean), *Prionocheilus cautleyensis* Ingham and species of *Ascetopeltis* (not known from pre-Ashgill strata), *Gravicalymene* and *Atractopyge*. Orthoid and strophomenoid brachiopods, corals, bryozoans and molluscs also occur.

The Yarlside Rhyolite, up to about 70 m thick, follows the Stile End Formation conformably and has thermally metamorphosed its highest beds. Its outcrop is similarly restricted to the area north-east of Kentmere. As the overlying Applethwaite Formation is now believed to rest conformably on the basal beds of the Coniston Limestone Group along much of the main outcrop, it seems likely that the area of the present outcrop of the Yarlside Rhyolite reflects the limited western extent of the lava outpourings. The basal Applethwaite Formation is locally conglomeratic east of Longsleddale, suggesting local erosion at the top of the rhyolite prior to further submergence. On the palaeogeographical reconstruction (Fig. 42b), the source volcano has been placed on the horst-like structure referred to earlier and nearby the present outcrop. Such a volcanic centre also could well have supplied the widespread rhyolitic ashes known from the later Ashgill in northern England.

The Applethwaite Formation is a persistent unit along much of the main outcrop although there are some striking thickness variations. The lithology is mainly one of blue-grey calcareous mudstone with nodules and beds of argillaceous limestone. The basal strata, consisting dominantly of limestone beds with minor interbeds of shale, form a clearly defined and widespread sub-unit. Bedded limestones and shales also characterize a level near the summit of the formation between Troutbeck and Kentmere. The thickest development of the Applethwaite Formation is seen in the Kentmere area where about 100 m of beds are known. Here, a conformable top is not seen owing to overstep by a higher Ashgill horizon. To the north-east of Kentmere, more upper Applethwaite strata are absent at this level and this is clearly a contributory factor to the much reduced thickness of the formation in Longsleddale (Fig. 44). The thickness of the Applethwaite Formation is again reduced in the central part of the Lake District outcrop but as a conformable top is known in this sector it is evident that the basal limestones are diachronous and probably younger here than in the Kentmere area. South-west of Torver, the formation begins to thicken but changes its character, an equivalent unit developing which consists of much siltier beds, not unlike those of the Stile End Formation.

The basal Applethwaite limestone unit persists and thickens towards the south-west also with shale horizons becoming less common: near Millom it reaches 28 m in thickness. About 2 km north of Millom, this horizon is developed as a purer, crystaline limestone which as yet has not yielded a fauna.

The typical Applethwaite Formation has yielded a large fauna of trilobites which enable a convincing correlation to be made with the Cautleyan Stage (Zone 2) of the type area. '*Diacalymene*' cf. *marginata* Shirley is the most common calymenid and, among others, species of *Stygina*, *Illaenus*, *Stenopareia*, *Pseudosphaerexochus* and *Tretaspis* also occur. The silty facies, developed above the basal limestones south-west of Torver, has yielded both *Errantencrinurus* and *Celtencrinurus* and thus not only

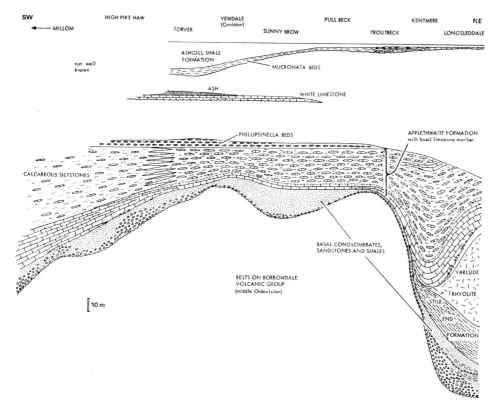

FIG. 44. Lateral variation within the Coniston Limestone Group (Ashgill Series) of the main outcrop in the southern Lake District; based on recent work by McNamara.

resembles the Stile End Formation lithologically but contains a virtually identical fauna.

The Phillipsinella Beds, first recognized by Marr (1916) as a distinct unit, overlies the Applethwaite Formation conformably but is only preserved south-west of Trout-beck: it reaches but 7 m in thickness and consists of blue argillaceous limestone containing a fauna which resembles that of the underlying Applethwaite Formation, except that calymenid trilobites are rare. *Calymene* cf. *prolata* Ingham is known, however, and this, together with the relative abundance of *Phillipsinella parabola aquilonia* Ingham, permits a correlation with Cautleyan Stage, Zone 3 of the Howgill Fells. South-west of High Pike Haw, the Phillipsinella Beds and higher strata of the Coniston Limestone Group are poorly exposed and consequently not well under-stood.

During late Cautleyan or early Rawtheyan times the area of the main Lake District outcrop became emergent and subject to erosion for the next unit of the Coniston Limestone Group, the White Limestone of Marr (1916), is of fairly late

Rawtheyan age and thus rests disconformably on the Phillipsinella Beds. This situation is paralleled in the Cross Fell Inlier (see below) but not at Cautley in the Howgill Fells and it reflects the temporary emergence of the southern Lake District horst-like structure already referred to.

The White Limestone is thin, not exceeding 3 m, and is a pale grey argillaceous limestone, brown weathering but commonly with a thin, white surface layer. It has yielded a sufficiently diagnostic fauna to place it at the summit of Zone 6 of the Rawtheyan Stage and, as at Cautley, it is followed by a rhyolitic ash. The latter, however, does not exceed 5 m in thickness and is rapidly cut out by overstepping late Ashgill strata towards the north-east. Nevertheless, the existence of a volcanic horizon at this level indicates the originally widespread development of the Cautley Volcanic Formation: it is even known as far east as the Austwick and Horton-in-Ribblesdale Inlier of the Craven district (Ingham and Rickards 1974, p. 35).

The Mucronata Beds (= Phacops mucronatus Beds of Marr, 1916) constitute the succeeding unit in the Lake District succession. This is believed to be of latest Rawtheyan age and thus rests disconformably on the rhyolitic ash near Torver. Further north-east it has overstepped on to the White Limestone and beyond Pull Beck, pre-Mucronata Beds erosion had cut down on to even earlier strata, the unit firstly transgressing the Phillipsinella Beds and finally, towards Longsleddale, resting on various levels within the Applethwaite Formation (Fig. 44). This stratigraphical break is more widespread than the one below the White Limestone and it represents a general emergence: it is recognized both at Cautley and in the Craven district (Ingham and Wright *in* Williams et al. 1972, figure 8; Ingham and Rickards 1974, figures 7, 8).

Near Torver, the Mucronata Beds comprise 4 m of greenish-grey calcareous shales and although the unit persists, as far as known, along the entire Lake District outcrop, the thickness and lithological characters vary somewhat. In Longsleddale, it consists of a thin, honeycomb-weathering argillaceous limestone but thickens south-westwards so that near Troutbeck both limestones and shales are represented in a 5 m thick succession, one of the limestones being crowded with endoceratids.

The trilobite fauna of the Mucronata Beds closely resembles that of the Cystoid Limestone at Cautley, being dominated by *Mucronaspis olini*. Other forms also indicate a correlation with the Upper Drummuck Group of Girvan. The brachiopod fauna is either like that from Cautley or, in the shaley facies, contains elements of a *Hirnantia* fauna. This suggests that the *Hirnantia* fauna is strongly facies controlled and may even occur in pre-Hirnantian strata in Britain; but this does not preclude Mucronata Beds being diachronous, partly of Rawtheyan and partly of Hirnantian age. *M. olini* is confined to the Mucronata Beds in the Lake District, suggesting their equivalence throughout the main outcrop but this may be misleading as *M. olini* occurs stratigraphically higher in the Cross Fell Inlier (see below).

As at Cautley, the highest beds of the Coniston Limestone Group in the Lake District constitute the Ashgill Shale Formation, lithologically almost identical with its Cautley counterpart, although it is the basal beds which are locally sandy. However, the Lake District development is much thinner, reaching a maximum of 20 m near Torver. The formation thins rapidly to the north-east so that to the east of Longsleddale, little more than 1 m is present. A shoreline for Ashgill Shale times was probably little different from that proposed for mid-Cautleyan times (Fig. 42b). The fauna is dominated by brachiopods of the *Hirnantia* assemblage and only one trilobite species, *Mucronaspis mucronata* is known, as at Cautley. The Ashgill Shale Formation is succeeded conformably by Silurian Stockdale Shales (Skelgill Beds).

The Ashgill Series in the Cross Fell Inlier

In the vicinity of Knock and Dufton, the youngest Dufton Shale Formation is known and beds of Pusgillian age and low Cautleyan (Zone 1) have been recognized. The highest Pusgillian and low Cautleyan strata near Dufton are developed in an atypical sandy facies and the faunas are perhaps more diverse than those from equivalent beds in the Howgill Fells. The earliest known representative of the genus *Phillipsinella* in northern England (*P. preclara* Bruton) is found in the highest Pusgillian strata of Billy's Beck, near Dufton.

In Swindale Beck, near Knock, the argillaceous and nodular Swindale Limestone, of mid Rawtheyan age rests unconformably on Pusgillian strata, yet in Billy's Beck low Cautleyan strata are preserved although the junction with the Swindale Limestone is faulted. This stratigraphical break is almost certainly the same one as that occurring beneath the White Limestone in the Lake District and it shows that the mid Ashgill emergence of the horst-like area extended at least as far as the Cross Fell Inlier.

The fauna of the Swindale Limestone is extensive and over thirty trilobite species have been found. A reliable correlation with a level at or near the Zone 5/Zone 6 boundary in the Rawtheyan Stage of the Howgill Fells is indicated.

The top of the Swindale Limestone is nowhere seen and the next youngest unit is the Ashgill Shale Formation, neither of whose boundaries is seen. In Swindale Beck, a more calcareous development within the Ashgill Shale Formation has yielded *Mucronaspis olini* in association with a *Hirnantia* fauna and cystoids, whereas the typical Ashgill Shale facies contains *M. mucronata* and the *Hirnantia* fauna, as elsewhere. This may have an important bearing on the dating of the Mucronata Beds in the Lake District, already noted.

In and around Keisley Quarry, east of Dufton, a fault-bounded area of Keisley Limestone is seen. This contains both massive crystalline and bedded limestones and represents a 'reef'-type facies similar to that of the Boda Limestone in central Sweden and the Chair of Kildare Limestone in Ireland. The fauna is clearly of Ashgill age but its precise position is not certain. It is possible that it represents a carbonate mound of Rawtheyan age, with the Swindale Limestone perhaps representing a flank facies but both lower and high Ashgill ages have also been argued (Ingham and Wright *in* Williams, et al. 1972, p. 47).

A *Hirnantia* fauna from thin decalcified limestones apparently overlying the Keisley Limestone and regarded as Llandovery in age by Temple (1968) are probably Hirnantian and these beds are thus likely to be correlatives of some part of the Ashgill Shales.

The occurrence of the Keisley facies in part of the Cross Fell Inlier may relate to the alignment and position of the southern Lake District horst-like structure (see Ingham and Rickards 1974, p. 36)

The graptolite zones of the Ashgill Series

Most of the evidence for correlation of late Ordovician graptolite zones with the type Ashgill Series lies mainly outside northern England (e.g. Girvan, in southwest Scotland). *Dicellograptus anceps* Zone faunas are not known from beds younger than those dated as Rawtheyan. Pusgillian correlatives contain the horizon of the *D. complanatus* fauna but also probably the higher part of the *Pleurograptus linearis* Zone (Ingham and Wright 1970).

SILURIAN

R. B. RICKARDS

LLANDOVERY

The Ordovician — Silurian Boundary

Since the choice of stratotype and horizon for this boundary is currently the subject of international debate, the horizon taken here is the traditional one in the graptolite facies; the base of the *Glyptograptus persculptus* Zone. The Lake District is of considerable interest in this respect, for east of Ambleside, and as far as the Howgill Fells, the *persculptus* Zone is unproven in several continuously exposed sequences, whereas west of Ambleside an excellent passage from the highest Ordovician rocks reveals a distinctive *persculptus* fauna followed by a black shale sequence from the *acuminatus* Zone upwards.

In the Howgill Fells an average of 1 m of impure shelly limestone and pale, pyritous mudstone is followed by 0.1 m of *acuminatus* Zone black shales (Rickards 1970, 1974). Below these 'Basal Beds' are the Ashgill Shales in type development, and whilst the former may be regarded as being approximately equivalent to the *persculptus* Zone elsewhere, this has not been proven. The nearest good Lake District section to the Howgill Fells is found in Browgill [497 060] a tributary of Stockdale Beck; both streams lending their names to the standard Browgill Beds and Stockdale Shales of the Lake District Llandovery. Here at least 1 m of black shales, yielding an *acuminatus* Zone fauna, directly and conformably overlies about 1 m of brown-weathering grey shales of certain Ordovician age. Thus in this section there is a non-sequence at the level of the *persculptus* Zone. The highest parts of the stream [500 062] have 0.051 m of bioturbated beds above a thin development of Ashgill Shales and below the *acuminatus* Zone strata (Hutt 1974, p. 5). Strike faults separate these two localities and strike slip movement may have brought into juxtaposition the rather different developments.

Further west, in the classic sections in Skelgill Beck [395 032], 'Basal Beds' (0.17 m) are again present overlying undoubted Ordovician strata, but here they are overlain by upper *atavus* Zone black shales (Hutt 1974) so that they are approximately equivalent to the *persculptus*, *acuminatus* and lower *atavus* zones so superbly developed only a few kilometres to the west of Ambleside. In the latter area the boundary is well seen at Yewdale Beck [305 986] where the Coniston Limestone is followed by rather less than 1 m of Ashgill Shales and then by 0.3 m of bluish mudstone with a mixed shelly and graptolitic fauna, the latter clearly indicating the *persculptus* Zone (Rickards and Hutt 1960, Hutt 1974). Thus the western Lake District would make a good stratotype for the Ordovician-Silurian boundary, possessing easily accessible, well-exposed sections with a mixed graptolite and shelly fauna at the crucial levels. However, recent work by the authors of this chapter in Dobb's Linn in the Southern Uplands indicates that the Scottish section may be equally useful and have a greater potential for international correlation. Nevertheless the sections in Ashgill Beck [268 953] and Yewdale Beck are highly instructive and well-exposed. Figure 45 summarizes the stratigraphy at this level from the Howgill Fells to Yewdale Beck. An interpretation of the depositional environment of the boundary beds follows in the next section.

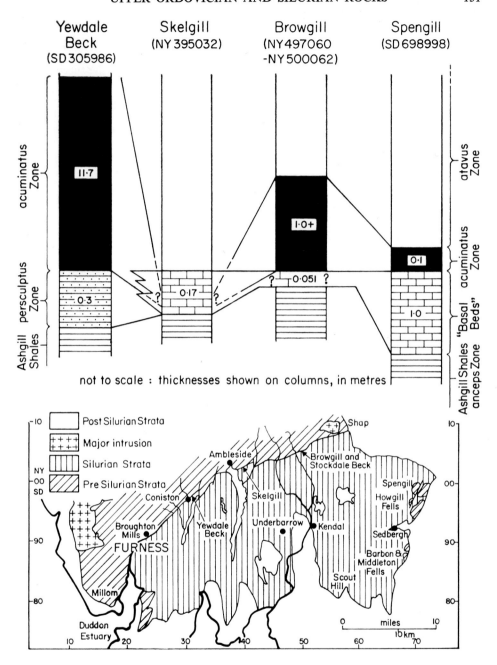

FIG. 45. Variations in rock types and thicknesses about the Ordovician-Silurian boundary across the Lake District: Ashgill Shales, striped; calcareous beds, blocked; *persculptus* mudstones, with dots; *acuminatus* mudstones, black; *atavus* mudstones, white: and general location map for places mentioned in the text.

Early Silurian Palaeogeography

The development of the early Silurian rocks is best considered as a gradually deepening sequence following the late Ordovician, shallow-water limestones and mudstones. There are few non-sequences above the 'basal beds' and the succession is largely pyritous black shales until at least the *magnus* and *leptotheca* zones. Figure 46 depicts the suggested development of the depositional environment.

Those areas of the Lake District which have the least complete Caradoc and Ashgill sequences also have a thin limestone at the base of the Silurian and condensed deposition representing about two and a half graptolite zones. The western Lakes, which already showed increased sedimentation in the Hirnantian (Ingham 1970, Fig. 8), continued with a relatively thick development of black shales (16 m) equivalent to 0.17 m of limestone at Skelgill. Hutt (1974, p. 5) considers the western deepening to be probably fault-controlled as shown in figure 46. Elsewhere, there is no direct indication of block faulting and the structure of the rises shown in figure 46 is thus largely speculative. The water depth at the time of deposition of these limestone horizons could have been of the order of 100–1000 m (K. D. Hemsley pers. comm.) and the depth of deposition of the graptolitic shales may have been considerably greater. The *persculptus* to *atavus* Zone shales of Yewdale Beck are richly pyritous with characteristic ribbed weathering. Much of the pryite is primary and contributes considerably to the thickness of sediment. A trough-like anaerobic environment is distinctly possible and is suggested both by narrow belts of thickening of the pyritous black shales in the Howgill Fells (see below), and the presence and position of the elevated areas established by Ingham and Rickards 1974 (Fig. 47).

The position of the Browgill section is difficult to envisage. There is no evidence to indicate that the non-sequence represented a subaerial rise and it is suggested here that the uplift was subaqueous. It is of interest that this position is on what must be the highest point of a north-east pointing Caradoc promontory. (Ingham and Rickards 1974 and Fig. 47).

Subsequently the rise in the Howgills and the Browgill region must have foundered, for both are overlain by a layer of black graptolitic shales, respectively 0.1 m and 1 m thick: the counterparts in the western Lakes (Fig. 45) are 11.7 m of conspicuously pyritous shale indicating continued relatively rapid subsidences and a strongly anaerobic environment. The region east of Ambleside, represented by the Skelgill section, remained a topographic high until half-way through the *atavus* Zone when it too became subject to black shale deposition, although sedimentation rates throughout the Skelgill Beds remained less than in the west (28 m compared with 40 m).

Sedimentation in the Howgill Fells, which, apart from the main Lake District section has the only other well exposed sections (Ingham and Rickards 1974, Fig. 10), totalled about 39 m for the Skelgill Beds (Rickards 1970), but is of further interest in that thick sequences are developed locally, as at Wards Intake [716 976] and Pickering Gill [689 966], and form a 1 km wide NE.–SW. trending zone of thickening. The black shales are richly pyritous, with pronounced pyrite ribs recalling those of Yewdale Beck, and the *atavus* to *cyphus* Zones (top not seen) are represented by 18.3 m, equivalent to 7.3 m elsewhere in the Howgill Fells. This is interpreted as a narrow trough in which highly anaerobic conditions persisted.

At this stage of the Llandovery, therefore, black shale deposition was established throughout the region, but the sea floor irregularities which characterized *persculptus* Zone times persisted in their effect, in some parts possibly into the highest Llandovery deposits as will be explained below.

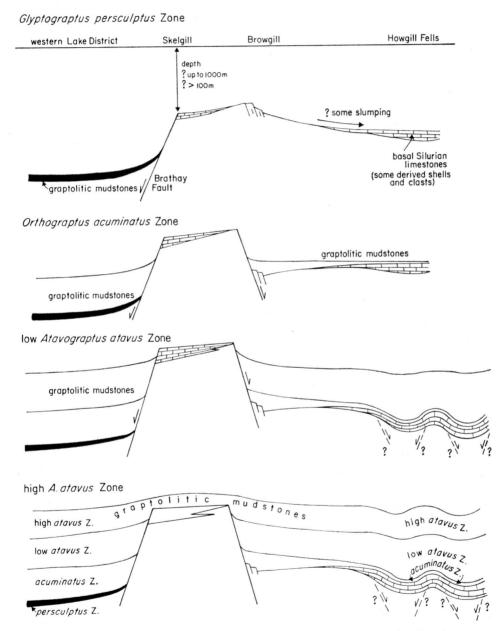

FIG. 46. Sections running East–West across the Lake District showing the stages in the development of the Early Silurian depositional environment.

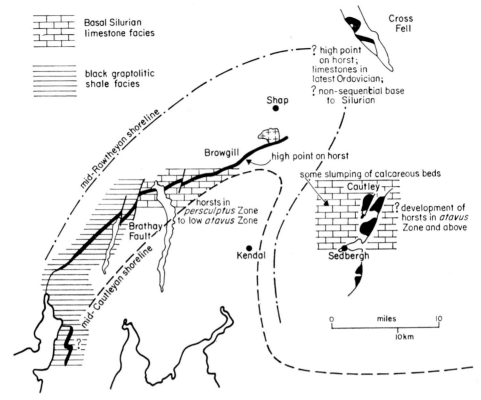

Fig. 47. Approximate minimal distribution of basal Silurian calcareous and black graptolitic shale facies. Outcrop of Coniston Limestone and Ashgill Shales is shown in black. Shap Granite intrusion is shown with crosses.

Llandovery stratigraphy

Figure 48 depicts the Llandovery stratigraphy, of the Lake District and the Howgill Fells. Cross Fell and Horton-in-Ribblesdale were considered by Ingham and Rickards (1974) and are not repeated here. Figure 49 summarizes the lithostratigraphy from west to east.

The environment of deposition of the Rhuddanian to Fronian black shales has already been described as a deepening sequence following the late Ordovician shallow seas. The widespread transgressions at the beginning of the Silurian may well have resulted from the melting of late Ordovician ice sheets (Allen 1975). Rickards (1964) has accounted for the anaerobic environment of the black shale deposition by influxes of carbonaceous matter (?algal) associated with plankton which sank upon death to the sea bed. When such influxes temporarily ceased, there was a slow reversal to aerobic bottom conditions, represented by grey-blue mudstones and thin nodular limestones (Fig. 49) which are typical of the Idwian and Fronian where they alternate with black graptolitic shales. Such grey-blue mudstones yield sparse trilobite and brachiopod faunas in the more calcareous levels; and these were utilized by Marr and

Nicholson (1888) in the recognition of their *Encrinurus punctatus, Phacops glaber* and *Ampyx aloniensis* horizons (Fig. 48).

The patterns of sedimentation caused by the alternation of shelly and graptolitic facies are maintained throughout the region (Fig. 49), and in some cases can be traced as far afield as Aberystwyth, whilst very similar patterns exist near Moffat in the Southern Uplands. It is not supposed that any shallowing took place during deposition of the grey-blue muds but rather that extensive influxes of planktonic matter did not occur. There is, in any case, independent evidence that the trilobite and brachiopod faunas of the calcareous nodule horizons were offshore forms living beyond the continental slope.

Series, stages & other divisions				Lake District Zones (Hutt 1974)	Lake District Zones (Marr & Nicholson 1888)		Howgill Fells Zones (Rickards 1970,1973)	Lithostratigraphy	
L L A N D O V E R Y S E R I E S	"upper"	Telechian	C6	?			crenulata	Grey Beds Red Beds	S T O C K D A L E S H A L E S
				?	? top crispus	Ba2	griestoniensis	BROWGILL BEDS	
			C5	crispus			crispus		
			C4	turriculatus (upper)	turriculatus	Ba1	turriculatus (upper)		
				maximus Subzone			maximus Subzone		
		Fronian C1-3		sedgwickii	Acidaspis erinaceus Ac4 spinigerus(=sedgwickii)Ac3		sedgwickii		
	"middle"	Idwian	B3	convolutus	Ab5 Ab6 Ac1 Ac2		convolutus		
			B2	argenteus(=leptotheca)	Phacops glaber argenteus	Ab3	argenteus (=leptotheca)		
			B1	magnus	Encrinurus punctatus Ab2		magnus	SKELGILL BEDS	
				triangulatus	fimbriatus	Ab1	triangulatus		
	"lower"	Rhuddanian	A4	cyphus	confertus	Aa2	cyphus		
			A3	acinaces			acinaces		
				atavus			atavus		
			A2	acuminatus	Atrypa flexuosa Band/acuminatus	Aa1	acuminatus		
			A1	persculptus			? persculptus	"Basal Beds"	

FIG. 48. The Llandovery lithostratigraphy, biostratigraphy and chronostratigraphy of the Lake District and Howgill Fells (after Marr and Nicholson 1888, Rickards 1970, 1973 and Hutt 1974).

The alternation of grey-blue mudstones with black shales is also seen on a small scale, particularly in the *convolutus* Zone. One mudstone band, the so-called Green Streak, is well known for its spectacular appearance throughout the Howgill Fells, Lake District and parts of west central Wales. In the Lake District, it is well displayed in Skelgill at the foot of the main cliff in the Lower Bridge section [395 034], as a 2 mm greyish mudstone in the middle of the 0.25 m *argenteus* Zone black shales. In sections where the *argenteus* bed itself is thinner as in the Howgills, the Green Streak is proportionately thinner, and vice versa. It is not a bentonite; this and other thin mudstone bands have the same mineralogy and geochemistry as the black shales except that the latter have in addition quantities of primary pyrite, graptolites, and carbonaceous matter (Rickards 1964, Spencer 1966). In the Lake District Rhuddanian to Fronian bentonites are largely absent, in contrast to the Moffat region where they are more common at this level than later in the Llandovery.

In the Fronian (*sedgwickii* Zone) a transition is seen from the typical Skelgill Beds with alternations of thick black shales and bluish-grey mudstones to the Browgill Beds in which black graptolitic shales are very thin in much greater thicknesses of fine-grained, hard, pale grey mudstone. Associated with the transition are thin coarse horizons, interestingly enough of the same age as parts of the Aberystwyth Grits and the *sedgwickii* Zone sandstone horizons at Balbriggan near Dublin. In the Howgill Fells there are more sandy horizons than in the Lake District and the immediately succeeding *maximus* level has evidence of penecontemporaneous erosion in that the black *maximus* bands contain shale pellets which themselves contain a *maximus* fauna. Thus for the first time since the *persculptus* Zone there is some evidence of more vigorous transport and deposition of sediment.

Whilst the rock types in the Telychian of the Lake District Browgill Beds closely resemble those of the Howgill Fells, a puzzling feature of the former is the paucity of black graptolitic shales (Fig. 49). In the main outcrop the *turriculatus* Zone is proven in only four very thin bands (Hutt 1974) and the *griestoniensis* and *crenulata* zones cannot be regarded as established. In the Howgills, by contrast, numerous such bands prove the presence of the *turriculatus*, *crispus*, *griestoniensis* and *crenulate* faunas (Rickards 1970, 1973). The Lake District *turriculatus* bands may all be referable to the *maximus* Subzone, so that the upper part of the *turriculatus* Zone is represented only by barren mudstones.

In the Skelgill and Browgill Beds graptolites are very abundant. Skelgill is a spectacular section in this respect. In the course of her (1974) work Hutt collected over 20,000 Llandovery graptolites, and the number of species available is equally striking. The highest number of species is reached in the *triangulatus* Zone (59), and the average number of species in each zone is about 30. Many aspects of the evolution of Llandovery graptoloids (Hutt and Rickards 1970; Rickards et al. 1977) can be observed by collecting from these Lake District sections, particularly from Yewdale Beck, Church Beck [299 978] and Skelgill.

Towards the top of the Browgill Beds, there is a distinct change of facies involving red beds and what have been termed Grey Beds (Rickards 1967, Ingham and Rickards 1974). The latter occur in all sections which span the Llandovery–Wenlock boundary, throughout the Howgill Fells and Lake District outcrop, and, indeed, as far afield as Balbriggan. They comprise 8 m of pale grey, hard mudstone, showing little variation in thickness over large areas. Their age is in doubt since they have yielded but one ostracod, but they can conveniently be taken as the top of the Llandovery, on present evidence, and are overlain conformably by a rich *centrifugus* Zone (Wenlock) fauna.

The red beds are discontinuously developed being absent in most sections. They are invariably mudstones, uniform in colour, with occasional paler coloured calcareous nodules yielding a sparse fauna of phacopid trilobites. No other benthonic fossils have been obtained but several thin dark bands have yielded a *crenulata* Zone grapto-lite fauna (Rickards 1973). The red colour is produced by an abundance of fine grains of haematite whereas in grey-green mudstones, which often alternate with the red mudstones at the upper and lower limits of the red bed sequences, the iron occurs in the form of pyrite. It has been shown by Wilson (1954) that in those Howgill Fells sections where red beds are absent, the rest of the sequence of both sediments and faunas is quite complete. A reasonable deduction for the geometry of the red beds

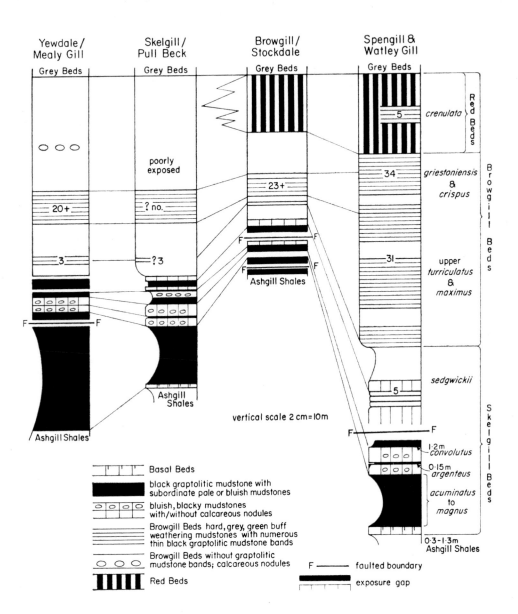

FIG. 49. Stratigraphical sections of the Llandovery strata running East–West across the Howgill Fells and Southern Lake District. Figures in the columns relate to the number of black graptolitic bands present for each zone.

lenses is shown in figure 51. It is almost certain that their local absence is caused by sea floor ridges, the red mud being deposited in hollows between the ridges. As the red beds thin out towards the ridges they are slightly more calcareous, suggesting increased aeration of the bottom at these points to a degree that benthos attains a tenuous hold.

A similar situation obtains in the main Lake District outcrop, but here, presumably because of the greater distance from the shoreline to the south-east (see Ingham and Rickards 1974, p. 39) the red beds are thinner and have only been established in Pull Beck, Hall Gill, Stockdale Beck and Browgill. In Stockdale Beck (Fig. 50a) there are 1.5 m of calcareous nodular red beds, yet only 400 m away on Browgill at least 11 m of non-calcareous red beds were discovered by the author and K. D. Hemsley in 1976. Browgill remains the only Lake District section where red beds are as thick as those in the Howgill Fells and Cross Fell.

Ziegler and McKerrow (1975) interpret the red mudstones as being derived from a red desert landscape possibly, and infilling hollows or gullies upon the sea floor. Absence of potentially reducing organic matter is considered a contributory factor to their preservation as haematite-rich red beds. In Hebblethwaite Hall Gill [6910 9320] the author and K. D. Hemsley have added to the discoveries of *crenulata* Zone graptolite bands (Rickards 1973) by the identification of a further four bands in

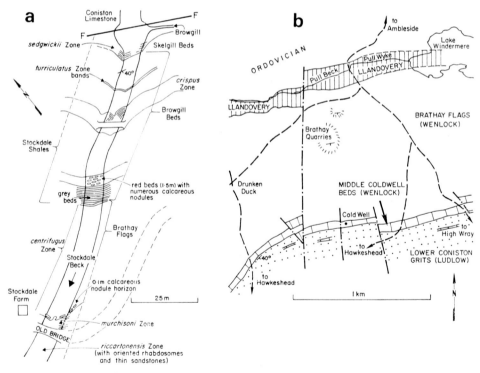

FIG. 50. a. Plan of Stockdale Beck showing distribution of the Upper Llandovery and Lower Wenlock Strata.

b. Distribution of Silurian strata to the West of Lake Windermere.

new exposures. These may well be of only local development for a nearby section [6911 9325] quite continuous through the whole of the red beds reveals a complete absence of black beds. It may be that the oxidizing conditions associated with the red mudstone deposition were sufficient generally to oxidize the small amount of carbon-rich planktonic debris reaching the bottom. The horizon of the red beds, as shown by these five graptolite bands and by a similar band in Cross Fell, is undoubtedly *crenulata* Zone, highest Llandovery.

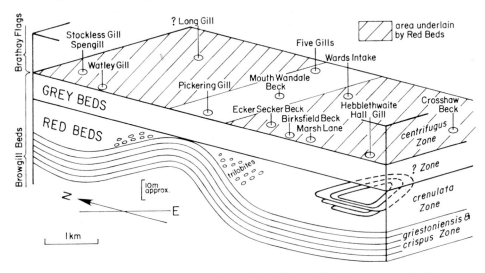

FIG. 51. Geometrical representation of the Upper Llandovery strata in the Howgill Fells.

The Stockdale Thrust

Present in all Lake District sections through the Skelgill Beds is a crush zone several metres wide largely affecting the *triangulatus* Zone and occasionally the *magnus* Zone. In terms of rock competancy, the crush is in black shales forming the weakest point in the Llandovery sequence. The break coincides roughly with bedding dipping about 40° to the south-east. The structure is absent in the Howgill Fells and probably in the outcrops near the Duddon Estuary. It is interpreted here as a bedding-plane thrust, but needs modern structural analysis.

WENLOCK

Wenlock stratigraphy

Figure 52 summarizes the Wenlock stratigraphy of the district. More deatiled accounts are given by Furness et al. (1967), Ingham and Rickards (1974), and Rickards (1969, 1970b) and include correlations with other sections in northern England. The most notable recent advances have been the recognition of a full suite of Wenlock graptolite zones and precise dating of the main formations, allowing easy correlation with the type Wenlock area (Cocks et al. 1971; Bassett et al. 1975). Interpretations of the depositional environments of the major formations were attempted by Rickards (1964) and Furness (1965).

The three formations of the Wenlock rocks of the Lake District are the Brathay Flags and the Lower and Middle Coldwell Beds. The age of the Lower Coldwell Beds was established by Rickards (1969) as *lundgreni* Zone, whilst the Middle Coldwell Beds were proved to be largely of *ludensis* Zone age by the same author (1970b). Prior to this Marr (1892) equated the Brathay Flags with the *C. murchisoni* Zone and the Lower Coldwell Beds doubtfully with the *M. nilssoni* Zone, whilst Blackie (1933) supported the first of these opinions but then identified three further zones in his rather broad view of the Middle Coldwell Beds. In fact the Brathay Flags have a rich and varied suite of graptolites enabling recognition of the *centrifugus* to *lundgreni* zones, and the Middle Coldwell Beds, up to the base of the upper graptolitic mudstone band, have a *ludensis* fauna (Rickards 1970b, Fig. 1). When compared with the type area for the Wenlock only the *nassa* Zone has not been established, thus making formal recognition of the Whitwell and Gleedon chronozones difficult (Fig. 52).

The type locality for the Brathay Flags is Brathay Quarry near Ambleside [357 016], which was actively worked until recently (Fig. 50b). Most of the formation is a hard, bluish-grey laminated mudstone with common calcareous nodules, rare unlaminated grey mudstones a few centimetres thick, and bentonite horizons which in the quarry may be up to 15 cm thick.

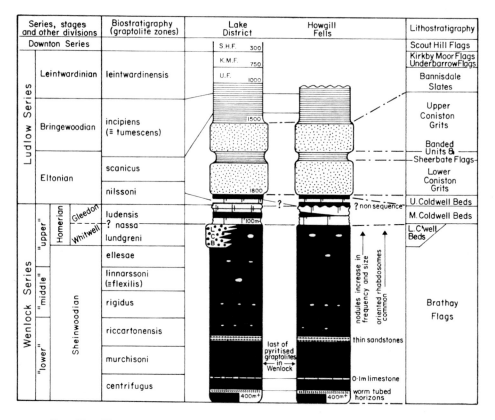

FIG. 52. The stratigraphy of the Wenlock and Ludlow strata of the Lake District and Howgill Fells.

The Lower Coldwell Beds, developed essentially west of Ambleside, are proximal turbidites of silt or sand grade, the only coarse sediments in the Wenlock of the Lake District. They exhibit graded bedding, flute casts and groove casts. Interbedded with them, at irregular intervals, are beds of the Brathay Flags laminated mudstone lithology, usually about 2–10 cm thick, containing graptolites used for dating. The most common graptolites are *Monograptus flemingii* (Salter) and *Pristiograptus pseudodubius* Bouček. The Lower Coldwell Beds thin to the east which, taken with the directions of sole markings, supports a derivation from the west or north-west.

The Middle Coldwell Beds, underlain in the west by the Lower Coldwell Beds and in the east by the Brathay Flags, form a conspicuous gully across the ridges of the southern Lake District. They consist largely of poorly calcareous, pale gray mudstones, often unlaminated, but equally commonly, strongly bioturbated with rust-coloured worm tubes. Occasional lenses of purer limestone have a rich trilobite and brachipod fauna, the former including species of *Delops, Struveria, Miraspis, Decoroproetus* and *Encrinurus*. In addition there are thinner but more continuous layers, sometimes up to 0.6 m, of the Brathay Flags lithology with numerous graptolites used for dating. The lower bands yield *Gothograptus nassa* (Holm), *Monograptus ludensis* (Murchison), *Pristograptus jaegeri* Holland et al., *P. dubius* (Suess) and *P. pseudodubius*, and the higher bands *Saetograptus* (*Colonograptus*) *varians* (Wood).

Although the Lower Coldwell Beds are restricted to the western Lake District, the Brathay Flags and Middle Coldwell Beds are widespread, and can be recognized in the Cross Fell Inlier, the Howgill Fells and at Horton-in-Ribblesdale.

Wenlock Palaeogeography and depositional environments

Furness (1965) suggested that Cockburnland acted as a barrier during Wenlock times which stopped the coarser south-easterly flowing turbidites of southern Scotland from reaching the Lakes, although the Lower Coldwell Beds suggest that the barrier was temporarily breached in latest Wenlock times. The Silurian ocean was certainly much narrower at this time than in the Llandovery (McKerrow and Cocks 1976) and Furness (op. cit.) convincingly argues a Scottish source for the low Ludlow proximal turbidites of the Barbon and Middleton Fells.

There is evidence of a shoreline to the east and south of the district in that both the Wharfe Conglomerate, reaching to the Howgill Fells, and the Wenlock Austwick Formation sandstones are derived from the south-east. The bulk of bottom current indications in the Brathay Flags, taken on oriented graptolite rhabdosomes (Ingham and Rickards 1974), indicate ENE. to WSW. currents, which may perhaps be due to bottom currents moving along the geosynclinal axis with the NW. and SE. derivation of coarser sediments (Lower Coldwell Beds and Austwick Formations) indicating down-slope bottom currents (but see Loubere 1977).

Depth of water is difficult to deduce, but at the onset of Brathay Flags deposition it was presumably greater than during the *crenulata* Zone red beds deposition. The *centrifugus* Zone is remarkably uniform in thickness over a large area, but thins very gradually from about 18 m in the Howgill Fells to about 9 m in the western Lakes (Ashgill Beck). Its transition from the underlying Grey Beds is marked by a distinct bluish hue. The lowest thin beds of laminated graptolitic mudstone are largely worked out by worms (Rickards 1964, pl. 16), a feature demonstrable on all sections across the Lake District, but as the supply of planktonic matter (algae, graptolites etc.) reaching the bottom increased, the benthos was overwhelmed and the worms, brachiopods and rare trilobites are replaced by graptolites preserved in an anaerobic environment rich in primary pyrite. Thus the bioturbated but unlaminated bluish mudstones are replaced about half-way through the zone by a preponderance of laminated Brathay Flags mudstones, and in the rest of the Wenlock succession this rock type predominates.

There is a remarkable uniformity of sediment type and thicknesses across the Lake District, seen also on a small scale. For example, at the top of the *centrifugus* Zone in the Howgill Fells there is a 0.1 m calcareous horizon which weathers distinctively, and can in fact be traced as far as Ashgill Beck in the west of the Lake District. West of Ambleside the bed tends to form large tabular nodules rather than a continuous bed, and is well seen in this form at the roadside at Pull Wyke [3595 0205].

The total thickness of the Brathay Flags is of the order of 400 m, perhaps more in the Howgill, Barton and Middleton Fells. Most is referable to the *lundgreni* Zone, and below this there is little lateral variation in thicknesses of each zone (Rickards 1967, 1969). In the western Lakes the Brathay Flags are thinner largely because much of the *lundgreni* Zone is represented by the Lower Coldwell Beds sandstones. The laminations running through the calcareous nodules indicate a reduction in thickness from the original mud of over 40 per cent.

Having emphasized the uniformity of Brathay Flags deposition, the changes in depositional environment towards the top of the sequence can be described. Half-way up the *murchisoni* Zone, the graptolites are preserved flattened, indicating a slight change from the strongly anaerobic, pyrite-producing conditions of the earlier strata. Within the *riccartonensis* Zone, the graptolite rhabdosomes are commonly current-oriented, and the first small calcareous nodules appear. These increase in frequency and size upwards culminating in the huge nodules seen just below the Middle Coldwell Beds calcareous horizons. Concomitant with this increase in lime, is a gradual increase in the number of benthonic elements (brachiopods, trilobites, bryozoa, corals, gastropods), again culminating at or about the Middle Coldwell Beds. From the *rigidus* Zone upwards there is a perceptible increase in grain size, and fine silt laminae are quite common; the *lundgreni* Zone of the eastern Lakes passes laterally into the sandstones of the western Lakes, and commonly shows thin sand and silt beds and small scale sedimentary structures. All these factors indicate a gradual aeration of the bottom environment by currents of increasing strength, possibly coupled with some shallowing at the time of the Middle Coldwell Beds.

However, the Middle Coldwell Beds are demonstrably of the same age as the (type) Much Wenlock Limestone, undoubtedly a shelf deposit. Since west of the shelf, through Long Mountain, there are unquestionable slope deposits of the same age, and since North Wales has calcareous basinal beds, again of the same age, the Middle Coldwell Beds must be regarded as offshore, relatively deep-water deposits. The benthonic elements found in the Middle Coldwell Beds are different from contemporary faunas in the shelf regions: *Delops* and *Struveria*, for example, are typical offshore genera. At about the time of deposition of the Middle Coldwell Beds are the first signs of instability in what has been a relatively uniform environment. These have been outlined by Rickards (1964) and include slumping of the calcareous muds, the appearance of the first sand grade turbidites (earlier in the western Lake District), and the fact that the Middle Coldwell Beds are overlain by early Ludlow deposits which indicate bottom currents from several directions. This instability and bottom current variation is taken as heralding the final stage in the geosynclinal development, namely its infilling by thicker and coarser beds of sand grade, and its final eclipse as a marine environment.

LUDLOW AND DOWNTON

Ludlow and Downton stratigraphy

Figure 52 depicts the Ludlow stratigraphy of the Lake District and the Howgill Fells and includes the six named formations. Recent accounts of the Ludlow rocks of the Lake District and adjacent regions can be found in Burgess et al. (1970) Furness (1965) Furness et al. (1967), Ingham and Rickards (1974), McCabe (1972), Rickards (1964, 1967, 1970b) and Shaw (1971a, 1971b).

The age of the Upper Coldwell Beds has been established as *nilssoni* Zone by Rickards (1970b) and it is known that in the Howgill Fells the *nilssoni-scanicus* fauna extends up to and includes the lowest Bannisdale Slates (Fig. 52). The remainder of the Bannisdale Slates are referable to the *incipiens* (approximately *tumescens*) and *leintwardinensis* Zones. Above this level no graptolites have been found but Shaw (1971a) has correlated the Kirkby Moor Flags and Scout Hill Flags with the Whitcliffian and Downtonian respectively. The Underbarrow Flags are transitional to the Kirkby Moor Flags.

The Upper Coldwell Beds revert to the Brathay Flags lithology, but with silt grade horizons, as laminae or beds, much more common than in the Wenlock. Small-scale sedimentary structures such as ripple-drift lamination are not uncommon. Pyritized graptolites occur in places amongst a rich graptolite fauna. This lithology persists as thin bands, from a few millimetres to 3–4 m, throughout the overlying Coniston Grits and Bannisdale Slates and provides the graptolite faunas which date these horizons. On faunal and lithological grounds the Upper Coldwell Beds can be correlated with the Horton Flags, unnamed mudstones below the Coniston Grits in the Howgill Fells, and the Stennerley Mudstones of Blawith.

The Coniston Grits are mostly fine sand-grade turbidites, occasionally coarse sand or conglomerate, often with a greywacke texture and composition. Variously complete Bouma sequences are commonly displayed, although slumped horizons are thin and uncommon. Above the lowest 50 m or so, the Coniston Grits show a dominant north-westerly derivation, but in the Broughton or Blawith districts Norman (1963) noted bottom currents from two directions: north-west and north-east, the latter being interpreted as axial to the geosyncline.

There are several mudstones within the Coniston Grits (Salthouse Mudstones; Sheerbate Flags; Longsleddale Siltstones) which Rickards (1964) termed the Banded Unit Facies. These are transitional from the Brathay Flags type of lithology to the Bannisdale Slates, and are typified by a stripy appearance due to rapid alternations of laminated mudstones, siltstone laminae, ripple drift laminae, bioturbated bluish mudstones and homogeneous bluish mudstones.

The Bannisdale Slates comprise at least 600 m of mud to silt grade rock, with sandstone beds at some horizons, but with graptolitic, laminated mudstones restricted to rare bands each only a few millimetres thick. Sole markings indicating bottom current directions are uncommon, but a north-westerly source of sediment is assumed. At some horizons, a rich *in situ* benthos of trilobites and bivalves is found, whilst at other levels lenses of transported shells occur. At their upper limit Shaw (1971a) has shown that the Bannisdale Slates, with a Leintwardinian fauna, alternate with the Kirkby Moor Flags lithology which contains its own distinctive fauna. This transitional sequence, with clear indication of facies control of the benthos, is the Underbarrow Flags, named after a region close to Kendal.

The Kirkby Moor Flags have a Whitcliffian fauna, and an abundance of benthonic species of bivalves, brachiopods, bryozoa, worms, cephalopods, tentaculitids, hyolithids and gastropods. Lithologically they comprise thick, sorted, homogeneous siltstones with characteristic de-watering structures, which contain most of the fossils in lenses, separated by flaggy siltstones and muddy siltstones. Slightly calcareous beds are common. Shaw (1971a) has concluded that the primary bottom current direction was from the north-west. The base of the formation is diachronous, appearing earlier in the north of its outcrop, at about the base of the Whitcliffian. Further south the Underbarrow Flags lithology persists into the low Whitcliffian.

The Scout Hill Flags (Fig. 52) differ from the underlying Kirkby Moor Flag not only faunally, but in having more flaggy horizons, and in the occurrence of red beds, in places up to 152 m thick. The presence of the ostracod *Frostiella groenvalliana* enables correlation of the strata with the Downtonian.

Ludlow and Downton depositional environments

The Ludlow is typified by more common coarse sedimentation, particularly in the Coniston Grits, than in the underlying Wenlock and Llandovery, and culminates in horizons with some thick red beds of silt grade, surely heralding the approaching Devonian non-marine environment. The dominant bottom current and source of sediments lie in the north-west, except toward the base of the Ludlow where oscillating current directions, from bed to bed, and some slumping indicate a sharp disturbance of the relatively tranquil conditions of deposition prevailing in the Wenlock. Perhaps the best general way of viewing the depositional environment is to consider the deposition of fine muds, including graptolite muds, took place in an offshore, relatively deep-water environment, subject to periodic influxes of coarse turbidites which increased in frequency up to and including the onset of Bannisdale Slate times. Most of these bottom currents came from the north-west, but there is some evidence of localized axial flow along the geosyncline. Thereafter the sequence gradually shallowed, and a rich benthos became established from the upper Bannisdale Slates into the Scout Hill Flags, which later Shaw (1971a) considered to have a declining population.

REFERENCES

ALLEN, P. 1975. Ordovician glacials in the central Sahara. *In* Wright, A. E. and Moseley, F. (Eds). *Ice ages ancient and modern*. Geol. J. special issue, 275–86.

BASSETT, M. G., COCKS, L. R. M., HOLLAND, C. H., RICKARDS, R. B., and WARREN, P. T. 1975. The Type Wenlock Series. *Rept. Inst. Geol. Sci. (GB)* 75/13, 19 pp.

BLACKIE, R. C. 1933. The Silurian Rocks of the Kentmere District, Westmorland. *Proc. Liverpool Geol. Soc.* 16, 88–105.

BURGESS, I. C., RICKARDS, R. B. and STRACHAN, I. 1970. The Silurian Strata of the Cross Fell area. *Bull. Geol. Surv. G.B.* 32, 167–82.

COCKS, L. R. M., HOLLAND, C. H., RICKARDS, R. B., and STRACHAN, I. 1971. A correlation of Silurian Rocks in the British Isles. *Q. Jl geol. Soc. Lond.* 127, 103–36.

DEAN, W. T. 1959. The stratigraphy of the Caradoc Series in the Cross Fell Inlier. *Proc. Yorks. geol. Soc.* 32, 185–228.

—— 1962. The trilobites of the Caradoc Series in the Cross Fell inlier of Northern England. *Bull. Br. Mus. nat. Hist., Geol.* 7, 65–134.

—— 1963. The Stile End Beds and Drygill Shales in the east and north of the English Lake District. *Bull. Br. Mus. nat. Hist., Geol.* 9, 47–65.

FURNESS, R. R., 1965. The geology of the Barbon and Middleton Fells near Kirkby Lonsdale, Westmorland. *Unpublished Ph.D. thesis, University of Hull.*

—— LLEWELLYN, P. G., NORMAN, T. N., and RICKARDS, R. B. 1967. A review of Wenlock and Ludlow stratigraphy and sedimentation in N.W. England. *Geol. Mag.* 104, 132–47.

HARKNESS, R. and NICHOLSON, H. A. 1877. On the strata and their fossil contents between the Borrowdale Series of the north of England and the Coniston Flags. *Q. Jl geol. Soc. Lond.* 33, 461–84.

HUTT, J. E. 1974. The Llandovery graptolites of the English Lake District Pt. 1. *Palaeontogr. Soc.* (*Monogr.*) **128**, 1.
—— and RICKARDS, R. B. 1970. The evolution of the earliest Llandovery monograptids. *Geol. Mag.* **107**, 67.
INGHAM, J. K. 1966. The Ordovician rocks in the Cautley and Dent districts of Westmorland and Yorkshire. *Proc. Yorks. geol. Soc.* **35**, 455–505.
—— 1970. A monograph of the Upper Ordovician trilobites from the Cautley and Dent districts of Westmorland and Yorkshire. *Palaeontogr. Soc. Monogr.* Pt. 1, 1–58.
—— 1974. Ibid. Pt. 2, 59–87.
—— 1977. Ibid. Pt. 3, 88–121.
—— and RICKARDS, R. B. 1974. Lower Palaeozoic Rocks. *In* Rayner, D. H. and Hemingway, J. E. (Eds). *The Geology and Mineral Resources of Yorkshire.* Yorkshire Geological Society, 29–44.
—— and WRIGHT, A. D. 1970. A revised classification of the Ashgill Series. *Lethaia.* **3**, 233–42.
LOUBERE, P. 1977. Orientation of Orthocones in the English Lake District based on field observations and experimental work in a flume. *Jour. Sed. Petrol.* **47**, 419–27.
MCCABE, P. J. 1972. The Wenlock and lower Ludlow strata of the Austwick and Horton-in-Ribblesdale inlier of north-west Yorkshire. *Proc. Yorks. geol. Soc.* **39**, 167–74.
MCKERROW, W. S., and COCKS, L. R. M., 1976. Progressive faunal migration across the Iapetus Ocean. *Nature.* **263**, 304–06.
MARR, J. E. 1892. On the Wenlock and Ludlow Strata of the Lake District. *Geol. Mag.* **9**, 534–41.
—— 1916. The Ashgillian succession in the tract to the west of Coniston Lake. *Q. Jl geol. Soc. Lond.* **71**, 189–204.
—— and NICHOLSON, H. A. 1888. The Stockdale Shales. *Q. Jl geol. Soc. Lond.* **44**, 654–732.
NORMAN, T. N. 1963. Silurian (Ludlovian) Palaeo-current Directions in the Lake District area of England. *Bull. Geol. Soc. Turkey.* **8**, 27–54.
PHILLIPS, W. E. A., STILLMAN, C. J. and MURPHY, T. 1976. A Caledonian Plate tectonic model. *Jl. geol. Soc. Lond.* **132**, 579–610.
RICKARDS, R. B. 1964. The graptolitic mudstone and associated facies in the Silurian strata of the Howgill Fells. *Geol. Mag.* **101**, 435–51.
—— 1967. The Wenlock and Ludlow succession in the Howgill Fells (north-west Yorkshire and Westmorland). *Q. Jl geol. Soc. Lond.* **123**, 215–51.
—— 1969. Wenlock graptolite zones in the English Lake District. *Proc. geol. Soc. Lond.* No. 1654, 61–65.
—— 1970a. The Llandovery (Silurian) graptolites of the Howgill Fells, northern England. *Palaeontogr. Soc. Monogr.* 1–108.
—— 1970b. Age of the Middle Coldwell Beds. *Proc. geol. Soc. Lond.* No. 1663, 111–14.
—— 1973. On some highest Llandovery red beds and graptolite assemblages in Britain and Eire. *Geol. Mag.* **110**, 70–72.
—— and HUTT, J. E. 1970. The earliest monograptid. *Proc. geol. Soc. Lond.* No. 1665, 115–19.
—— —— and BERRY, W. B. N. 1977. Evolution of the Silurian and Devonian Graptoloids. *Bull. Brit. Mus. (Nat. Hist.) Geol.* **23**, 1–20.
SHAW, R. W. L. 1971a. The faunal stratigraphy of the Kirkby Moor Flags of the type area near Kendal, Westmorland. *Geol. Jour.* **7**, 359–80.
—— 1971b. Ostracoda from the Underbarrow, Kirkby Moor and Scout Hill Flags (Silurian) near Kendal, Westmorland. *Palaeontology* **14**, 595–611.
SPENCER, D. 1966. Factors affecting element distribution in a Silurian graptolite band. *Chem. Geol.* **1**, 221–49.
TEMPLE, J. T. 1968. The Lower Llandovery (Silurian) brachiopods from Keisley, Westmorland. *Palaeontogr. Soc. Monogr.*, 1–58.
WILLIAMS, A., STRACHAN, I., BASSETT, D. A., DEAN, W. T., INGHAM, J. K., WRIGHT, A. D. and WHITTINGTON, H. B. 1972. a correlation of Ordovician rocks in the British Isles. *Spec. rep. geol. Soc. Lond.*, No. 3, 1–74.
WILSON, D. W. R. 1954. The Stratigraphy and Palaeontology of the Valentian Rocks of Cautley (Yorks. W. R.). *Unpublished Ph.D. Thesis University of Birmingham.*
ZIEGLER, A. M. and MCKERROW, W. S. 1975. Silurian Marine Red Beds. *Am. J. Sci.* **275**, 31–56.

J. K. INGHAM, PH.D.
Hunterian Museum, The University, Glasgow W2

K. J. MCNAMARA, PH.D.
Sedwick Museum, Downing Street, Cambridge

R. B. RICKARDS, PH.D.
Sedgwick Museum, Downing Street, Cambridge

10

Intrusions

R. J. FIRMAN

The outcrops of the Shap, Eskdale and Skiddaw Granites coincide with the minima of an extensive area of low Bouguer anomalies suggesting that they are the exposed parts of a large, composite, acid batholith underlying the central and northern Lake District (Bott 1974). Isotopic dating suggests that these granites are Devonian, but there are also older and younger intrusions, although considerable doubt exists about the age of many of them. Consequently in the following account they are speculatively classified as 'older, possibly Ordovician'; 'younger, probably Devonian', and 'post-Devonian' intrusions. A more rigorous chronology is impossible in the absence of Rb-Sr dates and may never be possible for the most intensely altered rocks, but see the comment following the section on the Eskdale Granite.

THE OLDER INTRUSIONS, AGE UNCERTAIN, POSSIBLY ORDOVICIAN

The Carrock Fell Complex

Gabbro, granophyre and diabase, apparently intruded successively and discordantly along the junction between the Skiddaw Slates and Eycott Volcanics, were subsequently thermally metamorphosed by the Lower Devonian Skiddaw Granite. The granophyre and diabase are younger than the gabbro but their relative ages are uncertain (Skillen 1973). Felsite, chemically related to the Skiddaw Granite, injects into, and intervenes between, the Caradocian Drygill Shales and the Carrock Granophyre (Fig. 53). The gabbro, granophyre and diabase are thus older than the Lower Devonian Skiddaw Granite and younger than the lower part of the Llanvirnian Eycott Volcanics.

The gabbros show a crude bilateral symmetry consisting of a central belt of quartz-hornblende-gabbro passing outwards into leucogabbro and marginal melagabbro. The southern junction is approximately vertical, apparently unfaulted, and parallel to the regional strike of the Skiddaw Slates and the northern junction is marked by a belt of steeply dipping hybrid melagranophyres. The gabbros and granophyres are probably terminated eastward by a N–S fault and westward the gabbros appear to be truncated by the diabase. Correlation with the westernmost gabbro is uncertain. The melagabbros are extremely variable and include hornblende, ilmenite and plagioclase rich varieties. Visible biotite often occurs; xenolithic, banded and pegmatitic gabbros are not uncommon. Plagioclase (An_{50-65}) is most sodic near the contact; augite, wholly or partly converted to hornblende, tends to envelop plagioclase, and opaque minerals, which constitute 2.5–20% of the rocks, include

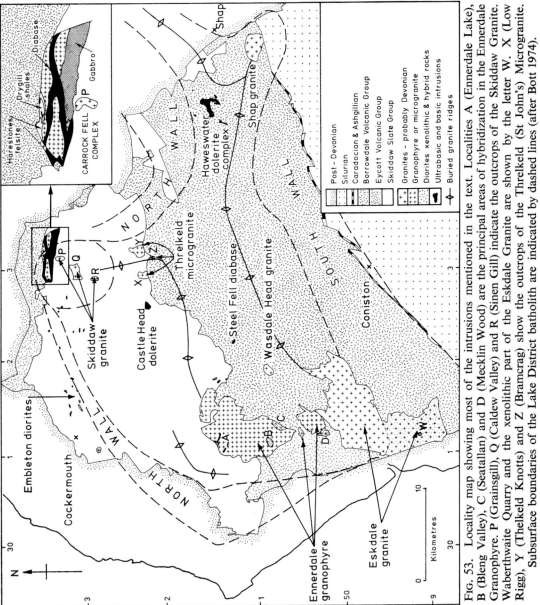

FIG. 53. Locality map showing most of the intrusions mentioned in the text. Localities A (Ennerdale Lake), B (Bleng Valley), C (Seatallan) and D (Mecklin Wood) are the principal areas of hybridization in the Ennerdale Granophyre. P (Grainsgill), Q (Caldew Valley) and R (Sinen Gill) indicate the outcrops of the Skiddaw Granite. Waberthwaite Quarry and the xenolithic part of the Eskdale Granite are shown by the letter W. X (Low Rigg), Y (Thelkeld Knotts) and Z (Bramcrag) show the outcrops of the Threlkeld (St John's) Microgranite. Subsurface boundaries of the Lake District batholith are indicated by dashed lines (after Bott 1974).

ilmenite and ilmenomagnetite. The leucogabbros contain more feldspar; augite partly replaced by hornblende; a little ilmenite and interstitial quartz and orthoclase often micrographically intergrown. The central gabbro has more hornblende, sometimes intergrown with biotite and interstitial quartz, perthite and a little apatite.

Lava inclusions were enriched in soda, depleted in silica, and converted into granulitic pyroxene-plagioclase rocks. Skiddaw Slate xenoliths, which are confined to the southern margin, are similar to the hornfelsed wall-rocks, containing porphyroblasts of biotite, albite, anorthoclase and spessartine-bearing almandine.

Eastwood et al. (1968) argue that neither Ward's (1876) hypothesis of metamorphic replacement nor Harker's (1894) application of the Soret principle explain the genesis of the gabbros. The rapid transition between the belts, and lack of igneous sedimentation, suggests separate intrusions rather than differentiation *in situ*. Eastwood et al. (1968) therefore suggest successive injections of crystal mushes derived from an unconsolidated intrusion at depth.

In the normal granophyre, albite-oligoclase phenocrysts zoned to albite are surrounded by fine-grained micropegmatite, sometimes crudely spherulitic. Soda is far in excess of that required for the plagioclase phenocrysts and must be an important component in the granophyric groundmass (Harker 1895). Scattered crystals of hedenbergite are accompanied by sporadic magnetite, apatite, late amphibole and chlorite. Near the gabbro narrow belts of basic and melagranophyre occur. The junction between these and the normal granophyre is usually sharp, suggesting separate intrusions. The melagranophyres are grey, mottled, plagioclase-hedenbergite rocks with a coarse micropegmatitic matrix containing minor zircon, opaque minerals, apatite and late biotite. The plagioclase is more calcic and hedenbergite more plentiful than in the normal granophyre. The basic granophyres are intermediate in composition between normal and melagranophyre. Contact phenomena, described by Harker (1895) and Eastwood et al. (1968) seem to indicate recrystallization of the gabbro and mingling with granophyre in the pegmatitic zone. Early concentration of hedenbergite is a possible alternative (Eastwood et al. 1968).

The fine-grained diabase appear to cut across the gabbro and granophyre; nevertheless Eastwood et al. (1969) claim that it contains a few granophyric veins and hence is older than the granophyre. Skillen (1973) tentatively reinterprets these veins as later felsite and favours Harker's (1895) view that the diabase is younger than the granophyre.

Late stage alteration is widespread and it is difficult to distinguish deuteric effects from metasomatism caused by the Skiddaw Granite (see later). Thus the anomalous K-Ar age of the biotite hornfels near the gabbro (356 $\pm$ 20 Ma Brown et al. 1964) may reflect late metasomatic events rather than intrusive phenomena.

Minor intrusions in the Skiddaw Slates, Eycott Volcanics and Carrock Fell Complex

Minor intrusions, mostly basic, abound between Carrock Fell and Cockermouth but few can be dated. They are restricted to the Lower Palaeozoic and are presumably pre-Carboniferous. Some of the acid intrusions, such as the Harestone Felsite, are chemically related to the Skiddaw Granite and are probably Lower Devonian. Other, more basic, dykes and sills have been thermally metamorphosed by the granite. Some diabase dykes cut the Drygill Shales and are thus post Caradocian; others could have been feeders to the earlier Eycott lavas. The majority have petrological features which link them with the Eycott Volcanic Group and Carrock Fell Complex rather than to the Borrowdale Volcanic Group or the Skiddaw Granite. As in the Carrock Fell Gabbro, hornblende is plentiful and opaque minerals, particularly ilmenite, are common. Acid varieties tend to be granophyric, both acid and basic types tending to have anomalously high soda (Fig. 54). The altered diorites around Embleton seem to show a tholeitic trend (Fig. 54), akin to the Eycott volcanics. Some of the smaller intrusions may have been involved in the deformation of the Skiddaw Slates: more massive intrusions sometimes show a close spaced jointing comparable to that in the Carrock Fell Granophyre. There is therefore, evidence suggesting that most of these intrusions are related to the Eycott volcanics and Carrock Fell Complex, are possibly Ordovician and certainly pre-Skiddaw Granite.

The more basic rocks crop out in small isolated exposures between Carrock Fell and Bassenthwaite (Fig. 53). Most are coarse hornblende-rich with calcic plagioclase, rare augite, minor orthoclase, quartz and chloritized biotite. Accessory apatite, ilmenite and magnetite are often abundant. Often the original minerals are replaced by calcite, chlorite, quartz and chloritized biotite. Phemister noted talcose pseudomorphs after olivine (Eastwood et al. 1968) and in the Little Knott 'picrite' the presence of two different types of pseudomorphs suggests that both enstatite (Bonney 1885) and olivine (Harker 1902) may have been present. Since the Skiddaw granite probably underlies these outcrops much of the hornblende and alteration of other minerals may be due to metamorphism and metasomatism caused by the granite.

Further west near Bassenthwaite (Postlethwaite 1893), around Embleton and near Eaglesfield (Eastward et al. 1928) sills of granophyric diorites of Markfield and Spessart type crop out. A similar intrusion at Castle Head near Keswick has been described as a volcanic neck (Ward 1876). These intrusions consist of albitized plagioclase laths; interstitial augite, partly or wholly replaced by amphibole; biotite usually accompanied by abundant accessory iron oxides and variable amounts of micropegmatite. Chemically the Embleton intrusion is characterized by low alumina, high ferrous iron and high alkalies, particularly soda, compared with other Markfield type diorites.

Of more local occurrence is the pink and green Sale Fell 'minette' which contains both plagioclase and orthoclase with chloritized biotite, quartz, a little colourless augite, and accessory ilmenite and apatite. Biotite, unusually, is ophitic towards plagioclase and the quartz frequently grades into orthoclase through a fringe of micropegmatite. Plagioclase is albitized and the rock considerably altered.

Further west, between the northern outcrop of the Ennerdale granophyre and the Carboniferous, numerous small intrusions occur in the Skiddaw Slates. The majority are dykes, ranging from acid keratophyres through andesites to basalts, which Eastwood et al. (1931) claim were intruded before the folded slates were cleaved. They trend between NW–SE and E–W, and appear to have formed at a relatively early stage. Other intrusions include granophyric and basic camptonite dykes. The age relationships have not been established and chemical details are not available.

Ennerdale Granophyre

The gravity data suggests that the volume of the Ennerdale Granophyre is much smaller than its outcrop implies. Bott (1974) commented that it 'might be a relatively thin marginal low density phase within the batholith'. Alternatively it might be an early intrusion which acted as a 'cap-rock' to the later Eskdale Granite magma.

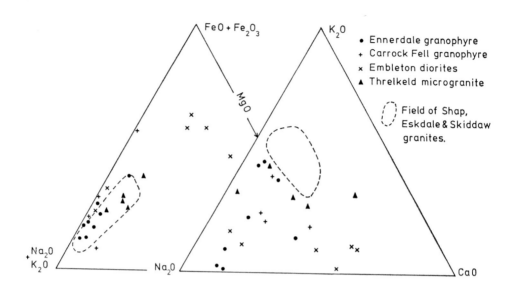

FIG. 54.   AFM and KCN diagrams of the older, possibly Ordovician, granophyres, microgranites and diorites. Analyses of the Threlkeld Microgranite are
those quoted by Hadfield and Whitesides (1936) supplemented by one XRF
analysis (University of Nottingham) of a sample from Bramcrag. Carrock Fell
Granophyre analyses are those quoted by Eastwood et al. (1968). Most of the
Embleton Diorite analyses are new but an analysis quoted by Eastwood et al.
(1968) is included. Ennerdale Granophyre analyses are new.

Its granophyric texture suggests that it was intruded at a higher structural level than
the coarsely crystalline Eskdale Granite; yet, like the granite, it penetrates only the
lowest andesites of the Borrowdale Volcanic Group. Possibly the granophyre was
intruded under a shallow cover of volcanic rocks in pre-Bala times whereas the
Eskdale Granite consolidated under a thicker cover during the Devonian. Chemical
variations within the granophyre and amphibolitization of the country rocks, attributed by Clark (1964) to metamorphism by the granophyre, may be due to the
underlying Eskdale Granite.

Green (1917) and Clark (1964) thought that the granophyre had been intruded
into the core of a major pre-Bala N–S anticline whereas Eastward et al. (1931) and
Simpson (1967) doubted the existence of such a fold. Rastall (1906) considered the
intrusion was a serious overlapping laccolites but, in the north, Eastward et al. (1931)
convincingly demonstrated a flat roof and vertical walls modified by vertical faults
and steeply dipping thrusts. In the south abrupt changes in altitude suggest faulting,
separate intrusions (Rastall 1906), or an undulating top to the granophyre.

The bulk of the outcrop consists of a fine-grained, pink, quartz-alkali feldspar
rock only rarely macroscopically porphyritic. Chlorite and patches of epidote are
sporadically distributed. Microscopically this 'normal' granophyre is very variable,
consisting of euhedral albite-oligoclase, sometimes perthite and rarely microcline set
in a granophyric matrix. Amongst the accessory minerals only chlorite, apparently
replacing biotite, and epidote are common: magnetite, apatite, sphene, zircon and

calcite have also been noted (Rastall 1906). Texturally the chief variations are in the size, distribution and proportion of phenocrysts to matrix, and in the amount and character of the micropegmatite. Marginal felsitic and micrographic varieties, and semi-marginal porphyritic varieties have been described (Trotter et al. 1937). Micropegmatite apparently increases, becomes finer and more perfectly developed towards the centre of the separate granophyre outcrops (Rastall 1906). Several different habits of micropegmatite occur including poorly developed spherulitic varieties and fringes around the plagioclase phenocrysts. Patches of micropegmatite become coarser outwards and the feldspar component is often in optical continuity with the central nuclei whether this is a plagioclase phenocryst or a fringe of orthoclase. Comparable features are ascribed by Dunham (1965) to magmatic crystallization and his suggestion that the felsic margin at Ennerdale represents a quickly cooled area which rapidly lost its volatiles has much to commend it. The Ennerdale Granophyre differs from most granophyres in the extreme variability of its potash/soda ratios which cannot satisfactorily be explained by variations in the feldspar phenocryst content. Evidently a substantial part and sometimes virtually all the feldspar in the micropegmatite is sodic. This variability could be due either to variations in the original sodium content of the magma, deuteric alteration or later metasomatism. Despite two unpublished Ph.D. theses (Govinda Rajulu 1959 and Clark 1963), no systemmatic study of the variations in alkalies has been attempted. In the writer's opinion the variations in the potash/soda ratios are too irregular and extreme to be other than a metasomatic effect; the timing and the source of soda being uncertain.

Basified and hybrid granophyres associated with dolerite are very common (Fig. 53). The dolerite appears older than the enclosing granophyre. However, hybrid rocks and granophyre gave K-Ar ages of 334 $\pm$ 18 Ma (334 $\pm$ 18 Ma and 370 $\pm$ 20 Ma respectively, Brown et al. 1964). The evidence needs to be reassessed, since basic magmas intruded into acid rocks often remobilize the acid material forming structures and textures which falsely make it appear as if the basic rock is older than the acid intrusion. Thus the dolerite and hybrid rocks could, as the K-Ar ages suggest, be Devonian; the slightly older date of the granophyre being due to isotopic overprinting during its remobilization. Alternatively the K-Ar ages could be due to loss of argon possibly caused by a metasomatic event connected with the underlying Eskdale Granite.

Threlkeld Microgranite

All three exposures of this prophyritic microgranite are sufficiently similar to suggest that they were formerly part of one irregular laccolite, intruded into Skiddaw Slates very close to the overlying Borrowdales. Moseley (in discussion of Wadge et al. 1975) claims that the Skiddaw Slates at Low Rigg were complexly folded before the microgranite was intruded. Similarly Rastall (1940) described a large Skiddaw Slate xenolith in the Threlkeld quarry which showed V-shaped folding. He argued that since the folding of the Skiddaw Slates was Caledonian the microgranite must be Calaedonian or younger. Conversely, now that the intrusion has yielded a Rb/Sr age of 445 $\pm$ 15 Ma (Wadge et al. 1975) it can be argued that the folding must be pre-Bala and might be pre-Borrowdale. Rastall's xenolith is thus relevant to the Skiddaw Slate/Borrowdale Volcanic junction controversy (Chapter 6) and it is unfortunate that he neither illustrated it nor noted whether it contained a cleavage.

The microgranite resembles the semi-marginal porphyritic variety of the Ennerdale Granophyre (Trotter et al. 1937) differing only in containing corroded quartz phenocrysts. Like the granophyre it contains albite and orthoclase micro-phenocrysts set in a felsic matrix with some bipyramidal quartz and chlorite. Published analyses (Ward 1876) Hadfield and Whiteside 1936) show considerable fluctuations in potash/soda ratios compared with the granites (Fig. 56). 6.4% calcite occurred in one sample and extensive metasomatic replacement of the microgranite by calcite was also described by Rastall (1940). Late stage patchy alterations involving both albit-ization and carbonitization with widespread chloritization seem characteristic. Much of the chemical variation could be due to the abundance of xenoliths, most of which seem to consist of Skiddaw Slate in various stages of assimilation. According to Rastall (1940) some xenoliths are derived from the Borrowdale Volcanics and he attributes the frequent solitary garnet crystals to the same source in spite of a lack of garnetiferous volcanics in the vicinity. Wadge et al. (1975) on the evidence of a calculated very low initial ratio of $^{87}Sr/^{86}Sr$ suggest that the intrusion is a mantle differentiate. If so, the almandine garnet might also have separated from this mantle differentiate in an analogous manner to that postualted by Fitton (1970) for garnets in the Borrowdales.

Older intrusions in the Borrowdale Volcanic Group

The Haweswater Complex is the largest basic intrusive within the Borrowdale Volcanic Group; yet was unnoticed during the initial Geological Survey mapping. Hancox (1934) noted that parts of the intrusion are cleaved and suggested that it is a pre-cleavage dolerite plug coeval with the Borrowdale volcanics. Banding in some of the coarser dolerites (Nutt 1966) suggests vertical flow within the intrusion. Hancox (1934) states that the main original constituents of the dolerite are plagioclase, a diopsidic clinopyroxene, orthoclase and quartz with accessory ilmenite and apatite. Secondary minerals include albite, sericite, actinolite, chlorite, sphene, epidote, biotite and tourmaline.

Numerous minor intrusions exist in the Borrowdale Volcanic Group and many more are shown on Geological Survey six inch maps. Some of these are altered lavas or the massive centres of lava flows. No comprehensive map of the minor intrusions in the Borrowdales has been published; nor is it possible to distinguish between 'early' and 'late' intrusions.

YOUNGER INTRUSIONS, PROBABLY DEVONIAN

Eskdale Granite

The outcrop consists of well exposed perthite-muscovite granite, coinciding with the lowest Bouguer anomaly in the Lake District, and a poorly exposed southern area with granodiorite and xenolithic granite. The shape of the outcrop and the gravity contours (Bott 1974) are consistent with intrusion into the core of a Caledonian anticline superimposed on a N–S trending pre-Bala anticline (Clark 1964). The granite is in contact with the Skiddaw Slates near Muncaster and Devoke Water and elsewhere penetrates the Borrowdale Volcanic Group. In the Boot (Dwerryhouse 1909) and Muncaster (Trotter et al. 1937) areas, the roof is relatively flat. However, contacts elsewhere suggest an undulating roof with distinct cupolas. The gravity contours seem to confirm that the main outcrop is contiguous with the small outcrop near Wasdale Head and also that the granite probably passes north under the Enner-dale Granophyre. A comparatively narrow aureole and close spaced gravity contours, parallel to the southern margin of the perthite granite, strongly suggest a steeply dipping south wall comparable to that postulated for the NW junction near Muncaster (Trotter et al. 1937). Bott (1974 and Chapter 3) estimated the depth of the batholith to between 6 and 18 km. Thus the perthite granite is more likely to be partially unroofed stock (cf. Trotter et al. 1937) than a laccolith (Dwerryhouse 1909, Simpson 1934). Structural studies (Firman 1960) suggest that the perthite granite was affected by wrench faulting before it had completely consolidated; minor thrusting and later vertical movements further modifying the shape of the intrusion. The southern granodiorite and xenolithic granite are apparently above the southern wall of the batholith (Bott 1974) and might therefore be relatively superficial phenomena.

The northern mass consists predominately of coarse pink perthite-muscovite granite, sometimes containing small amounts of biotite. Petrographic descriptions by Dwerryhouse (1909), Simpson (1934) and Trotter et al. (1937) differ in detail probably because of variations within the granite. Both strain free (Dwerryhouse 1909) and strained quartz (Simpson 1934) occur. The larger quartz crystals tend to be anhedral but small euhedral quartz enclosed in feldspar and fine quartz feldspar intergrowths occur (Trotter et al. 1937).

Perthite is the dominant feldspar, with albite (Simpson 1934) or oligoclase (Dwerryhouse 1909) lamellae. Plagioclase ranges in composition from albite to An_{25} and two generations of orthoclase occur. Simpson argued that the patchy distribution of muscovite in spherulitic aggregates indicated that it was due to 'some later phenomena' whereas Rose (in Trotter et al. 1937) thought that muscovite was usually primary. Biotite, when present, is often chloritized or replaced by haematite.

Tourmaline occurs sporadically both as joint coatings (Jones 1915) and as a replacement of feldspar (Rose op. cit.). Accessory minerals are rarer than in other intrusions (Rastall and Wilcockson 1915) Despite intense chloritization, sericitization and haematite staining, the perthite granites show little chemical variation.

Marginal rocks include normal perthite granites, compact granular, slightly porphyritic microgranites, greisen and injection breccias of quartz with very little muscovite. Acid rocks occur patchily in the main granite mass. Topaz, andalusite and fluorite have been noted in the Devoke Water Greisen and sericitized feldspar, epidote, apatite, rutile and sphene in the Eskdale Moor injection breccia (Simpson 1934).

The poorly exposed southern area is more varied. In the Waberthwaite quarries Simpson (1934) recorded xenolithic granite overlying a grey granite with white aplite veins containing large nests of tourmaline and garnet. This 'grey granite' apparently corresponds to the biotite-almandine granodiorite now exposed in the deeper levels. Perthite and quartz are less abundant and biotite, orthoclase and plagioclase are more abundant than in the perthite muscovite granites. Almandine garnets occur every 20–50 mm in most specimens, and range up to 5 mm in diameter. They often appear to be broken fragments of euhedral crystals. Zoned and altered plagioclase with epidote cores and sericitized margins are common; perthite is intensely sericitized and most of the biotite is chloritized. Quartz is always anhedral and intensely strained.

Beyond the quarry garnets are rare and xenoliths abundant. The xenoliths are angular or subangular, up to 300 mm in diameter, and often have ill defined boundaries. Usually they consist of biotite and plagioclase with a little interstitial quartz. Often hornblende is present and the plagioclases are zoned and altered. All stages between

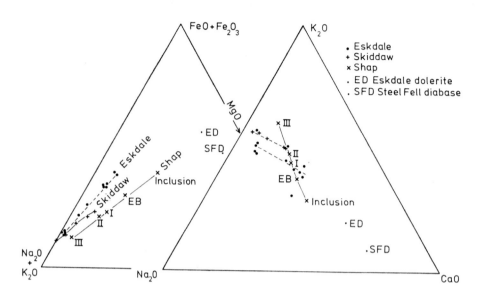

FIG. 55. AFM and KCN diagrams of Lower Devonian intrusions. Two analyses quoted by Trotter et al. (1937) are included otherwise all the analyses from the Eskdale area are new. Shap analyses (Grantham 1928) consist of an andesitic inclusion, early basic and stages I, II and III granites. Skiddaw Granite analyses are those quoted by Hitchen (1934).

well defined xenoliths and vague patches containing slightly higher concentrations of biotite have been observed. The xenoliths do not seem to increase towards the granite margin but the enclosing granite often becomes fine grained and granophyric (Rose op. cit.).

The abundance of xenoliths and the presence of apparently xenocrystal zoned plagioclase in quartz rich perthitic rocks led Rose (op. cit.) to suggest hybridization between the perthite granite magma and intermediate rocks such as andesites or diorites. On balance he favoured assimilation of Borrowdale Volcanic andesites. Recent gravity data (Bott 1974) suggests that the area is underlain by rocks which increase in density southward, so diorites may occur at depth. Moreover, the andesites hereabouts are not garnetiferous and, unlike the xenoliths, are not granitized at the granite contacts. Skiddaw Slate xenoliths have not been observed and hence it is tentatively suggested that the area is underlain by diorites which are in places garnetiferous.

Miller (1961) obtained an average K-Ar total volume age of 383 $\pm$ 2 Ma on biotite extracted from grey, biotite rich samples, collected near Bootle. Contrary to the impression given by Brown et al. (1964) and Bott (1974) the main pink perthite granite has not been dated.

Since this article went to press recent Rb-Sr and K-Ar age determinations have been made available by C. C. Rundle (in press). They have shown that the Eskdale, Ennerdale and Embleton intrusions are Ordovician, that the Carrock gabbros are older than the granophyres; but both are Ordovician as is the Harestones Felsite.

Skiddaw Granite

Although the gravity data suggests that the Skiddaw Granite is a large, homogeneous ovoid body, approximately 9×12 km and 6 km deep (Bott 1974), only three small exposures occur (Fig. 53). The southernmost has been interpreted as a dyke by Hitchen (1934) and as part of the main granite by Eastwood et al. (1968). The Caldew valley outcrop appears to be a flattish portion of the roof and the Grainsgill mass is probably a steep sided cupola, elongated N-S (Eastwood et al. 1968). The gravity data is not sufficiently detailed to permit precise definition of the undulations in the roof of the granite. However, the shape of the aureole suggests a buried cupola north of the Carrock Fell Complex, beneath the Harestones Felsite (Fig. 53).

The central outcrop consists largely of biotite-granite, with oligoclase up to 1.5 mm long enclosed in orthoclase perthite up to 5 mm across. Quartz blebs occur in both kinds of feldspar and interstitially. The brown biotite contains zircon and is partially converted to chlorite and sphene. A little smokey apatite and magnetite is present. Eastwood et al. (1968) also noted that biotite was less abundant and probably less iron-rich in cores from boreholes at the northern end of the outcrop. Here they also recorded a little greisen and noted oligoclase zoned outwards from An_{14} to An_8 Aplitic and pegmatitic varieties occur, and Hitchen (1934) claimed that biotite progressively diminishes towards the contact, being replaced by white mica, haematite, kaolin, epidote, etc. The exposures in the Caldew valley are, thus, very variable, nevertheless, differences between the three outcrops are greater than variations within outcrops.

The Sinen Gill exposure differs in having more biotite, and larger phenocrysts than either of the other outcrops. Microcline occurs here and nowhere else in the Skiddaw Granite. Hitchen (1934) described this rock as a coarse, porphyritic biotite-granite consisting of quartz, various feldspars, muscovite, biotite and a few accessory minerals. The large phenocrysts, up to 50 mm long are coarse perthites enclosing plagioclase, quartz, muscovite, biotite, zircon and apatite. They seem to be either contemporaneous (Rastall 1910) or later than the other constituents. The matrix consists of early formed biotite, plagioclase, orthoclase, microcline, perthite, occasional muscovite, rare micropegmatite, and late quartz. Accessory minerals include pyrrhotite, ilmenite, zircon, apatite, epidote, sphene, rutile, anatase, andalusite, and relatively abundant tourmaline (Rastall and Wilcockson 1915).

The Grainsgill outcrop consists of normal granite exhibiting various stages of greisenization leading to the formation of a coarse grained rock consisting of about two-third quartz and one third muscovite with accessory tourmaline, apatite, arseno-pyrite, pyrite, molybdenite, pyrrhotite, epidote, fluorite, rare bismuth sulpho-salts and other minerals associated with the epigenetic veins. Pioneer petrographic studies by Harker (1895) were supplemented with chemical analyses by Hitchen (1934) and Eastwood et al. (1968). These combine to suggest that, 'in the metasomatism of granite to form greisen, much of the potash is presumably derived from orthoclase, some from biotite, but probably a little is removed. The addition of some silica, water and fluorine is indicated, and removal of soda possibly with some lime and iron' (Dunham in Eastwood et al. 1968). Recently oxygen isotope and fluid inclusion studies have indicated that this process was probably caused by moderately saline NaCl brines being drawn into the Grainsgill cupola as the granite cooled (Shepherd et al. 1975). These authors estimate that greisenization took place between 250–300°C. at 800 bars. K-Ar determinations on muscovite date the greisenization at 385 ± 4 Ma.

The Skiddaw Granite has an extensive gentle dome-shaped aureole in Skiddaw Slates which Eastwood et al. (1968) estimated to be about 750 m thick. The Carrock Fell Complex, Eycott Volcanic Group and Drygill Shales have also been metamorphosed but it is difficult to distinguish between the metasomatic effects caused by the Skiddaw Granite and deuteric changes within the Carrock Fell Complex. If, for instance, hornblendization is due to fluids emanating from the granite, then all the Carrock Gabbro might have been affected. Alteration is most intense near mineral veins (Holmes 1917) and it is possible that much of the metasomatism in the granite is related to much younger epigenetic mineralization (cf. Chapter 16).

In the Skiddaw Slates metamorphic effects can easily be distinguished from narrow metasomatic zones adjacent to mineral veins. Chemical analyses (Eastwood et al. 1968) suggest that the thermal metamorphism was essentially an isochemical process. Eastwood et al. (1968) show that in the outermost zone spotted slates are frequent and tiny flakes of biotite are developed at the expense of chlorite in some bands. The spots consist of either white mica with a chloritic margin or chlorite with a micaceous margin. They believed the former to be the precursor of andalusite and the latter the forerunner of cordierite. In the intermediate zone visible andalusite and cordierite porphyroblasts have developed. The andalusites commonly are chiastolite varieties. The matrix usually consists of fine white mica and chlorite with reddish brown biotite, carbonaceous matter, black oxides and occasional pyrite. Cordierite-biotite-hornfelses without andalusite also occur and retrogressive metamorphism has been noted. Inner zone hornfelses are similar but are completely metamorphosed. Grain sizes are larger and the white mica has recrystallized.

Minor dolerite intrusions within the aureole are amphibolitized in a similar manner to the Carrock Fell Gabbro and because of this Eastwood et al. (1968) refer the hornfels to the Amphibolite Facies rather than the Pyroxene-Hornfels Facies.

No direct isotopic dating of the aureole rocks has been attempted. As shown by Barratt (in discussion of Brown et al. 1964), the chiastolite is younger than the slaty cleavage and is affected by a later crenulation cleavage. Soper and Roberts (1971) suggest that the granite was emplaced towards the end of the D_2 (see Chapter 5) deformation, but before D_3. The mean of five K-Ar 'total volume' ages on biotites from the Sinen Gill Granite outcrop indicate an age of cooling of 399 $\pm$ Ma (Miller 1961), which may be compared with a mean age of 392 $\pm$ 4 Ma obtained more recently by Shepherd et al. (1976) on two biotites from the same locality. Thus the final emplacement of the granite appears to have been 'end-Silurian' or early Devonian, the latter being the most likely since, as pointed out by Dodson et al. (1961) 'total volume' techniques tend to give anomalously older apparent ages.

Shap granite

The 8 km² outcrop, is just north of the minimum of an extensive area of low gravity values which probably are caused by a subsurface granite of variable density. According to Bott (1974) a less dense phase probably occurs beneath the Silurian and may form an incomplete subsurface ring around the outcrop with denser varieties occurring further north. The outcrop appears to be a small boss emerging from a larger composite batholith about 10 × 12 km and 8 km deep; suggestions of a cedar-tree laccolith (Harker and Marr 1891) are not substantiated by the gravimetric evidence.

The granite reached a higher stratigraphical level than any other major Lake District intrusion. The country rocks which range from the Borrowdale Volcanic Group to Coniston Grits were folded, faulted and cleaved before being thermally metamorphosed (Harker and Marr 1891). The main sub-vertical cleavage was deflected around the granite, presumably due to forcible intrusion (Boulter and Soper 1973). Outside the aureole in Silurian strata this main cleavage is often folded by kink-bands. These have not been seen in the hornfelses, probably because the kink-bands formed after thermal metamorphism, and the hornfelses were resistant to kinking (P. Redfern personal communication). Thus the Shap Granite, like the Skiddaw intrusion, appears to have been intruded before the main D_2 structures were refolded. The approximate contemporaneity of the Shap and Skiddaw Granites is further demonstrated by the isotopic dating (Brown et al. 1964 and Shepherd et al. 1976).

Grantham (1928) has shown that the granite consists largely of a coarse porphyritic adamellite (Stage II) containing xenoliths of earlier 'Stage I' and 'early basic' granites. Angular Borrowdale Volcanic and Coniston xenoliths are rare, now that the quarry face is further from the granite margin.

Stage II granite contains 20–30% orthoclase perthite phenocrysts up to 50 mm long, set in a moderately coarsely crystalline matrix of quartz, orthoclase and oligoclase, in roughly equal proportions, and biotite. Oligoclase is euhedral in the groundmass and also occurs as irregular apparently corroded inclusions in the phenocrysts. Groundmass orthoclase is subhedral, apparently later than the plagioclase, slightly altered, often cryptoperthitic, and sometimes surrounded by an albitic margin. Some late orthoclase occurs. Biotite tends to be paler and less abundant than in the earlier stages. The large subhedral late quartz crystals are one of the distinctive features of this stage. Primary accessory minerals are abundant, apparently formed before the major components. Euhedral sphene, apatite and magnetite are the most prolific with lesser amounts of zircon, fluorite, monzonite, amphibole and pyrite. Grantham (1928) regarded some minerals as either secondary or due to hydrothermal infiltration. These include chlorite, calcite, pyrite, anatase, fluorite, tourmaline, muscovite, rutile and bismuthinite. Spears (1961) has shown that 95% of the alpha radioactivity comes from the 2% accessory minerals and Farrand's (1960) trace element analyses show high lithium contents.

Xenoliths of stage I granite range from a few centimetres to a raft at least 36 × 30 m and 6 m deep (Grantham 1928). Phenocrysts are fewer and about the same size as in stage II, often having an albitic rim. Feldspars occur in the groundmass and accessory minerals are as in stage II except that the plagioclase is more frequently zoned with altered cores and clear margins. Biotite is more abundant, darker and euhedral. Quartz is earlier than the groundmass orthoclase.

The 'early basic' xenoliths occur in stage I and stage II granites. Frequently they have a coarse granitic margin containing quartz, oligoclase and biotite, passing inwards to very dark fine-grained biotite hornfels. Orthoclase perthite phenocrysts constitute 6–8% of the rock. They have albitic margins and are smaller and more rounded than in later granite stages. Occasionally phenocrysts occur partly in the surrounding granite and partly in the xenoliths. Grantham (1928) regarded these xenoliths as part of an early peripheral intrusion, disrupted and partly absorbed by the stage I and stage II granites. Alternatively, they may represent partially granitized and felspathized fragments of Borrowdale rocks, the rounded shapes suggesting considerable incorporation of Borrowdale material by the granite magma.

The 'phenocrysts' formed late but before final consolidation. Fluxion structures are common and, particularly in the stage II granite, the phenocrysts are often disrupted and veined with later minerals (Grantham 1928). Harker and Marr (1891) noted that, whereas biotite and plagioclase inclusions in the phenocrysts were common, quartz is rare, and tends to be concentrated in the outer margin. They therefore ascribed the formation of phenocrysts to a period later than the crystallization of plagioclase and biotite and before the formation of the bulk of the quartz. Vistelius (1960) in a statistical study of the textures of grey and pink granites (presumably equivalent to Grantham's stage I and II) emphasized the anomalous paucity of potassic feldspars in the groundmass. He suggested that late fluids soaking through the granite dissolved some of the groundmass orthoclase and recrystallized others into phenocrysts in the granite and its xenoliths, attributing the albitic fringe to an abrupt change in pressure. Textural evidence suggests that the solution and recrystallization was not limited to potash feldspars, as implied by Vistelius (op. cit.), but also involved the plagioclase feldspars. Redistribution of potash also occurred in the later granite stages such as Grantham's 'stage III' and in some of the associated dykes in the aureole (Redfern personal communication).

Reddening of the phenocrysts, and subsequent reddening of zones adjacent to master joints in the granite, could be due to the iron released during the partial replacement and alteration of biotite. This may also be the source of iron for the formation of pyrometasomatic andradite garnet, magnetite and haematite in the aureole.

The metamorphic aureole has been described in detail by (Harker and Marr 1891, 1893). Their statement 'that no transference of material has taken place within the mass of rocks except between closely adjacent points', needs modification. Analyses by Firman (1957) and Farrand (1961) indicate potash metasomatism in contact hornfelses. Extensive fissure metasomatism occurs along pre-granite fractures and contact hornfelses. Extensive fissure metasomatism occurs along pre-granite fractures and cleavages both in Borrowdales (Firman 1957) and the Silurian. Phyllic alteration was observed in biotite hornfels 400 m north of the granite (Firman 1957), in volcanic rocks on the fringe of the aureole in Wet Sleddale (under the dam) and in a dyke near Shap Wells. Transference of potash, albeit along restricted channels, was therefore more extensive than envisaged by Harker and Marr (1893). The Ca^{2+} needed for the formation of epidote-andradite seems likely to have been derived from pre-granite calcite-quartz-chlorite veins (Firman 1972).

Within the aureole, chloritized andesites, tuffs and Silurian siltstones were metamorphosed into biotite hornfels with quartz and plagioclase. Sillimanite, andalusite and cordierite occur in the more aluminous rocks, and hornblende in metamorphosed chlorite-rich rocks such as the vesicular andesites. Idocrase and grossularite, in the lower part, and diopside, anorthite, tremolite and wollastonite, in the upper part, are the commonest minerals in the metamorphosed Coniston Limestone. Similar assemblages of lime silicates occur in calcareous bands within the Silurian.

Minor intrusions

The Steel Fell Diabase has been shown by K-Ar dating to be Devonian (Mitchell and Ineson 1975) but the age of most other minor intrusions into pre-Silurian rocks is uncertain. Dykes in the Silurian are almost certainly Devonian since no comparable dykes have been found in the Carboniferous. They have been found as far west as Lake Coniston and east to Sedbergh. Similar intrusions have been reported from the Eskdale Granite, the Cross Fell Inlier and the Ingletonian. In the Duddon Valley and the Silurian between Coniston and Kentmere the minor intrusions are parallel to the gravity contours, which hereabouts are parallel to the regional strike. In the vicinity of the Shap intrusions the dykes tend to strike either at right angles or parallel to the isogals and cut across the regional strike. A swarm of dykes trends SSE from the Shap Granite and also extends 3 or 4 km north of the granite outcrop. This swarm appears to be developed above the crest of the Shap intrusion. The Lake District batholith thus appears to have controlled the distribution of the later minor intrusion in the southern area and probably throughout the whole region.

Bonney and Houghton (1878) classified the minor intrusions into the Silurian as felsite, kersantites and minettes. Morrison (1918) added acid intrusions with Shap type phenocrysts. Published analyses are over 100 years old (Bonney and Houghton 1878) and probably unreliable. They seem to indicate potash metasomatism similar to many of the dykes near to the Shap Granite (Redfern personal communication). Detailed petrography and a suggested sequence are given by Morrison (1918) but a modern study is long overdue

POST-DEVONIAN INTRUSIONS

A fresh olivine dolerite, intruded into the Mell Fell Conglomerate, is the only intrusion in the Lake District which penetrates post-Silurian strata. It is associated with tuffs, a vent and a basalt lava which Capewell (1954) thought most likely to be contemporaneous with the Cockermouth Lavas. Petrologically similar dykes intruded into the Borrowdale Volcanic Group crop out in the Scafell region (Dwerry-house 1909, p. 78) and near Melmerby in the Cross Fell Inlier. The latter have yielded a mean K-Ar age of 296 $\pm$ 8 Ma which suggests that they are contemporaneous with the Whin Sill (Wadge et al. 1972). Thus, the Lake District olivine dolerites seem most likely to be either Lower Carboniferous or Stephanian/Westphalian in age. A Tertiary age has been suggested (Dwerryhouse 1909) but seems less likely since the undoubted Tertiary Armathwaite dyke does not contain olivine.

CONCLUSIONS, SPECULATIONS AND SUGGESTIONS FOR FURTHER RESEARCH

The intrusions are classified here into 'older', 'younger' and 'post-Devonian' suites. The Carrock Fell Complex is grouped with the 'older, possibly Ordovician' intrusions, otherwise the classification is the same as Harker (1902). Of the older suite, only the Llandeilo/Caradocian Threlkeld Microgranite appears to be reasonably dated (Wadge et al. 1975). There is need for Rb-Sr and K-Ar dating to clarify, if possible, the age of emplacement and subsequent metasomatic events affecting the Carrock Fell Complex. Ennerdale Granophyre and the minor intrusions. Isotopic dating suggests that the Shap and Skiddaw Granites are contemporaneous end-Silurian or early Devonian intrusions. The Eskdale Granite seems to be a little younger (Brown et al. 1964). The dates quoted for the Eskdale and Skiddaw Granites have not been confirmed by Rb-Sr methods but are reasonably in accord with the structural evidence (Firman 1960, Soper and Roberts 1971). Shepherd et al. (1976) have confirmed that the Grainsgill Greisen is only slightly younger than the main Skiddaw Granite, but greisen in the Eskdale Granite, Ennerdale Granophyre and Threlkeld Microgranite and other metasomatic phenomena have not been dated.

A review of published petrochemical data, supplemented by analyses prepared for this chapter, reveals that metasomatism in the older intrusions was far more extensive than previously suspected. Figures 54, 55 and 56 suffer from the disadvantage that the analyses were done in different laboratories, often using different techniques. The results, therefore, are not strictly comparable. Nevertheless the variation is usually greater than that attributable to analytical error. Acid rocks and diorites of the older group have, with one exception K_2O/Na_2O ratios less than unity. On the AFM diagram (Fig. 54) the granophyres and microgranites occupy a similar field to the later granites, and, with the diorites, appear to show a tholeitic differentiation trend. When K_2O and Na_2O are plotted separately, as in the KCN diagram (Fig. 54), all semblance of a magmatic differentiation trend is lost. The simplest explanation of the extreme variability of the KCN diagram, contrasted with the comparative orderliness of the AFM diagram (Fig. 54), is that there has been widespread, patchy replacement of potash by soda. This soda metasomatism is particularly marked in the Ennerdale Granophyre, but is also exhibited by some specimens of the other older intrusions. Several analyses of the Ennerdale Granophyre and Embleton Diorites lie outside the 'igneous spectrum' as defined by Hughes (1972) whilst other have K_2O/total alkali ratios more appropriate to intermediate and basic rocks than to acid intrusives.

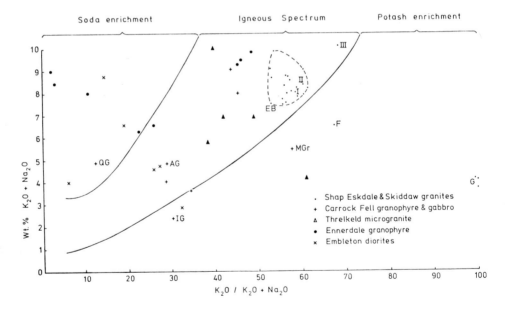

FIG. 56. $K_2O \pm Na_2O$ versus $K_2O/K_2O \pm Na_2O$ (after Hughes 1972). Analyses plotted include those on Figs. 54 and 55 plus greisen associated with the Skiddaw Granite (Hitchen 1934) and Carrock Fell gabbros and felsite (quoted by Eastwood et al. (1968) QG — quartz gabbro, IG — ilmenite gabbro, AG — albitic granophyre, MG — melagranophyre, F — felsite, G — greisen.

If Hughes (1972) is correct, the majority of the older intrusions have been metasomatized, including the Carrock Fell quartz and ilmenite gabbros both of which lie outside the 'igneous spectrum' (Fig. 56). The one carbonate rich rock to have been analysed (Threlkeld Microgranite with 6.4% calcite, Hadfield and Whiteside 1936) is depleted in total alkalies and enriched in K_2O relative to Na_2O (Fig. 56). It is tentatively suggested that this is the result of soda metasomatism followed by replacement of the Na plagioclase by calcite. The extent, timing and mechanism of the soda metasomatism of the older intrusions is unknown. One possible mechanism is that if these high level intrusions were intruded into wet sediments, NaCl rich connate waters may have been drawn into the cooling igneous masses resulting in the replacement of potash feldspars by albite. Such a process would enhance the expelled formation water in KCl and would be expected to release barium and lead from the feldspars. This would provide a possible source for mineralizing brines which later precipitated baryte and galena (cf. Chapter 16). This is but one of many models which need to be tested by comprehensive chemical, fluid inclusion and isotopic studies.

In contrast, the Shap, Eskdale and Skiddaw Granites are far less variable chemically. On the AFM and KCN diagrams (Fig. 55) they occupy small, well defined fields, and seem to show distinct differentiation trends. In spite of considerable sericitization in the Eskdale Granite (Simpson 1934), greisenization in the 'normal' Skiddaw Granite (Hitchen 1934), and extensive transference of potash in the Shap intrusion (Vistelius 1969) all granite analyses lie within Hughes igneous spectrum (Fig. 56). Nevertheless minor intrusions associated with the Skiddaw and Shap Granites sometimes have alkali contents outside Hughes igneous spectrum and appear to have suffered considerable potash metasomatism. The various stages of the Shap Granite show a direct lineage from the andesitic inclusion (Grantham 1928) with a pronounced enrichment in both total alkalies and K_2O in the latter stages (Figs. 55 and 56) suggesting incorporation of considerable amounts of Borrowdale Volcanic material and late potash metasomatism. There are close chemical similarities between the Eskdale and Skiddaw Granites although the Skiddaw Granite tends to be slightly more potassic and has more magnesia. Both granites have been likened to the Weardale Granite (Dunham et al. 1950) the closest analogy being with the Skiddaw Granite. Like the Weardale Granite, the Skiddaw Granite is surrounded and penetrated by a very varied set of epigenetic base metal mineral veins of which most, if not all, are much younger than the granite. The problems of epigenetic mineralization are discussed more fully in Chapter 16 but it may be noted here that the role of the Lake District intrusions as possible sources, heat engines and conduits for mineralizing brines remains one of the most promising fields of Lake District research.

REFERENCES

BONNY, T. G. 1885. On the so-called diorite of Little Knott (Cumberland) with further remarks on the occurrence of picrites in Wales. Q. Jl geol. Soc. Lond. **41**, 511–22.
—— and HOUGHTON, F. T. S. 1897. On some mica-traps from the Kendal and Sedbergh districts. Q. Jl geol. Soc. Lond. **35**, 165–80.
BOTT, M. H. P. 1974. The geological interpretation of a gravity survey of the English Lake District and the Vale of Eden. Jl geol. Soc. Lond. **130**, 309–31.
BOULTER, C. A. and SOPER, N. J. 1973. Structural Relationships of the Shap Granite. Proc. Yorks. geol. Soc. **39**, 365–69.
BROWN, P. E., MILLER, J. A. and SOPER, N. J. 1964. Age of the principal intrusions of the Lake District. Proc. Yorks. geol. Soc. **34**, 331–42.
CAPEWELL, J. G. 1954. The basic intrusions and an associated vent near Little Mell Fell, Cumberland. Trans. Leeds geol. Assoc. **6**, 243–48.

CLARK, L. 1964. The Borrowdale Volcanic Series between Buttermere and Wasdale, Cumberland. *Proc. Yorks. geol. Soc.* **34**, 343–56.

DOBSON, M. H., MILLER, J. A. and YORK, D. 1961. Potassium-argon ages of the Dartmoor and Shap Granites using the total volume and isotopic dilution techniques of argon measurement. *Nature.* **190**, 800–82.

DUNHAM, A. C. 1965. The nature and origin of groundmass textures in felsites and granophyres from Rhum, Inverness-shire. *Geol. Mag.* **102**, 8–23.

DUNHAM, K. C., DUNHAM, A. C., HODGER, B. L. and JOHNSON, G. A. L. 1965. Granite beneath Visean sediments with mineralization at Rookhope, northern Pennines. *Q. Jl geol. Soc. Lond.* **121**, 383–417.

DWERRYHOUSE, A. R. 1909. Intrusive rocks in the neighbourhood of Eskdale. *Q. Jl geol. Soc. Lond.* **65**, 55–80.

EASTWOOD, T., DIXON, E. E. L., HOLLINGWORTH, S. E. and SMITH, B. 1931. The Geology of the Whitehaven and Workington District. *Mem. geol. Surv.,* **304** + xviii pp. H.M.S.O. London.

——, HOLLINGWORTH, S. E., ROSE, W. C. C. and TROTTER, F. M. 1968. Geology of the Country around Cockermouth and Caldbeck. *Mem. geol. Surv.,* x + 298 pp. H.M.S.O. London.

FARRAND, M. G. 1960. The distribution of some elements across four xenoliths. *Geol. Mag.* **97**, 488–93.

FIRMAN, R. J. 1957. The Borrowdale Volcanic Series between Wastwater and Duddon valley, Cumberland. *Proc. Yorks. geol. Soc.* **31**, 39–64.

—— 1957a. Fissure metasomatism in volcanic rocks adjacent to the Shap Granite, Westmorland. *Q. Jl geol. Soc. Lond.* **133**, 205–22.

—— 1960. The relationship between the joints and fault patterns in the Eskdale Granite (Cumberland) and the adjacent Borrowdale Volcanic Series. *Q. Jl geol. Soc. Lond.* **166**, 317–47.

—— 1972. The Shap area. *Proc. Geol. Assoc.* **83**, 499–50

FITTON, J. G. 1972. The genetic significance of almandine-pyrope phenocrysts in the calc-alkaline Borrowdale Volcanic Group, Northern England. *Contr. Mineral. and Petrol.* **36**, 231–48.

GOVINDA RAJULU, B. V. S. 1959. The structural and petrological relations of the Ennerdale Granophyre (Cumberland) to the adjoining rocks. Unpublished Ph.D. thesis, University College, London.

GRANTHAM, D. R. 1928. The Petrology of the Shap Granite. *Proc. Geol. Assoc.* **39**, 299–331.

GREEN, J. F. N. 1917. The age of the chief intrusions of the Lake District. *Proc. Geol. Assoc.* **28**, 1–30.

—— 1921. Long excursion to the Lake District. *Proc. Geol. Assoc.* **32**, 123–38.

HADFIELD, G. S. and WHITESIDE, H. C. M. 1936. The Borrowdale Volcanic Series of High Rigg and the adjoining Low Rigg Microgranite. *Proc. Geol. Assoc.* **47**, 42–64.

HANCOX, E. G. 1934. The Haweswater Dolerite, Westmorland. *Proc. Liverpool Geol. Soc.* **16**, 173–97.

HARKER, A. 1894. Carrock Fell. A study in the variation of igneous rock masses: Part I. The Gabbro. *Q. Jl geol. Soc. Lond.* **50**, 311–17.

—— 1895. Carrock Fell. Part II. The Carrock Fell Granophyre. Part III. The Grainsgill Griesen. *Q. Jl geol. Soc. Lond.* **51**, 125–48.

—— 1902. Notes on the igneous rocks of the English Lake District. *Proc. Yorks. geol. Soc.* **14**, 487–93.

—— and MARR, J. E. 1891. The Shap Granite and Associated rocks. *Q. Jl geol. Soc. Lond.* **47**, 266–328.

—— and MARR, J. E. 1893. Supplementary notes on the Metamorphic rocks around the Shap Granite. *Q. Jl geol. Soc. Lond.* **49**, 359–71.

HITCHEN, C. S. 1934. The Skiddaw Granite and its resdiual products. *Q. Jl geol. Soc. Lond.* **90**, 158–200.

HOLMES, A. 1917. Albite-granophyre and quartz-porphyry from Brandy Gill, Carrock Fell. *Geol. Mag.* **4**, 403–407.

HUGHES, C. J. 1972. Spilites, keratophyres, and the igneous spectrum. *Geol. Mag.* **109**, 513–27.

MILLER, J. A. 1961. The Potassium-argon ages of the Skiddaw and Eskdale Granites. *Geophys. Journ.* **6**, 391–93.

MITCHELL, J. G. and INESON, P. R. 1975. Potassium-argon ages from the graphite deposits and related rocks of Seathwaite, Cumbria. *Proc. Yorks. geol. Soc.* **40**, 413–18.

MORRISON, J. 1919. The Shap minor intrusions. *Q. Jl geol. Soc. Lond.* **74**, 116–44.

NUTT, M. J. C. 1966. Field meeting in the Haweswater area. *Proc. Yorks. geol. Soc.* **35**, 429–33.

POSTLETHWAITE, J. 1893. An intrusive sheet of diabase and associated rocks at Robin Hood, near Bassenthwaite. *Q. Jl geol. Soc. Lond.* **49**, 531–35.

RASTALL, R. H. 1906. The Buttermere and Ennerdale Granophyre. *Q. Jl geol. Soc. Lond.* **62**, 253–74.

—— 1910. The Skiddaw Granite and its Metamorphism. *Q. Jl geol. Soc. Lond.* **66**, 116–40.

—— 1940. Xenoliths at Threlkeld, Cumberland. *Proc. Yorks. geol. Soc.* **24**, 223–32.

—— 1940a. A carbonate rock at Threlkeld, Cumberland. *Proc. Yorks. Geol. Soc.* **24**, 232–34.

RASTALL, R. H. and WILCOCKSON, W. H. 1915. Accessory minerals of the Granitic rocks of the Lake District. *Q. Jl geol. Soc. Lond.* **71**, 592–622.

SHEPHERD, T. J., BECKINSALE, R. D., RUNDLE, R. D., and DURHAM, J. 1976. Genesis of the Carrock Fell tungsten deposits, Cumbria: fluid inclusion and isotopic study. *Trans. Inst. Min. Met.* **85**, B63–73.

SIMPSON, A. 1967. The stratigraphy and tectonics of the Skiddaw Slates and the relationship of the overlying Borrowdale Volcanic Series in part of the Lake District. *Geol. J.* **5**, 391–87.

SIMPSON, B. 1934. The petrology of the Eskdale (Cumberland) Granite. *Proc. Geol. Assoc.* **45**, 23–38.

SKILLEN, I. E. 1973. The Igneous complex of Carrock Fell. *Proc. Cumb. geol. Soc.* **3**, 363–86.

SOPER, N. J. and ROBERTS, D. E. 1971. Age of cleavage in the Skiddaw Slates in relation to the Skiddaw aureole. *Geol. Mag.* **108**, 293–302.

SPEARS, D. A. 1961. The distribution of Alpha Radioactivity in a specimen of Shap Granite. *Geol. Mag.* **98**, 483–87.

STRENS, R. J. G. 1965. The graphite deposits of Seathwaite in Borrowdale, Cumberland. *Geol. Mag.* **102**, 393–406.

VISTELIUS, A. B. 1969. O granitakh Shep (Westmorland Angliya). (The Shap Granite, Westmorland). *Akad. Nauk. SSR., Dokl.* **187**, 391–94, (In Russian–English Translation).

WADGE, A. J., HARRISON, R. K. and SNELLING, N. J. 1972. Olivine-dolerite intrusions near Melmerby, Cumberland, and their age-determination by the potassium-argon Method. *Proc. Yorks. geol. Soc.* **39**, 59–70.

WADGE, A. J., HARDING, R. R. and DERBYSHIRE, D. P. F. 1974. The rubidium-strontium age and field relations of the Threlkeld Microgranite. *Proc. Yorks. geol. Soc.* **40**, 211–22.

WARD, J. C. 1876. The geology of the northern part of the Lake District. *Mem. Geol. Surv.* 12 + 132 pp.

R. J. FIRMAN, PH.D.
Department of Geology, The University, Nottingham

II

Devonian

A. J. WADGE

The Mell Fell and Polygenetic conglomerates occur as pockets of coarse detritus beneath the Carboniferous sequences of the eastern Lake District and the Vale of Eden. The Mell Fell Conglomerate underlies Great Mell Fell, Little Mell Fell and the rounded hills at the foot of Ullswater, and the Polygenetic Conglomerate crops out at three localities in the Cross Fell Inlier (Fig. 57). The two formations are probably Devonian in age, and may even date from early Devonian times but they are barren of fossils and conclusive evidence of their age is lacking. They are, however, distinguished from the clastic sediments forming the lowest beds of the Lower Carboniferous sequence by their stratigraphical relations as described below.

The conglomerates post-date the end-Silurian earth movements and rest unconformably upon the Caledonian land-surface. The period of time between the deposition of the conglomerates and the Carboniferous rocks seems to have been considerable. Certainly further extensive erosion of the land surface took place during this time as the Caledonian mountains were worn down to the topography of low relief that was eventually transgressed by the Carboniferous seas. It is not known however whether further earth movements occurred during this period. An angular unconformity separates the weathered top of the Polygenetic Conglomerate from the overlying Carboniferous rocks and a similar relationship has been suggested (Capewell 1955) at the top of the Mell Fell Conglomerate, although this junction is not exposed. Whether the steeper dips in the conglomerates are tectonic in origin, implying earth movements, or depositional, is still in doubt.

Polygenetic Conglomerate

The deposit was first differentiated from the Carboniferous rocks by Marr (1899) and was further described by Shotton (1935). Its name derives from its distinctive pebbles which are more varied in lithology than those of the Carboniferous Basement Beds. It is exposed at three localities (Fig. 57a), lying to the east of Gamblesby, Melmerby and Ousby Townhead respectively (Burgess and Wadge 1974). The thicknesses of the Conglomerate is variable. In the best exposure, along the fell-track east of Melmerby, 35 m of the Conglomerate pinch out entirely along the crop within 250 m. The base of the deposit rests with marked unconformity upon rocks of the Skiddaw Group [6280 3705] and its weathered top is also exposed nearby [6287 3704] beneath the unconformable Carboniferous Basement Beds. The Conglomerate is crudely stratified, and dips are generally to the east, exceeding those in the Basement Beds by up to 40 degrees.

The clasts range in shape from sub-angular to well-rounded and are set in a purple-red sandy, matrix. They are up to 1 m across and are commonly coated with a thin layer of hematite, resembling the desert varnish of modern desert pebbles. They can generally be matched with local Lower Palaeozoic lithologies. Skiddaw Group mudstones, siltstones and greywackes predominate and Eycott lavas and acid porphyries are also common. Rare boulders of biotite-granite found in the Ousby outcrops are of unknown provenance.

The deposit seems to represent several coalescent fans of coarse detritus laid down in arid conditions. The steep depositional dips suggest derivation from higher ground close to the west (Burgess and Wadge 1974) and it is possible that this was an east-facing scarp defined by the Deep Slack and Fellside faults (Fig. 57a).

Mell Fell Conglomerate

The most comprehensive study of the deposit, including an analysis of the heavy minerals, was by Capewell (1955). The earliest accounts of the lithologies and outcrops were given by Ward (1876) and Dakyns et al. (1897), and Oldham (1900) was first to suggest that the conglomerate accumulated as a fan of torrent detritus in an arid climate whilst Green (1918) deduced the nature of the source areas from the lithologies of the phenoclasts.

The rocks are best seen on the southern and western slopes of the Mell Fells, by the roadside near Pooley Bridge, and in Dacre Beck and its tributaries. Recent boreholes on the line of the A66 trunk road show that the western margin of the conglomerate lies beneath the drift-covered ground around Tarn Moss [399 277], about 1.5 km farther west than is shown on the published maps.

The commonest lithology is cobble-conglomerate, generally with sub-angular clasts about 0.1 to 0.3 m across but with occasional boulders up to 1.0 m in diameter. The purple-red, sandy matrix has a calcite or iron oxide cement. The flatter clasts are aligned to give a crude stratification, and local channelling and infilling is common. The sediments become finer in grade to the north-east.

The thickness of the conglomerate has been variously estimated at about 275 m (Dakyns et al. 1897) and at 1500 m (Capewell 1955). In this account the lower figure is favoured for the reasons given below. The wide variation in these estimates arises from the different views taken of the nature and overall shape of the deposit. Thus, the horizontal section drawn by Ward (Aveline et al. 1881) shows that the conglomerate is more or less constant in thickness and that the bedding is largely cross-stratified. In contrast, Capewell contends that the interbedding of flaggy sandstones with the conglomerates around Dacre confirms that the dips throughout the deposit are mainly tectonic. By adding together all the exposed beds, he arrives at the much higher total thickness. Similar reasoning led Green (1918) to an initial estimate that more than 1000 m were present to the south of Penruddock, but he saw serious difficulties in accepting this figure and the still greater thicknesses necessary around Dacre, and preferred therefore to regard the deposit as the truncated remnant of a cone of detritus dipping outwards from the mouth of a torrent. In this view, the coarser sediments around the Mell Fells are more likely to pass laterally into the finer sandstones around Dacre than to pass directly beneath them. Whilst this latter view seems sensible, the picture of a single cone of detritus is not supported by the dips, which do not radiate towards the edges of the conglomerate but generally converge towards the north-east and east. Instead, the deposit is envisaged here as a series of coalescing

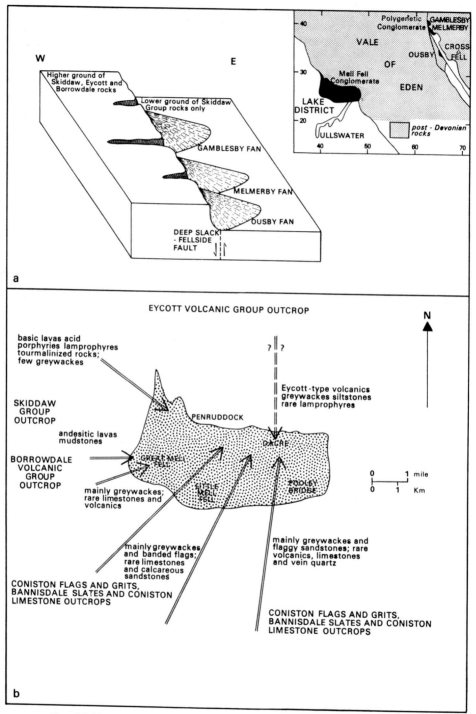

FIG. 57 (a) The mode of accumulation of the Polygenetic Conglomerate.
         (b) The Mell Fell Conglomerate outcrop showing the clasts
               derived from nearby outcrops in Devonian times.

fans derived from higher ground close to the south, west, north-west and possibly the north. The fan-conglomerates probably pass laterally eastwards into the well-bedded sandstones around Dacre, which are fluvial in character with upward-fining sedimentary cycles.

The nature of the rocks forming the surrounding mountains has been deduced by detailed analysis of the clasts by Green (1918) and Capewell (1955), summarized in figure 57b. In the west, the Skiddaw and Borrowdale Volcanic Groups cropped out as at present but yielded only minor amounts of detritus. Farther north, the Eycott Group was probably exposed not only in the Caldbeck Fells as at present but also eastwards towards Penrith, where it is now covered by younger rocks. The nature of these source areas in Devonian times might reasonably have been anticipated from present outcrops, but those farther south are more surprising. Instead of the Borrowdale Volcanic Group rocks, which now form the mountains in this direction, Silurian lithologies predominate. These rocks are now found only in the southern Lake District, but it is probable that during the Devonian they entirely covered the ground around Ullswater and Haweswater. The predominance of greywacke pebbles, and the record (Capewell 1955, p. 26) of a Lower Ludlovian monograptid, '*Monograptus colonus*', from a greywacke clast on Great Mell Fell, shows that the Coniston Grits formed much of the outcrop, although Bannisdale Slates and Coniston Limestone lithologies are also present in subordinate amounts.

Acknowledgement

This chapter is published by permission of the Director, Institute of Geological Services.

REFERENCES

ARTHURTON, R. S. and WADGE, A. J. *in press*. Geology of the country around Penrith. *Mem. geol. Surv.*

AVELINE, W. T., DAKYNS, J. R. and WARD, J. C. 1881. Horizontal Sections Sheet 119. *Geol. Surv. England and Wales.*

BURGESS, I. C. and WADGE, A. J. 1974. The Geology of the Cross Fell area. H.M.S.O. London.

CAPEWELL, J. G. 1955. The post-Silurian pre-marine Carboniferous sedimentary rocks of the eastern side of the English Lake District. *Q. Jl geol. Soc. Lond.* **111**, 23–46.

DAKYNS, J. R., TIDDEMAN, R. H. and GOODCHILD, J. G. 1897. The Geology of the country between Appleby, Ullwsater and Haweswater. *Mem. Geol. Surv.*

GREEN, J. F. N. 1918. The Mell Fell Conglomerate. *Proc. Geol. Ass.* **29**, 117–25.

MARR, J. 1899. Note on a Conglomerate at Melmerby. *Q. Jl geol. Soc. Lond.* **55**, 11–15.

OLDHAM, R. D. 1900. The basal Conglomerate of Ullswater. *Geol. Mag.* **7**, 564.

SHOTTON, F. W. 1935. The stratigraphy and tectonics of the Cross Fell Inlier. *Q. Jl geol. Soc. Lond.* **91**, 639–701.

WARD, J. C. 1876. The geology of the northern part of the English Lake District. *Mem. geol. Surv.*

A. J. WADGE, M.A.
Institute of Geological Sciences, Ring Road, Halton, Leeds LS15 8TQ.

12

Carboniferous

M. MITCHELL, B. J. TAYLOR and W. H. C. RAMSBOTTOM

By late Devonian — early Carboniferous times the Caledonian mountains had been reduced to a gently-domed land surface which was remarkably even over much of the northern Lake District, but was of moderate relief in southern and eastern areas. The early Carboniferous deposits preserve a record of the encroachment of the sea onto this landmass, and before the close of the Lower Carboniferous (Dinantian) the Lake District was covered. Upper Carboniferous rocks too, formerly extended over the whole of the area. Because of Tertiary uplift Carboniferous rocks now crop out only in an incomplete collar around the Lake District (Plate 13) and in the central area they have been removed by erosion.

The Lake District formed part of a stable area covering much of northern England between the Northumberland Trough and the Craven Basin. The Stainmore Trough separated the Alston Block from the Askrigg Block and spread westwards to the flanks of the Lakes. The northern margin of this trough was an active fault line but the poorly defined southern margin was formed by the gradual northward tilting of the Askrigg Block and the southern part of the Lake District Block (Johnson 1967, p. 184; Burgess and Mitchell 1976, p. 620). Carboniferous sedimentation was closely controlled by these structural units of the pre-Carboniferous basement.

The Carboniferous rocks are discussed here under their major time divisions; Dinantian (broadly the Carboniferous Limestone); Namurian (the Millstone Grit generally, but here the lower boundary is taken at the base of the First Limestone), and the Westphalian (Coal Measures).

DINANTIAN

M. MITCHELL

The Carboniferous seas gradually advanced across the uneven ancient platform of Lower Palaeozoics, and the Dinantian rocks rest with striking onlapping unconformity on these older beds. There are few surface exposures of the contact in the Lake District, but the unconformity has been seen during mining operations in West Cumbria and Furness, and in trenches at Holker (Rose and Dunham 1978, p. 52). The top of the pre-Carboniferous surface is generally red-stained suggesting that it was exposed to arid or semi-arid conditions before being submerged. The engulfing of the Lake District massif occurred intermittently, rather than as a continuous process (Fig. 58).

The works of Garwood (1907, 1913, 1916) were of major importance to the study of the Dinantian of the Lake District and his zones have been applied to all the areas fringing the District. A summary of the Dinantian classification and of the correlations in the area are given in figures 59 and 60. The Dinantian crops out in three main areas. In the south, between Millom and Kendal, the limestones form prominent fault blocks and this area includes the extensively mined hematite deposits of Furness. In the east, the Stainmore Trough outcrops between Ravenstonedale and Shap were taken by Garwood (1913, p. 484) as the type district for his North-West England study. In the north, the outcrops extend from north-west of Penrith to Egremont, and, south of Cockermouth, include the West Cumbria hematite area. South of Egremont, Dinantian is present beneath the New Red Sandstone, but no Carboniferous is known between Calder Bridge and Silecroft (Smith 1924, p. 20) but it may be present to the west.

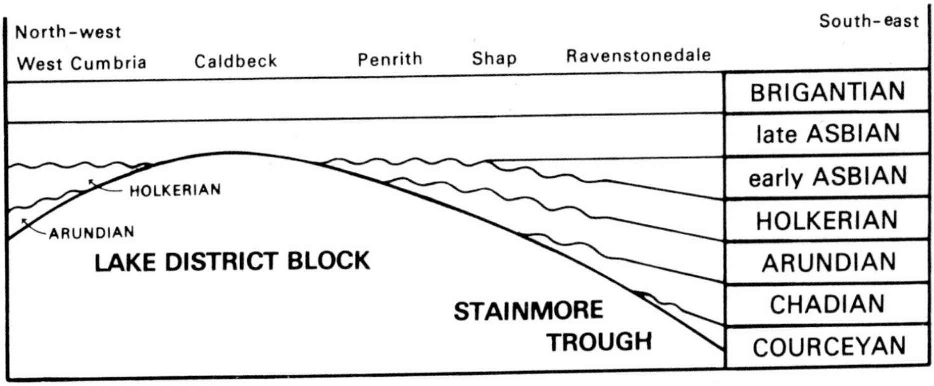

FIG. 58. Diagrammatic section across the Lake District from north-west to south-east showing the progressive onlap of the Dinantian stages onto the pre-Carboniferous land surface. Vertical scale is exaggerated.

Cyclic deposits

Above the Basement Beds the Dinantian is predominantly of limestone facies. The main limestone masses were deposited in relatively shallow clear-water seas but the earliest beds often show evidence of restricted environments or near shore deposition. The limestones contain abundant foraminifera and rich coral and brachiopod faunas which provide the basis of the present correlations.

Cycles of deposition can be detected through much of the limestone sequence. Ramsbottom (1973) has described this cyclicity as evidence for eustatic changes of sea level with transgressions of the sea marked by marine bioclastic limestones and regressions by algal, dolomitic and fine-grained limestones, shales and sandstones and with emergence frequently indicated in shelf areas by breccia beds, unconformities, non-sequences and karstic horizons. Ramsbottom grouped these cycles (or cyclothems) into a series of major cycles of deposition or mesothems (Ramsbottom 1977, p. 282) each recognized by a distinctive fossil assemblage, and each mesothem is bounded above and below by unconformities in shelf areas. The cyclic nature of the sequence becomes pronounced in the Asbian which consists of distal cycles composed of limestone with thin clastics, and in the Brigantian typical Yoredale facies cyclothems are developed (Dunham 1950, Johnson 1960, Ramsbottom 1973, p. 591, fig. 7), with an upward sequence from limestone through shale and sandstone to seat-earth and coal which are not always present.

Basement Beds

The Basement Beds are the local accumulations of debris eroded from the Lower Palaeozoic floor and swept up to fill the irregularities in this ancient platform. The material is terriginous and often coarsely conglomeratic especially where the source was from an area of igneous rocks. Sandstones, mudstones and occasional thin limestones are also present, and away from outcrop, boreholes in Furness have proved thin beds of gypsum. The lithologies grade into one another and the predominant colour is reddish-brown.

The thickest development is at Millom in the Duddon Valley where Rose and Dunham (1978, p. 27) recorded up to 240 m and this area must have been a considerable pre-Carboniferous hollow or major river valley. Few macrofossils have been recorded from the Basement Beds and much work remains to be done in attempting

STAGES George et al. 1976	MAJOR CYCLES Ramsbottom 1973	RAVENSTONE-DALE STRATIGRAPHY	GARWOOD'S 1913 ZONES AND SUBZONES	
BRIGANTIAN	Sixth Group of Minor Cyles	Upper Alston Group Peghorn Lst	*Dibunophyllum* Zone	*Dibunophyllum muirheadi* Subzone
				Lonsdaleia floriformis Subzone
ASBIAN late / early	Fifth Group of Minor Cycles	Robinson Lst Knipe Scar Lst Potts Beck Lst		*Cyathophyllum murchisoni* Subzone
HOLKERIAN	④	Ashfell Lst	*Productus corrugato-hemisphericus* Zone	*Nematophyllum minus* Subzone
				Cyrtina carbonaria Subzone
ARUNDIAN	③	Ashfell Sst Michelinia grandis Beds Brownber Pebble Beds	*Michelinia grandis* Zone	Gastropod Beds
				Chonetes carinata Subzone
				Camarophoria isorhyncha Sz
CHADIAN late / early	②	Scandal Beck Lst	*Athyris glabristria* Zone	*Seminula gregaria* Subzone
		Coldbeck Lst Stone Gill Lst		*Solenopora* Subzone / Algal Band
COURCEYAN	①	unexposed Shap Conglomerate Pinskey Gill Beds non - sequence	no zones	

Fig. 59. Stratigraphy and classification of the Dinantian of the Lake District with the Ravenstonedale succession (after Taylor et al. 1971, pl. 5E, Ramsbottom 1974, fig. 16 and George et al. 1976, fig. 11D) given as the most complete sequence. Garwood's 1913 zonal scheme is included only to facilitate reference to previous literature. Ramsbottom (1973, fig. 2) placed the base of Major Cycle 3 at the base of the M. grandis Beds. Identification of Archaediscids typical of the Arundian in the top of the Scandal Beck Limestone (George et al. 1976) shows this to have been in error.

to date these rocks especially the thin limestone bands. However the spore assemblages recovered from mudstones near Dalton (Owens, in Rose and Dunham 1978, p. 23) are of late Courceyan age. The Basement Beds thin rapidly northwards from Millom with about 20 m at Kirksanton and eastwards, with 50–100 m near Dalton, but they are absent in Cartmel.

At Shap the Basement Beds (about 10 m) consist of irregularly bedded shaly sandstones and conglomerates containing Shap Granite debris (the Shap Conglomerate of Garwood 1913). These beds thicken (up to 36 m) southwards (Butterfield 1921, p. 5). In Ravenstonedale the Basement Beds consist of the Feldspathic Conglomerate and the underlying Pinskey Gill Beds; calcareous mudstones with thin sandstones and dolomitic limestones probably deposited in hollows in the basement floor. A few undiagnostic marine fossils have been found in the Pinskey Gill Beds (Garwood 1913, p. 496) but they have yielded spores of Courceyan age (Johnson and Marshall 1971). In the Kendal area basal conglomerates are thin, but they are thickly developed, filling another river valley, near Sedburgh (Butterfield 1920, p. 249). In the north the Basement Beds are thickest (about 30 m) where they rest on volcanic ground, but on the softer Skiddaw Slates they are thin or absent. In West Cumbria they rarely exceed 3–4 m.

At Cockermouth, the Cockermouth Lavas are present within the Basement Beds (Eastwood 1928, Eastwood et al. 1968, p. 152). The lavas are olivine-basalts and occur as 4 or 5 flows, the tops and bottoms of which are vesicular, and reach a maximum thickness of about 90 m. They crop out from the Derwent valley west of Cockermouth to Bothel a distance of about 13 km. The age of the Cockermouth Lavas is indicated by the recovery (C. E. Butcher pers. comm.) of CM Zone spore assemblages of Courceyan age from the Basement Beds below the lavas at Redmain and above the lava at Blindcrake.

By the close of the deposition of the Basement Beds, most of the irregularities of the Lower Palaeozoic floor had been filled in and the resultant even surface was gradually flooded by the advancing Carboniferous seas.

Chadian

The Algal Band of Garwood (1913, p. 462) at the top of the Coldbeck Beds of Ravenstonedale is now thought (Ramsbottom 1977, p. 283) to correspond with a mid-Chadian regression rather than with the regression at the close of the Courceyan as indicated by George et al. (1976, p. 38, fig. 11; see also Ramsbottom 1973, p. 574). In Ravenstonedale, the Stone Gill Beds with Chadian foraminifera (Ramsbottom 1977, p. 283) and the Coldbeck Beds are of early Chadian age and most of the Scandal Beck Limestone with *Thysanophyllum pseudovermiculare* is late Chadian. Traced northwards to Shap these beds thin markedly especially in the lower part. In Furness and Grange, the mid-Chadian regression is correlated with the algal beds with a well-developed desiccation surface near the base of the Martin Limestone at Meathop (Mitchell in Rose and Dunham 1978, p. 33), the basal part of the formation being of early Chadian age and the upper part with *T. pseudovermiculare* being late Chadian.

The Chadian contains much dolomitic, shaly and fine grained limestone with algal bands. No rugose corals have been recognized in the early Chadian where the shelly fossils are concentrated in bands crowded with individuals of relatively few species. These characters suggest deposition in a restricted near-shore environment.

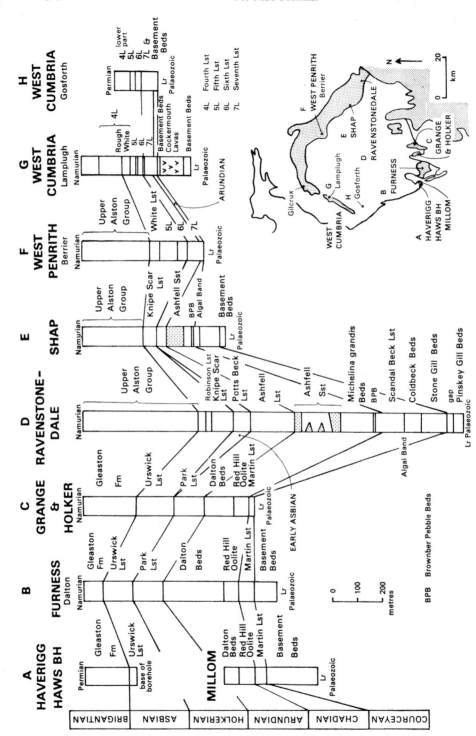

FIG. 60. Correlation of the Dinantian of the Lake District. Column A, after Rose and Dunham 1978, pp. 40, 125; B, after Rose and Dunham 1978, p. 45; C, after Rose and Dunham 1977, p. 52; D, after Taylor et al. 1971, pl. 5:E, Ramsbottom 1974, fig. 16 and George et al. 1976, fig. 11:D; E, after Garwood 1913, p. 486 and Taylor et al. 1971, pl. 5:F; F, after Edmonds 1922, p. 77, Eastwood et al. 1931, p. 62, and H, after Trotter et al. 1937, p. 64.

The presence of rugose corals (including *Carcinophyllum simplex, Koninckophyllum sp.* and *T. pseudovermiculare*) in the late Chadian suggests some connection with open-sea conditions.

The Chadian is not recorded from northern areas, and in the south and east the shore line was not far to the Lake District side of the present outcrops.

Arundian

The stage is well developed in the Furness–Kendal area and in the Stainmore Trough. As in the Chadian, the faunal and lithological similarities between the rocks in these two areas are again so striking that it seems certain that the beds were formed in a once laterally continuous shelf environment. The transgressive part of the stage is represented by the upper part of Scandal Beck Limestone with Arundian foraminifera (George et al. 1976, p. 39, fig. 11:4), the Brownber Pebble Beds (sandstones, oolites and layers of quartz pebbles) and the Michelinia grandis Beds in the Stainmore Trough, and by the Red Hill Oolite and lower part of the Dalton Beds in Furness. The brecciated limestones at the base of the Red Hill Oolite of Holker and of Meathop indicate a non-sequence before the major transgression. This break, with the lower beds of the stage missing, is confirmed by the forminiferal sequence. The Red Hill Oolite is a pale grey massive, pelletly limestone and the Dalton Beds are dark grey, well-bedded, crinoidal limestones with calcareous shale partings. The latter formation thickens rapidly from 110–120 m east of Ulverston to a maximum of 255 m near Dalton, indicating an area of greater subsidence, possibly the sagging margins of the broad shelf area on the southern borders of the Lake District.

In Ravenstonedale, the M. grandis Beds thin northwards to Shap where they are poorly developed, resting on pale grey massive pelletly limestone comparable with the Red Hill Oolite.

The close of the Arundian is marked by a major development of regressive phase sandy beds (the Ashfell Sandstone) in the Stainmore Trough. These beds thin northwards as they are traced onto the stable Lake District Block and the sandstones enter at earlier horizons until at Shap much of the stage is composed of barren sandstones and sandy limestones. In the Kendal–Furness area the upper part of the Dalton Beds is sandy, partly dolomitic, massively bedded limestone (the Gastropod Beds of Garwood 1913, p. 520). The stage is missing from much of the northern Lakes but thin Arundian is present in the lowest beds of the Seventh Limestone in at least two small areas. Trotter et al (1937, pp. 73, 75) recorded *Lonsdaleia spp.* from two boreholes at Beckermet and Eastwood et al. (1968, p. 247, locs. 17 and 19) noted *Lonsdaleia duplicata* from two exposures near Berrier north-west of Penrith. The specimens have been reidentified as *Thysanophyllum pseudovermiculare* which indicates late Chadian or early Arundian, and Ramsbottom has identified archaediscids characteristic of the Arundian among the associated foraminifera. The basal Seventh Limestone is therefore in places older than the S_2 Zone (= Holkerian) age previously assigned to it. The cross-bedded sandy limestones in the lower part of the Seventh Limestone indicate shallow water conditions and may represent the regressive phase of the Arundian.

The Arundian shoreline was probably close to the present outcrop in northern areas but in the south and east was some way towards the central Lake District. The faunas of the Red Hill Oolite contain *Michelinia megastoma, Palaeosmilia murchisoni* and *Stenoscisma isorhyncha*. The Dalton Beds contain the Arnside Fauna of

Garwood (1913, pp. 467, 508), one of the most distinctive and widespread faunas of the Western European Dinantian. *Delepinea carinata* is the diagnostic brachiopod, with the rich coral fauna including *Caninia subibicina*, *Caninia cylindrica* (Arnside form: Garwood 1916, pl. 14, fig. 5), *Lithostrotion martini*, *M. megastoma* and *P. murchisoni*.

Holkerian

The Holkerian transgression is the most striking Carboniferous transgression in northern England and much of the remaining pre-Carboniferous land surface of the Lake District was flooded by it. In the southern and eastern areas the stage is represented by the Park and Ashfell limestones which are poorly-bedded, pale grey limestones, with dark crinoidal fragments, and they often weather into thin platy layers. The Ashfell Limestone on Ashfell Edge, Ravenstonedale, shows evidence of cyclicity with sandy shales interbedded with limestone (Ramsbottom 1974, p. 58). The Holkerian stratotype (George et al. 1976, p. 9, fig. 2) is at the base of the Park Limestone at Barker Scar, Holker. A thin sandy, red-stained Park Limestone resting on Basement Beds is present near Kirksanton (Dunham and Rose 1941, p. 5).

In northern areas, the Seventh Limestone except for the few records of the basal parts being older (see above) is of Holkerian age. It shows shallow water near shore characters with calcite mudstones and calcareous sandstones. Llewellyn et al. (1968) recorded nodular anhydrite from a borehole 2.2 km west of Egremont in the upper part of the Seventh Limestone, and they concluded that the anhydrite and associated dolomites were of supratidal sabkha origin. The small outlier of sandy Seventh Limestone with *Lithostrotion minus* at Wilton 4 km east of Egremont (Eastwood et al. 1931, pp. 65, 92) shows that the Holkerian extended some distance to the east of the present outcrop. In southern and eastern areas the clear water limestones indicate a shoreline towards the centre of the Lake District.

The faunas, with *Lithostrotion minus* and *Davidsonina carbonaria*, are characteristic of the Holkerian but fossils are not common except in bands crowded with individuals of relatively few species, a distribution suggesting a slightly restricted environment.

Asbian

The Asbian stratotype is at Little Asby Scar in Ravenstonedale at the base of the Potts Beck Limestone (George et al. 1976, p. 40, fig. 11:D) and the stage comprises two major cycles or mesothems (Ramsbottom 1977, p. 283) here referred to as early and late Asbian. The early Asbian forms a distinct and separate major cycle or mesothem (D5a of Ramsbottom 1977, fig. 10). In this area it is restricted to the Stainmore Trough (Ramsbottom 1974, p. 59, fig. 20) where it forms the Potts Beck Limestone. These beds comprise several cycles of dark grey limestone becoming paler upwards, with *Daviesiella llangollensis* and *Dibunophyllum bourtonense*, and are thickest in the centre of the trough, thinning rapidly towards the surrounding block areas where they are not present. The regressive phase is marked by a shale-ferruginous sandstone horizon in the area of Little Asby Scar. The late Asbian rocks form the most spectacular limestone scenery of the Lake District with widespread karst topography of clints and grikes (Waltham 1974).

The Urswick Limestone is not preserved at Millom but the top of the formation was proved in the Haverigg Hawes Borehole [1473 7853]. Northwards towards Kirksanton thin (26 m) Asbian limestones, containing marly-layers, rest on Basement Beds and represent the feather edge of the Dinantian limestones (Dunham and Rose 1941, p. 6). The Urswick Limestone in Furness and Grange appears to rest conformably on the Holkerian Park Limestone, but the early Asbian has not been proved and the base of the Urswick Limestone which is a gently undulating surface on a thin shale parting, marks a non-sequence.

In Ravenstonedale the latest Asbian, the Robinson Limestone, is separated from the main scar forming Knipe Scar Limestone by a shale band and followed by a sandstone. These clastic intercalations are the forerunners of the Yoredale facies typical of the Brigantian rocks.

Between Penrith and Egremont the Sixth, Fifth and White limestones are of late Asbian age. The same formations are proved in the concealed area south of Egremont where they are the highest limestones below the New Red Sandstone unconformity.

The late Asbian rocks consist of a series of rhythmic thick-bedded, pale grey limestones with thin shale bands usually developed above potholed or palaeokarst surfaces (master bedding planes) indicating emergent conditions (Walkden 1974). The rhythms are typical of distal limestone cyclothems (Ramsbottom 1973, fig. 7) and give rise to typical terraced topography. The rhythmic character of the Fifth and White limestones and to a lesser extent the Sixth Limestone in West Cumbria was noted by Eastwood et al. (1968, p. 156). The more shaly, sandy nature and darker colour of the lower part indicates that these beds may have been laid down either nearer to a shoreline or in an area more affected by the entry of clastics (Eastwood et al. 1931, p. 66). The shale bands are rarely seen at surface (cf. Waltham 1971) but G. M. Walkden (pers. comm.) has established bentonitic characteristics in some of them and regards these as representing non-local volcanic-rich atmospheric dusts, accumulated and weathered to a soil during emergent phases (cf. Walkden 1972, 1974).

The basal and upper parts of the late Asbian are notable for pseudobrecciated limestones (the 'spotted beds' and 'stick beds' of Garwood 1913, p. 475). Pseudobreccias are very conspicuous with darker grey irregularly outlined sphaerical patches set in a matrix of pale grey limestone, often weathering to give a rubbly or brecciated texture. This lithology was possibly caused by bioturbation, the darker areas being animal burrows, and is present in the Asbian over a large area of northern England which must have been a thriving habitat for a large population of burrowing animals. This distinctive lithology is almost confined to the late Asbian in southern and eastern areas, but in the Haverigg Hawes Borehole and in West Cumbria it is also present in the Brigantian.

Much of the Asbian limestone is apparently barren of macrofossils but in the Settle area, Schwarzacher (1958, p. 139) noted concentrations in bands just below or just above the master bedding planes. The bands are rich in fossils with *Carcinophyllum vaughani*, *Dibunophyllum bourtonense*, *Lithostrotion junceum*, *L. martini*, *L. pauciradiale* and *Palaeosmilia murchisoni* forming the typical Asbian coral fauna. With the exception of the addition of *Davidsonina septosa* in the upper part of the late Asbian, the fauna shows little change through the stage, although *Linoprotonia hemisphaerica* occurs abundantly in the Robinson Limestone but is not confined to it.

No Asbian reef limestones are exposed in the Lake District but the records of reefs in the Carnforth area indicate that the Settle–Cracoe reef belt (Ramsbottom 1974, fig. 21) between the Askrigg Block and the Craven Basin, extends westwards towards the southern fringe of the Lakes (Ramsbottom 1969). Brigantian lithologies suggests that this block/basin margin crosses the southern part of the Furness peninsula to the south of the Carboniferous outcrops.

With the exception that small islands, not submerged until the Brigantian, may have remained in the north, the Lake District was probably flooded by the close of the late Asbian.

Brigantian

The close of the Asbian was marked by a major change in deposition and fauna with massif limestones giving way upwards to the onset of rocks of Yoredale facies. The Brigantian stratatype is at the base of the Peghorn Limestone at Janny Wood, and this boundary can be traced over a wide area of northern England (Burgess and Mitchell 1976). Most of the Lake District Brigantian is of Yoredale facies deposited in a broad shallow sea with a dominant source of clastics from the north-east. The persistent and widespread limestones are separated by variable thicknesses of clastics, thinly developed in the west but thickening in a north-easterly direction, and the distribution and proportions of limestones and clastics has been summarized by Rowley (1969, fig. 3).

In the north-east and east between Caldbeck and Ravenstonedale clastics are strongly developed and typical Yoredale facies cyclothems are present. Yoredale facies extended to the Holker area west of Grange but has not been preserved in the intervening ground. Traced westwards from Caldbeck, the amount of clastics decreases until in West Cumbria (Eastwood et al. 1931; 1968) from Cockermouth to Egremont the sequence is essentially one of limestone cycles with little clastic material. The cycles in this area are similar to those present in the Asbian, much of the limestone being pale grey but with increased evidence of shallow water or emergence conditions. Karstic surfaces are developed, and thin coals, calcite mudstones, dolomite horizons, rubbly limestones and non-sequences are present. The Brigantian has been removed by the New Red Sandstone overlap in the Beckermet–Calder Bridge area, but is again present to the south-west of Millom, and the Haverigg Hawes Borehole (Rose and Dunham 1978, p. 125) proved a thick sequence of limestones with subordinate clastics. This limestone development probably extends into the faulted and subdrift outcrops west of Dalton (Rose and Dunham 1978, p. 31).

In Furness the Brigantian is represented by the poorly exposed Gleaston Formation and details are from two recent boreholes (Rose and Dunham 1978, pp. 30, 134, 151). At Gleaston the sequence consists of thin mudstones and limestones with impersistent sandstones. At Roosecote to the south of the exposed Carboniferous, a thick sequence of dark cherty limestones with thin mudstones (the Roosecote Limestones) with poor faunas of zaphrentoid corals was proved. These lithologies indicate the approach to the open sea with sediments derived possibly from a southerly source, the Gleaston Formation of Furness being a marginal facies between the Lake District shelf sea and the Craven Basin.

Unlike the minor cycles in the Asbian which have not yet been correlated from one area to another, the faunas and distinctive lithologies of the Brigantian cyclothems enable them to be correlated over much of the area (Fig. 61). The faunas are rich and varied including clisiophylloids, *Lithostrotion spp.*, *Lonsdaleia spp.*, *Orionastraea spp.*, *Palaeosmilia regia* and *Gigantoproductus spp.* Several faunal horizons

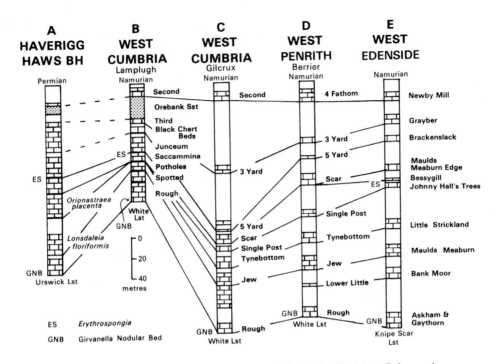

FIG. 61. Correlation of Brigantian rocks of the Lake District. Column A, after Rose and Dunham 1977, p. 125; B, after Edmonds 1922, pp. 77, 81; C, after Eastwood et al. 1968, fig. 14; D, after Eastwood et al. 1968, fig. 14 and Arthurton and Wadge (in preparation); E, after Rowley 1969. The West Edenside limestone names are included but correlation with the Alston Block limestones is sufficiently precise that use of this separate nomenclature should be discontinued.

are especially valuable in detailed correlation. A distinctive algal horizon, the Girvanella Nodular Bed (Garwood 1913, p. 482), is present just above the base of the stage in dark grey limestones with Brigantian faunas, and is used to fix the base of the stage with precision. Hudson (1929) recognized a thin lagoonal deposit with the sponge *Erythrospongia lithodes* at the top of the Single Post Limestone of the Askrigg Block. He traced this horizon through the Pennine area, to the dolomitic top of the Johnny Hall's Trees Limestone of Shap and found it again in West Cumbria between the Saccammina and Junceum beds. This confirmed the correlations of Edmonds (1922), and that the Cockleshell Limestone of the Alston Block had been cut out by unconformity in parts of West Cumbria (Eastwood et al. 1931, p. 77).

Most of the Yoredale limestones are darkish grey in colour but in West Cumbria and in the Haverigg Hawes Borehole, pseudobreccias which are restricted to the Asbian in southern and eastern areas persist into the Brigantian. This suggests that some of the limestones in West Cumbria were deposited in clearer, more open seas with less detritus being washed in.

NAMURIAN

W. H. C. RAMSBOTTOM

It has recently been shown that the Namurian, when complete, comprises the rocks of eleven major transgressions and regressions (mesothems N 1 to N 11, see Ramsbottom 1977). Of these only the earliest (N 1) and the latest (N 11) reached the Lake District area itself though in the Vale of Eden to the north-east some of the beds between are present (Fig. 62). This is taken to indicate the continued positive nature of the Lake District massif in the Namurian with only intermittent subsidence.

Namurian rocks crop out both to the north of the Lake District (Fig. 63) and to the south in the Furness area, but in the intervening ground, around Gosforth, the Series is probably absent. The rocks in the northern area are of shelf facies with a

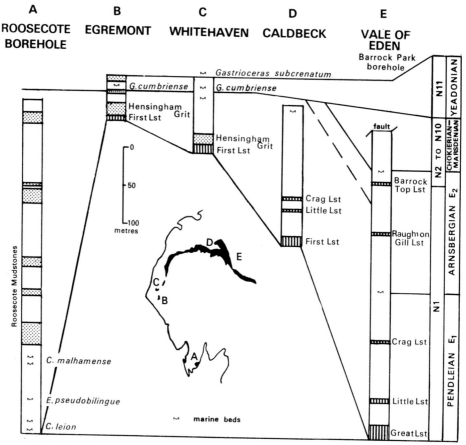

FIG. 62. Namurian sections around the Lake District. A, after Rose and Dunham (1978); B and C, after Eastwood et al. (1931); D, after Eastwood et al. (1968); E, after Arthurton and Wadge (in preparation). The Namurian stages and the mesothem classification of Ramsbottom (1977) are shown on the right.

varied fauna of brachiopods and mollusca comparable with similar beds of modified Yoredale facies on the Alston and Askrigg Blocks. In the south the facies is different and consists of shales and mudstones with goniatite faunas at some levels, and this type of deposition is comparable with that in the Craven Basin to the south of the Askrigg Block. It thus seems likely that the southern boundary of the Lake District massif which passes through Furness is a continuation of the southern boundary of the Askrigg Block and marks in a similar way the passage from shelf to basinal deposition.

Furness District

Little was known about the Namurian in this area until two recent I.G.S. boreholes at Roosecote (2304 6866) and at Gleaston Castle [2549 7185] allowed a reinterpretation of the succession (Rose and Dunham 1978, p. 32). At Roosecote over 400 m of mainly mudstone rocks (Roosecote Mudstones of Rose and Dunham) with subordinate sandstones and a few conglomeratic and slumped beds in the upper part are known. Near the base the mudstones contain the *Cravenoceras leion, Eumorphoceras pseudobilingue* and *Cravenoceras malhamense* marine bands of E_1, Pendleian, age. It is probable, indeed, that all these beds are of E_1 age, for higher faunas have not been found, and that the higher, more sandy beds are the lateral equivalents of the Pendle Grit of the Craven Basin. Two old boreholes in this area contain records of 'greenstone' taken to refer to beds of dolerite which are probably sills.

Northern area

The succession begins here with the First Limestone, correlated with the Great Limestone of the Alston Block and of similar character, being a bioclastic limestone with an abundant fauna of corals and brachiopods. It rests conformably on the underlying Dinantian.

All the Namurian beds above the First Limestone are referred to the Hensingham Group. Westwards from around Cockermouth the basal beds form a coarse sandstone, the Hensingham Grit, which is probably a near-coastal sandstone for it is thickest where the Namurian is thinnest. The overlying beds consist mainly of mudstones with a few thin beds of sandstone, some discontinuous limestone bands and a few thin coals. All but the top few metres of these beds are of E_1 and E_{2a} age but the uppermost beds contain the goniatite *Gastrioceras cumbriense* (its type locality is at Bigrigg, near Hensingham) and are of G_{1b} age. This substantial non-sequence cannot be detected except by means of the fossils. In the eastern area, near Caldbeck the G_1 beds are absent and there the Westphalian rests directly on red-stained Hensingham Group mudstones of E_{2a} age. This zonal position is suggested in particular by the brachiopods, some of which can only be matched in the Corbridge Limestone of the Northumberland Trough and its correlative the Orchard Limestone of the Midland Valley of Scotland. The position of these limestones is now known to be E_{2a} (Ramsbottom 1978). Records in the Hensingham Group of the nautiloid *Tylonautilus nodiferus* and of the goniatite *Anthracoceras*, both formerly believed to indicate the E_{2b} Zone, are now known to occur in E_{2a} as well. Moreover, spores studied by Dr B. Owens from these beds also support the suggestion that no beds higher than E_{2a} (i.e. the N 1 mesothem, not N 2 as stated in Ramsbottom 1977, p. 281) are present. Since the limestones of the Hensingham Group appear to be laterally impersistent, correlation of the individual beds is difficult, though east of Cockermouth the

Little and Crag Limestones of the Alston Block succession have been identified (Eastward et al. 1968).

Further east the Barrock Park Borehole in the Vale of Eden (Taylor et al. 1971, pl. vii) has shown that higher beds come in between the E_{2a} and the G_1 beds, but a detailed correlation has not yet been established, mainly because the higher beds of the Namurian are red-stained. It seems likely that the succession there is similar to that on the Alston Block, and that several major non-sequences are present (Ramsbottom 1977).

The Hensingham Group thickens markedly when traced northwards towards the basinal area of the Solway, and over 500 m are present in boreholes north of Maryport. It is not known whether this indicates that individual beds have thickened, or that strata cut out by the major non-sequence at outcrop are here present, and possibly both these things are happening.

WESTPHALIAN

B. J. TAYLOR

Outcrops of the Westphalian rocks are confined to the north-west, north and north-east flanks of the Lake District, where they form the outermost layer of Carboniferous sediments. Their absence from the southern sectors is due to erosion in Hercynian times rather than non-deposition, for such outliers as the Ingleton, Stainmore and Midgeholme coalfields bear witness to the former continuity of the Coal Measures.

The measures are best known in the West Cumbrian Coalfield between Whitehaven and Maryport, and are thickest around Workington and Maryport and in the offshore workings of Risehow, St Helens and Solway collieries. Here the interval from the *Gastrioceras subcrenatum* Marine Band at the base of the Westphalian to the St Helens (Top) Marine Band (i.e. the Lower and Middle Coal Measures) is some 450 m; the Upper Coal Measures are of unknown thickness. The sequence is highly condensed but all the Westphalian bivalve zones are present.

A gentle dip lowers the beds seawards beneath the offshore outcrop of the Permian. Coal workings extend to more than 5 km beyond the coast in Haig Colliery off Whitehaven, and to an average of 2 km in the Harrington, Solway, St Helens and Risehow collieries (Fig. 63). The coalfield is intensely faulted and working the seams, particularly offshore, has been difficult. Only the main faults are shown in figure 63. These have a north-west trend and some have throws greater than 300 m. The highest measures can be expected on the seaward side of the outcrop, but in places large faults have caused high measures to be preserved well inland.

In the northern sector, from Dearham eastwards through Oughterside and Aspatria to Mealsgate, over 300 m of coal-bearing Lower and Middle Coal Measures are present. Sandstones increase eastwards at the expense of marine and other shales in the lower part of the sequence. The higher measures are comparable with those in the coastal belt. Bores near Aspatria from which *Anthraconauta tenuis* Zone fossils have been identified suggest than an extensive area of Upper Coal Measures may lie concealed beneath the Permo-Triassic rocks of the Solway Plain, down-dip from the northern outcrop, although reliable estimates of thickness are impossible.

East of Caldbeck secondary reddening of strata descends abruptly down through the Coal Measures and into the underlying Namurian rocks. Here the *A. lenisulcata*

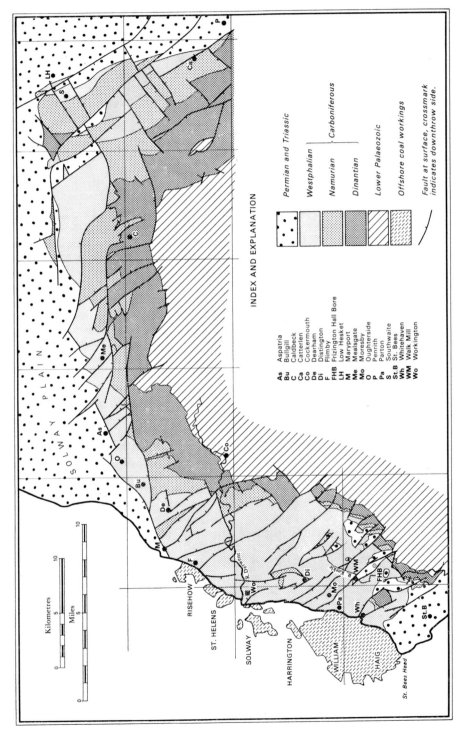

Fig. 63. Sketch map of Carboniferous outcrops around the northern fringe of the Lake District.

and *C. communis* zones are missing, and reddened rocks with *A. modiolaris* Zone fossils occur only some 3 m above beds with a Hensingham Group fauna. The period of non-deposition which elsewhere in Cumbria caused the absence of much of the Namurian here lasted through *communis* and *lenisulcata* times also.

In the north-eastern sector of the outcrop, north of Penrith, the reddening of strata persists; and so also does the non-sequence, for Lower *similis-pulchra* Zone bivalves are found close above Namurian strata. Westphalian rocks, possibly coal-bearing, are thought to extend widely beneath the Permian in the Vale of Eden east and north of Penrith.

History of Research and Classification

The original one-inch to one mile Geological Survey maps of the coalfield by Russell (1892) showed the succession as 'Coal Measures' with an upper division of red and purple 'Whitehaven Sandstone'. Much earlier, Sedgwick (1832) had considered the 'Whitehaven Sandstone' to be Permian and this view was not countered until the discovery by Brockbank (1891) of *Spirorbis* limestones in the group.

Kendall (1896) believed that his 'Whitehaven Sandstone Series' was a distinct group of mainly arenaceous red and purple Carboniferous rocks resting unconformably on the Coal Measures, though Russell, according to the 1895 edition of the map, regarded them as reddened Coal Measures. The resurvey of 1921–32 did not resolve the question, and Eastwood et al. (1931) and Eastwood (1930) leaned towards Kendall's interpretation. Trotter (1953) put forward criteria for distinguishing primary red beds in northern England from grey beds which had been reddened by oxidation in situ after deposition. Taylor (1961) applied these criteria to the Cumberland Coalfield and showed that much of the 'Whitehaven Sandstone Series', was Middle Coal Measures reddened in late Carboniferous or early Permian times, and the term was therefore abandoned.

Eastwood et al. (1931) and Eastwood (1930) made only limited use of fossils mainly because of a lack of material. Exploration by the National Coal Board, begun about 1947, provided the bulk of the fossil collections on which classification is now based. Taylor (1961) summarized the stratigraphical results of this exploration and Calver in the same work divided the measures into bivalve zones, and provided inter-coalfield correlations on the basis of the marine bands. Calver's (1968) study of the Westphalian marine faunas of northern England located the Westphalian of Cumbria relative to the margins and depositional centre of the sedimentary basin.

Study of Westphalian miospores was begun in Cumbria by Tomlinson (1940). Spores occur in both coals and fine grained sediments but most of the work in Cumbria has been on spores from coal seams (Smith and Butterworth 1967, pp. 28–30).

Sedimentary environment

The Cumberland Coalfield forms part of the Pennine Province of Coal Measures deposition lying between the Southern Uplands of Scotland and the Wales-Brabant Island. The granite-cored block of the Lake District was probably submerged throughout the Westphalian. Despite non-sequences caused by pauses in subsidence, or even emergence, there are no signs of the type of marginal detritus that might be expected from an exposed Lake District core. Evidence from the marine bands indicates that the periodic marine incursions came from an open sea area to the west (Calver 1969, p. 253).

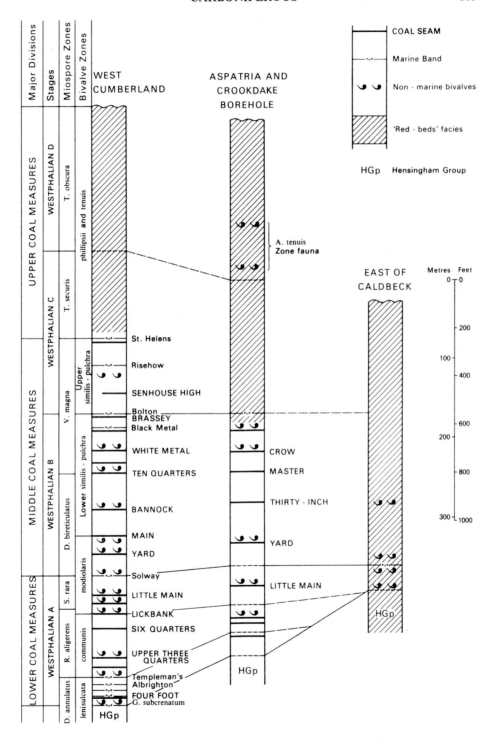

FIG. 64. Comparative vertical sections in the Westphalian.

A remarkable uniformity and continuity of sedimentation obtained throughout the depositional basin during the main coal-forming part of the Westphalian. Some marine bands, and horizons with characteristic non-marine fossil assemblages, can be correlated throughout the coalfields of this Province. The strata below the Upper Three Quarters seam show exceptional lateral constancy over wide areas in the coastal part of the coalfield, the main variation being found in the sandstones. This uniformity, however, is not maintained in those sectors of the Westphalian outcrop to the north and north-east of the Lake District. Traced round from the coast along the southern side of the Solway Plain the coals successively fail and a major non-sequence develops. In West Cumbria the Coal Measures thin towards the Lake District massif; evidence that this ancient block was still exerting an influence upon sedimentation.

The most marked characteristic of these deposits, like those of the older Carboniferous, is their cyclic nature. Typically, the most complete Coal Measures cyclothem consists of a marine shale, overlain by non-marine shale or mudstone, then sandstone, a rootlet bed or seat-earth and finally a coal seam. Each cyclothem records the building up of the sediments to water level after each negative change in the relative level of the delta top, and the coal represents forest-covered delta-swamps, just above sea level, which lasted for varying periods that are reflected by the seam thicknesses. Most cyclothems are incomplete and lack one or more elements. Individual cyclothems vary in thickness from 1 to 30 m or more, the thinnest commonly being mere alternations of rootlet beds and mudstone. A further variation in the West Cumbrian Coalfield is that sandstones may occupy the whole of a cyclothem, lying close above a coal. In places the coal may be removed by erosion and its place occupied by sandstone. The rhythmic pattern of sedimentation gives an impression of a jerky or pulsating subsidence, but the same result would be equally effectively and more plausibly arrived at by a fairly steady subsidence, accompanied by epeiric rises and falls in sea level.

Lithology

Lithologies range from coarse sandstones to the finest of mudstone. Argillaceous members account for at least two thirds of the strata up to the Bolton Marine Band; above this horizon sandstones in places become dominant. Argillaceous sediments with an appreciable sand content have attracted local names in the coalfield. Blocky varieties have been known as 'blaes' or 'caum' and fissile interlaminations of sandstone and shale as 'faikes'. Shales or mudstones with finely divided carbonaceous material form a continuous suite with carbonaceous black shale at one end and near cannel or 'rattler' at the other. Other lithologies include various types of coal, and the associated seat-earths, fireclays and ganisters.

Very large amounts of iron are present, mainly as the carbonates ankerite and siderite, either disseminated through the rocks or concentrated in nodules and thin bands. Siderite commonly occurs in association with beds of fossil bivalves, and replacing plant roots in seat-earths beneath coals.

Sandstones represent the deposition of coarser sediment from shifting distributaries, usually by building-out of successive foreset beds. In many cases the streams carrying the sediment began by strongly eroding the pre-existing sediments and then the sandstone has a minor unconformity at its base, which may cut down through argillaceous and mixed strata to rest upon a coal. The roof of a coal may be of sandstone over quite wide areas, and the Harrington Four Foot, the Six-Quarters or Lickbank and the Main Band have extensive though not ubiquitous sandstone roofs.

In the main coalfield, sandstones which form a stable roof to worked seams generally take their name from the coal. Thus we have the Four Foot Rock above the Harrington Four Foot seam, though this is known near Harrington to occupy all the interval from between its named seam and the Lower Three Quarters — elsewhere represented by seven distinct cyclothems. Sandstones, being laterally impersistent, are the main cause of local variations in the thickness of cyclothems, since they are subject to much less reduction in volume during lithification and dehydration than the argillaceous members.

Coals

Seams that have been of economic importance are all contained in a group of measures some 300 to 400 m thick, the lowest coal being the Harrington Four Foot and the highest the Senhouse High Band. They are bituminous coals, and have been used for coking, gas-making, steam raising and household purposes. Volatile content on a dry ash-free basis varies from 32 to 39 per cent, and the main variations in quality are due to differing proportions of included adventitious matter such as pyrites, phosphorus and inorganic sedimentary material.

The Harrington Four Foot, thin in the Whitehaven area, attains 1.3 m at Harrington and its greatest thickness, 2 m, in the St Helens area, north of which it decreases. Nowhere of high quality, the seam contains much pyrite. The Lower and Upper Three Quarters are thin around Whitehaven but improve northwards around Flimby, Risehow and Dearham. They were worked as far north-east as Aspatria. One of the Three Quarters seams was worked as the Micklam Fireclay Seam at Harrington Colliery and in the Micklam, Gillhead and Camerton districts in association with the underlying seat-earth which was used for making refractory bricks.

The group of measures from the Six Quarters to the Ten Quarters, inclusive, embraces the most valuable coals of West Cumbria. Both the Six Quarters and the Lickbank are in many places double seams, the two parts being separated by shale. The Little Main, or Two Foot, is the most consistent seam in the coalfield, maintaining a thickness between 0.7 and 1 m of good quality coal, though rather high in phosphorus. It rests upon a fireclay which has occasionally been worked. The Yard band is workable only north of the Derwent, and at Aspatria and Oughterside is the most important seam. The Main Band, the most celebrated Cumbrian coal, is everywhere a composite seam. The three elements are close together in the Whitehaven area and can be worked as one. The seam splits north of the Derwent, the lower section being known as the Cannel Band and the upper as the Metal Band. The top part of the latter is the Crow Coal, and in the Dearham, Bullgill and Oughterside areas this in turn is a workable seam separated from the underlying Metal Band. North-east of here there is a deterioration in all three coals.

The Bannock Band is at its best in the southern part of the main coalfield; northwards it becomes thinner, though the upper leaf, under the name Rattler, was mined in the Flimby–Oughterside area. The Ten Quarters seam is of workable thickness and quality from the Workington area north-eastward. It is the Five Foot of the Cleator Moor area east of Whitehaven. Higher seams of importance, chiefly in the northern part of the western outcrop, are the Upper Yard and White Metal Band, the Black Metal Band, the Brassy, a pyritic seam which was rendered valuable by the fireclay seat-earth on which it rests, and the Senhouse High Band.

Non-marine fossils

Among the non-marine fossils the most obvious and stratigraphically important are the bivalves or 'mussels'. In addition, *Spirorbis*, ostracods and fish remains are almost invariably present in the richer bands. *Estheria* occurs in isolated bands. In Westphalian A and B there is an abundance of 'mussel'-beds (Fig. 64), many of which can be correlated with similar beds in the main Pennine coalfields. In Westphalian C, bivalves are much rarer and the variety of fauna is also reduced. Important changes in the non-marine faunas (and the floras) take place at some of the major marine incursions, showing that these episodes had a profound effect on the total environment.

Of the 'mussel' genera, *Carbonicola* was dominant in Westphalian A, but did not survive the Solway Marine Band incursion after which it was replaced by *Anthracosia*. *Naiadites* was common throughout Westphalian A and B, but became extinct at about the time of the St Helens (Top) Marine Band incursion in Westphalian C, and was replaced by *Anthraconauta* which was important in the later part of Westphalian C and early Westphalian D. The ostracod *Geisina* is abundant in Westphalian A and again in the Upper *similis-pulchra* Zone.

At least two thirds of the coals in Westphalian A and B have 'mussel'-beds above them. Many of the bivalve species have a limited stratigraphical range, and this makes it possible to identify minor subdivisions within the main zones (Calver 1956).

Marine Bands

Marine fossils in the Coal Measures are contained in relatively thin layers of sediment, each usually close above a coal seam, and at the base of a cyclothem. The lithology is dark grey to black mudstone or shale, occasionally with harder 'canky' calcareous or ankeritic layers. These bands are the result of short-lived marine incursions, and some of them can be traced throughout all the British coalfields as well as those in adjacent continental Europe. Of the nineteen known marine bands of the British sequence possibly twelve occur in Cumbria.

Throughout the Lower and Middle Coal Measures, Cumbria lay to the north of the depositional centre in the southern Pennines, at times perhaps close to the 'shoreline' or limit of the incursion, so that the marine bands are thinner and their faunal content less varied. In only two bands are goniatites found; the rest contain a restricted assemblage chiefly of *Lingula*, foraminifera and fish remains.

As in the other British coalfields, marine bands are confined to the Lower and Middle Coal Measures, the last incursion, that of the St Helens (Top) Marine Band, marking the top of the Middle Coal Measures. In the case of the marine bands which lack a diagnostic goniatite, the correlation with marine bands in other coalfields can in most cases be established with reasonable certainty by their positions in the zonal sequences of bivalves or miospores. The faunal facies of the marine bands and their regional correlations were described by Calver (1969, pp. 249–53).

Reddening

Parts of the normally grey Westphalian sequence have been reddened by oxidation which converted ferrous iron compounds to ferric oxide, totally removed carbon including coal, and oxidized the sulphides, mainly pyrite, to sulphates such as gypsum.

The zone of reddening is related to the pre-Permian land surface, but varies considerably in thickness. Deep oxidation occurred when the climate was arid, the water-table lay well below the surface and air occupied the pore-spaces. The reaction, not unlike that which commonly takes place in colliery pit heaps, must have released a considerable amount of heat.

In the north-western sector the base of the reddened zone is between the Ten Quarters and Black Metal seams near the eastern side of the outcrop. Close to the coast the base of the zone is above the St Helens Marine Band. North of the Maryport Fault in Risehow Colliery, the base of reddening is some distance below the St Helens Marine Band. Along the northern outcrop from Dearham eastwards, the base of the zone is maintained fairly consistently a short distance above the Ten Quarters as far as the longitude of Caldbeck. Here it descends sharply along a roughly north-south line, east of which virtually the whole Westphalian sequence is reddened and the coals destroyed. Deep oxidation persists eastwards into the Southwaite and Low Hesket area where the outcrop flanks the Penrith Sandstone, and throughout the strip of Westphalian rocks extending from Braithwaite to Catterlen some 4 km north-west of Penrith. Between Caldbeck and Penrith reddening extends well down into the Dinantian close to the flanks of the Lake District massif.

The simplest explanation of the facts outlined above is that Hercynian earth movements raised the Lake District block, with its Carboniferous cover, so that the water-table fell away and oxidation set in. There was at the same time much erosion, and lower stratigraphical levels were reddened as the land surface was lowered. The deeper penetration on the east side of the Lake District may indicate a westward tilt of the block. The anomalously low position of the reddened zone at Risehow Colliery may be connected with Hercynian folding along the Maryport Fault line.

REFERENCES

ARTHURTON, R. S. and WADGE, A. J. (in preparation). Geology of the country around Penrith. *Mem. geol. Surv. Great Britain.*

BROCKBANK, W. 1891. On the occurrence of the Permian, Spirorbis limestones and Upper Coal Measures at Frizington Hall in the Whitehaven district. *Mem. Proc. Manchr. lit. phil. Soc.* **4**, 418–26.

BURGESS, I. C. and MITCHELL, M. 1976. Viséan lower Yordale limestones on the Alston and Askrigg blocks, and the base of the D_2 Zone in northern England. *Proc. Yorks. geol. Soc.* **40**, 613–30.

BUTTERFIELD, J. A. 1920, 1921. The conglomerates underlying the Carboniferous Limestone in the NW of England. *Naturalist* (for 1920), 249–52, 281–84; (for 1921), 5–8, 169–72, 205–207.

CALVER, M. A. 1956. Die stratigraphische Verbreitung der nichtmarinen Muscheln in den penninischen Kohlenfeldern Englands. *Z. deutsch. geol. Ges.* **107**, 26–39.

—— 1968. Distribution of Westphalian marine faunas in northern England and adjoining areas. *Proc. Yorks. geol. Soc.* **37**, 1–72.

—— 1969. Westphalian of Britain. *C.R. Congr. Avanc. Etud. Stratigr. Géol. Carbonif.* Sheffield 1967, **1**, 233–54.

DUNHAM, K. C. 1950. Lower Carboniferous sedimentation in the northern Pennines (England). *Rep. XVIII intern. geol. Congress, London*, Part IV, 46–63.

—— and ROSE, W. C. C. 1941. Geology of the iron-ore field of south Cumberland and Furness. *Wartime Pamph. geol. Surv. Great Britain.* **16**, 26 pp.

EASTWOOD, T. 1928. The Cockermouth Lavas, Cumberland — a Carboniferous volcanic episode. *Summ. Prog. geol. Surv. Great Britain.* Pt. 2, 15–22.

—— 1930. The geology of the Maryport district. *Mem. geol. Surv. Great Britain.*

—— DIXON, E. E. L., HOLLINGWORTH, S. E. and SMITH, B. 1931. The geology of the Whitehaven and Workington District. *Mem. geol. Surv. Great Britain.*

—— HOLLINGWORTH, S. E., ROSE, W. C. C. and TROTTER, F. M. 1968. Geology of the country around Cockermouth and Caldbeck. *Mem. geol. Surv. Great Britain.*

EDMONDS, C. 1922. The Carboniferous Limestone Series of West Cumberland. *Geol. Mag.* **59**, 74–83, 117–31.

GARWOOD, E. J. 1907. Faunal succession in the Carboniferous Limestone of Westmoreland, etc. *Geol. Mag.* **44,** 70–74.

—— 1913. The Lower Carboniferous succession in the north-west of England. *Q. Jl geol. Soc. Lond.* **68,** 449–586.

—— 1916. The faunal succession in the Lower Carboniferous rocks of Westmorland and north Lancashire. *Proc. geol. Assoc.* **27,** 1–43.

GEORGE, T. N., JOHNSON, G. A. L., MITCHELL, M., PRENTICE, J. E., RAMSBOTTOM, W. H. C., SEVASTO-PULO, G. D. and WILSON, R. B. 1976. A correlation of Dinantian rocks in the British Isles. *Geol. Soc. Lond., Spec. Rep.* No. 7, 87 pp.

HUDSON, R. G. S. 1929. A Carboniferous lagoon deposit with sponges. *Proc. Yorks. geol. Soc.* **21,** 181–96.

JOHNSON, G. A. L. 1960. Palaeogeography of the northern Pennines and part of north eastern England during the deposition of Carboniferous cyclothemic deposits. *Rep. XXI intern. geol. Congress, Norden,* Part XII, 118–28.

—— 1967. Basement control of Carboniferous sedimentation in northern England. *Proc. Yorks. geol. Soc.* **36,** 175–94.

—— and MARSHALL, A. E. 1971. Tournaisian Beds in Ravenstonedale, Westmorland. *Proc. Yorks. geol. Soc.* **38,** 261–80.

KENDALL, J. D. 1896. The Whitehaven Sandstone Series. *Trans. Instn. min. Engrs,* **10,** 204–24.

LLEWELLYN, P. G., MAHMOUD, S. A. and STABBINS, R. 1968. Nodular anhydrite in Carboniferous Limestone, west Cumberland. *Trans. Instn Min. Metall.* (*Sect. B: Appl. earth sci.*) **77,** B18–25.

RAMSBOTTOM, W. H. C. 1969. Reef distribution in the British Lower Carboniferous. *Nature, Lond.* **222,** 765–66.

—— 1973. Transgressions and regressions in the Dinantian: a new synthesis of British Dinantian stratigraphy. *Proc. Yorks. geol. Soc.* **39,** 567–607.

—— 1974. Dinantian. *In* Rayner, D. H. and Hemingway, J. E. (Eds), *The Geology and mineral resources of Yorkshire.* Leeds. ix + 405 pp.

—— 1977. Major Cycles of transgression and regression in the Namurian. *Proc. Yorks. geol. Soc.* **41,** 261–91.

—— 1978. Correlation of the Scottish Upper Limestone Group (Namurian) with that of the North of England. *Scott. J. Geol.* **13,** 327–30.

ROSE, W. C. C. and DUNHAM, K. C. 1978. Geology and hematite deposits of south Cumbria. *Mem. geol. Surv. Great Britain.*

ROWLEY, C. R. 1969. The stratigraphy of the Carboniferous Middle Limestone Group of West Edenside, Westmorland. *Proc. Yorks. geol. Soc.* **37,** 329–50.

SCHWARZACHER, W. 1958. The stratification of the Great Scar Limestone in the Settle District of Yorkshire. *L'pool Manchr. geol. J.* **2,** 124–42.

SEDGWICK, A. 1832. On the deposits overlying the Carboniferous Series in the valley of the Eden, etc. *Proc. geol. Soc. Lond.* **1,** 343–45.

SMITH, A. H. V. and BUTTERWORTH, M. A. 1967. Miospores in the coal seams of the Carboniferous of Great Britain. *Spec. Pap. Palaeont.* **1,** 324 pp.

SMITH, B. 1924. Special reports on the mineral resources of Great Britain. Vol. VIII, Iron ores — haematites of West Cumberland, Lancashire and the Lake District. Second Edition. *Mem. geol. Surv. Great Britain.*

TAYLOR, B. J. 1961. The stratigraphy of exploratory boreholes in the west Cumberland coalfield. *Bull. geol. Surv. Great Britain.* **17,** 1–74.

—— BURGESS, I. C., LAND, D. H., MILLS, D. A. C., SMITH, D. B. and WARREN, P. T. 1971. Northern England. *Br. reg. Geol.* (4th Edition). H.M.S.O. London, x + 121 pp.

TOMLINSON, T. E. 1940. Microspores of the coal seams of the Solway No. 1 shaft. *Rep. of Coal Survey Lab.,* Newcastle upon Tyne.

TROTTER, F. M. 1953. Reddened beds of Carboniferous age in north-west England and their origin. *Proc. Yorks. geol. Soc.* **29,** 1–20.

—— HOLLINGWORTH, S. E., EASTWOOD, T. and ROSE, W. C. C. 1937. Gosforth District. *Mem. geol. Surv. Great Britain.*

WALKDEN, G. M. 1972. The mineralogy and origin of interbedded clay wayboards in the Lower Carboniferous of the Derbyshire Dome. *Geol. J.* **8,** 143–59.

—— 1974. Palaeokarstic surfaces in upper Viséan (Carboniferous) limestones of the Derbyshire Block, England. *J. Sed. Pet.* **44,** 1232–47.

WALTHAM, A. C. 1971. Shale units in the Great Scar Limestone of the southern Askrigg Block. *Proc. Yorks. geol. Soc.* **38,** 285–92.

—— 1974. *The limestones and caves of north-west England.* Newton Abbot. 477 pp.

M. MITCHELL, M.A., B. J. TAYLOR, B.SC., and W. H. C. RAMSBOTTOM, PH.D.
Institute of Geological Sciences, Ring Road Halton, Leeds LS15 8TQ

13

Permian and Triassic

R. S. ARTHURTON, I. C. BURGESS and D. W. HOLLIDAY

The earliest 'Permian' sediments of the Lake District contain no fossils and their precise age cannot be ascertained. However, it is apparent that, in common with most of the British landmass, there is in this area an hiatus in the sedimentary record extending from mid-Westphalian times to an undetermined point in the Lower Permian. During this interval continental conditions were established over most of Europe, and the region suffered substantial upheaval associated with the Hercynian orogeny. The Lake District-Howgills area was uplifted and a ridge arose along the Pennine and Dent faults where major down-east monoclines were formed. This period of earth movement was closely followed by the intrusion of the Whin Sill into the Carboniferous strata of northern Pennines to the east.

The upheaval resulted in the widespread erosion of hundreds of metres of Carboniferous rocks, and Lower Palaeozoic rocks were exposed locally. However, Upper Carboniferous strata including some of Westphalian age were preserved in contemporaneous downwarps around the Lake District, notably along its north-western margin, in the Vale of Eden and Stainmore, and in Morecambe Bay. During this period the rocks were reddened (Trotter 1939, 1953) and limestones dolomitized to great depths in places. According to Dunham (*in* Rose and Dunham in press) it was such reddened rocks that, on subsequent erosion, provided the red detritus characteristic of the Permian and Triassic sequence. An alternative view that the red coloration largely results from diagenetic alteration of ferromagnesian minerals as in some modern deserts (Waugh 1973, Walker 1976) also is widely held.

Later several coalescing sedimentary basins were established around the Lake District. Once initiated the downwarping of these basins continued throughout the remainder of Permian and Triassic time, leading to the accumulation of thick sedimentary sequences. Lateral variations in the thickness and lithology of the Permian sediments show that the Lake District–Howgills area must have been mountainous at that time. The northern Pennines were also an area of non-deposition, but probably of lower relief, extending eastwards from an escarpment along the Pennine line to the North Sea basin. The Triassic sediments appear to have overlapped the Permian throughout, and progressively buried the contemporary mountainous terrain. In the view of Eastwood et al. (1931) the process of burial by these sediments was complete (see also Burgess and Wadge 1974, Fig. 26).

The Permian and Triassic rocks were first described by Sedgwick (1832, 1836) and later in more detail by Binney (1855, 1857), Harkness (1862) (who introduced the term Penrith Sandstone for the red sandstones at the base of the sequence in the Vale of Eden) and Goodchild (1893). The classification was revised for west Cumbria and the Carlisle Plain by Smith (1924), and his revision, in which he introduced several new formation names including St Bees Shales, was incorporated in the local memoirs of the Geological Survey published between 1926 to 1937. Important contributions to the regional stratigraphy of the Upper Permian rocks were made by Hollingworth (1942) and Meyer (1965), while the Permian and Triassic stratigraphy of the poorly exposed south Cumbria district was documented largely from borehole data by Dunham and Rose (1949). Accounts of the Permo-Triassic rocks of the Penrith (by Arthurton and Wadge) and Brough (by Burgess and Holliday) districts are at the time of writing in press.

CHRONOSTRATIGRAPHY

The Permian and Triassic sediments are mostly red beds lacking fossils useful in chronostratigraphical classification. Only the grey sediments and associated dolomites which occur in the lower part of the sequence in west Cumbria, and are intercalated with red beds in the Vale of Eden, have yielded fossils of value and these are all of Upper Permian age. A Lower Permian age has been inferred for breccias and aeolian sandstones at the base of the sequence (including the Penrith Sandstone) by comparison with the sequences of NE England and elsewhere in Europe (Mykura 1965, Smith 1972, Smith et al. 1974).

In common with other parts of Britain, there is no recognizable chronostratigraphical boundary for the base of the Triassic, and it has been customary to accept an arbitrary position fixed to a lithostratigraphical boundary of regional importance, even though such a line is likely to be diachronous. In the North Sea basin, current practice places the boundary at the top of the Z4 evaporite member of the Upper Permian Zechstein sequence, thus classifying the overlying red mudstones and sandstones of the Bacton Group as Triassic (Rhys 1974). In the basins of northwest Britain, however, the EZ4 evaporite member is unknown, and the older EZ3 member is believed to be restricted to the southern part of the Vale of Eden (Burgess 1965), so it is not feasible to apply the North Sea practice to Cumbria. Instead, it is convenient to regard the base of the St Bees Sandstone (part of the Sherwood Sandstone Group) as the arbitrary boundary, and this classification has been adopted by the Institute of Geological Sciences on their recently published Penrith, Brough and Barrow in Furness 1 : 50 000 geological sheets.

Only indirect evidence for the age of the Triassic sediments is available, and this comes from lithostratigraphical correlations with the more fully known sequence in the Fylde district of Lancashire. From this is seems possible that the youngest preserved strata in south Cumbria are of Ladinian age. Rocks of still younger age, including the Rhaetian, may be present under the proved Liassic sediments of the Carlisle Plain.

LOWER PERMIAN

Continental deposits of presumed Lower Permian age crop out in the Vale of Eden. The basal breccias of the west Cumbrian coast at Saltom Bay, and those proved in borings in south Cumbria may also be partly of this age. Everywhere they rest with marked unconformity on reddened Carboniferous rocks, in places on Dinantian limestones and elsewhere on sandstones and shales of Namurian or Westphalian age.

The deposits are thickest in the Vale of Eden where they form the Penrith Sandstone (Fig. 65). This has a broad, continuous outcrop on the western limb of the Vale of Eden Syncline from Kirkby Stephen to Armathwaite, and is seen also in a number of small outcrops on the steeper eastern limb, adjoining the Pennine Fault. Other small outcrops are known to the north-west of the Vale, around Ivegill. At least 400 m are present in the Appleby–Hilton area, though in the east, where the formation rests on an irregular surface of Carboniferous rocks, it reaches a maximum of about 100 m (Burgess and Wadge 1974).

The dominant lithologies are medium-grained and coarse-grained reddish brown sandstones, in which well-rounded, frosted quartz grains abound. Pronounced large-scale trough-cross-bedding is a characteristic feature, and this, together with the grain roundness, has led to the belief that the sands accumulated in a desert sand sea as the foresets of barchan dunes. Measurement of the dip directions of the foresets has shown that the prevailing wind blew from between east and south-east (Waugh 1970a). Footprints recorded from quarries in the Penrith Sandstone, are the only fossils known from the Lower Permian of the region (Smith 1884, Hickling 1909).

For the most part the component sand grains are poorly cemented, and make up a permeable, soft and easily weathered rock. However, in places the grains are cemented by secondary silica, as overgrowths in optical continuity with individual grains (Waugh 1970a, b), and it is the resulting quartz prism facets that give a lustre to the rock in sunshine (Murchison and Harkness 1864). The silicified rock locally has a patchy distribution, but in general the upper part of the formation in the main crop between Cliburn and Armathwaite is silicified, and gives a persistent escarpment best seen on Lazonby Fell. The silicification has been ascribed by Waugh (1970a) to the precipitation of silica from desert alkaline groundwaters.

The silicified rock has been quarried extensively in the past for building purposes, the product being known locally as Lazonby Stone. Today the principal economic value of the sandstones as a whole is as an aquifer, and water feeding the mains supply is pumped from a number of boreholes in the northern part of the outcrop.

In the extreme northern part of the outcrop, near Holmwrangle, where the sequence is thin compared with that farther south, dune-bedded sandstones are underlain by red water-laid sandstones, mudstones and breccias (Trotter and Hollingworth 1932), formed apparently by the spasmodic encroachment of alluvial fans onto a playa lake, perhaps marginal to the sand sea of the more central part of the basin. Thicker wedges of breccia, known as 'brockrams', are present between Appleby and Kirkby Stephen, and to a lesser extent in some of the small outcrops adjoining the Pennine Fault (Fig. 65).

In the Appleby–Hilton Beck area coarse breccia composed mainly of Carboniferous limestone clasts, commonly dolomitized, forms the lowest 150 m of the sequence. This is overlain by a similar thickness of dune-bedded sandstone, which is itself in turn overlain by about 100 m of water-lain sandstone with lenses of breccia with, at the top, more dune-bedded sandstones (Versey 1939, Burgess and Wadge 1974). In George Gill, near Appleby, these upper breccias have yielded pebbles of decomposed dolerite, probably derived from the Whin Sill (Holmes and Harwood 1928, Dunham 1932). Pebbles of Lower Palaeozoic rocks have also been recorded from the upper breccias (Kendall 1902), as they have from breccias near Ranbeck Farm, to the north of Appleby (Burgess and Wadge 1974). Pebbles of Roman Fell Quartzite in brockrams near Hilton (Kendall 1902) point to local derivation from the present Pennine escarpment area. Towards the basin margin the large-scale cross-bedded dune sandstones pass laterally into water-lain sandstones, with smaller-scale cross-bedding and planer lamination, interbedded with bands of breccia. At the southern end of the Vale, west of Kirkby Stephen, the entire formation is composed of breccia (Burgess 1965). Like those at the northern end of the basin, these breccias and their associated water-laid sandstones are considered to have been deposited on alluvial fans extending into the sand sea, the detritus being transported by periodic flash floods from mountains which flanked the basin, at least to the east and south. In the southern part of the Vale of Eden, water-laid sandstones and breccias at the top of the Penrith Sandstone appear to be the lateral equivalents of the Hilton Plant Beds at the base of the Upper Permian Eden Shales (Figs 65 and 67). Thus locally the top of the Penrith Sandstone is diachronous.

In west and south Cumbria a thin and impersistent sheet of breccia separating proved Upper Permian from reddened Carboniferous rocks may be partly of Lower Permian age (Fig. 66). In west Cumbria this deposit was termed Lower Brockram by Smith (1924), and Basal Breccia by Arthurton and Hemingway (1972). It has been proved in the Whitehaven–St Bees area in thicknesses up to 3 m, although exceptionally, in a boring at St Bees, 20.4 m were proved. The Basal Breccia is well exposed in the coasts section at Saltom Bay (Eastwood et al. 1931).

In south Cumbria the Basal Breccia (Rose and Dunham in press) has a maximum recorded thickness of 26 m in a boring at Haverigg Haws, near Millom, though generally it is much thinner and locally absent.

These thin, but widespread, basal breccias of west and south Cumbria contrast with the thick breccia wedges of the Vale of Eden, and may be residual piedmont gravels rather than alluvial fan deposits (cf. Smith 1972). They are basin-margin equivalents of thicker Lower Permian sequences offshore in the NE Irish Sea Basin (Colter and Barr 1975, Fletcher and Ransome, Chapter 16).

UPPER PERMIAN

The Upper Permian sediments around the Lake District are composed largely of red, brown and grey mudstones and siltstones with marginal breccias, and, like the Lower Permian sandstones and breccias which they overlap, were deposited by wind and water in a continental, largely desert environment. However, the red beds are associated with marine deposits including limestone and dolomite, and also layers of anhydrite which are taken to have accumulated in a coastal environment; where the latter approach outcrop they are represented by gypsum or by solution breccias and residues. These associated deposits have made possible detailed lithostratigraphical correlation, not only between the various sections around the Lake District, but also between Cumbria and the North Sea Zechstein basin.

facing page

PLATE 10. Steeply inclined Silurian greywackes, transitional between the Coniston Grits and Bannisdale Slates, A6 road cutting, Shap [555 055] — Vide Fig. 29. Steep 'main phase' cleavage is confined to the finer units. In thin section the cleavage is seen to consist of pressure solution seams more closely spaced in the mudstones orientation in the siltstones.   *Photograph: F. Moseley*

A

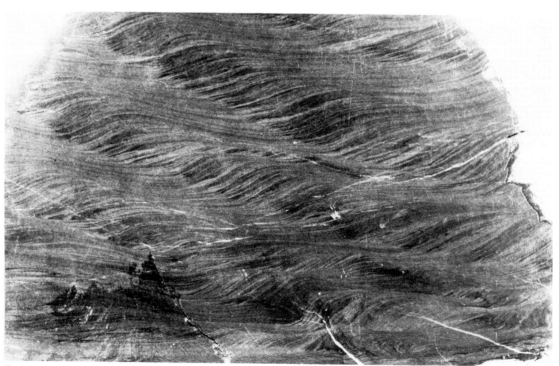

B

The position of the marine connexion between the Upper Permian sea of north-west England (*Bakevellia* Sea of Smith 1970a) and the Zechstein Sea is not known. Burgess (1965) proposed a direct marine link between the southern part of the Vale of Eden and the Zechstein by way of the present Stainmore Depression, during the deposition of the Belah Dolomite. However, this transgression from the east was of limited extent and there is no indication that a marine connexion existed through Stainmore during any other part of Upper Permian time. Smith (1970a) and Pattison (1970) have suggested that the north-west basins were connected with the Zechstein sea by way of a link around the north of Scotland rather than by a direct east–west strait across northern England.

As in the Lower Permian, the Vale of Eden was an intermontane basin, flanked on the east by high ground along the Pennine Line and on the west by the Lake District uplands. In this basin, a thick sequence of largely non-marine sediments, with only intermittent marine intercalations, accumulated (Fig. 65). In west and south Cumbria, marine deposits are more fully represented in a sequence lapping on to the western edge of the Lake District uplands (Fig. 66). Three cycles of marine transgression and regression are recorded in these beds.

Vale of Eden and Carlisle Plain

Upper Permian strata are preserved in the Vale of Eden, and are known at the margins of the Carlisle Plain at Riddings and Longtown in the north and Cocklakes in the south. Knowledge of their stratigraphy comes partly from natural sections, notably Hilton Beck near Appleby, but mainly from boreholes drilled either in exploration for gypsum and anhydrite (Sherlock and Hollingworth 1938, Hollingworth 1942, Meyer 1965), or by the Institute of Geological Sciences (Ann. Rept. Inst. geol. Sci. for 1967).

In each area the Upper Permian rocks are classified as one formation, those of the Carlisle Plain are known as the St Bees Shales (Dixon et al. 1926, Trotter and Hollingworth 1932), and their stratigraphy has been summarized by Meyer (1965); those of the Vale of Eden, although formerly called the St Bees Shales, are now termed the Eden Shales (Fig. 65), a formation defined locally in terms of its occurrence in the Hilton Borehole (Burgess and Holliday 1974).

The Eden Shales rest mainly on the Penrith Sandstone. In the extreme south, at Kirkby Stephen, they overlap on to Carboniferous rocks (Burgess 1965), while in the north they are continuous with the St Bees Shales of the Carlisle Plain. They consist mostly of red and subordinate grey terrigenous sediments, but also include several laterally persistent beds of anhydrite, present as gypsum near outcrop, of which the four thickest are known by the letters A to D. The marine Belah Dolomite, is also present in the southern part of the Vale. The formation is at its thickest (184 m) near

facing page

PLATE 11. A. Crag exposing the Applethwaite Formation of the Coniston Limestone Group (Ashgill Series, Cautleyan Stage, Zone 2) showing typical nodular, bioclastic limestone with interbeds of calcareous mudstone, cut by a strong cleavage, probably of pressure solution origin; east of Tarn Hows and north-east of Coniston [336 005].
B. Coniston Grit, Shap [555 055]. Cross lamination is well seen in this specimen.

Photographs: F. Moseley

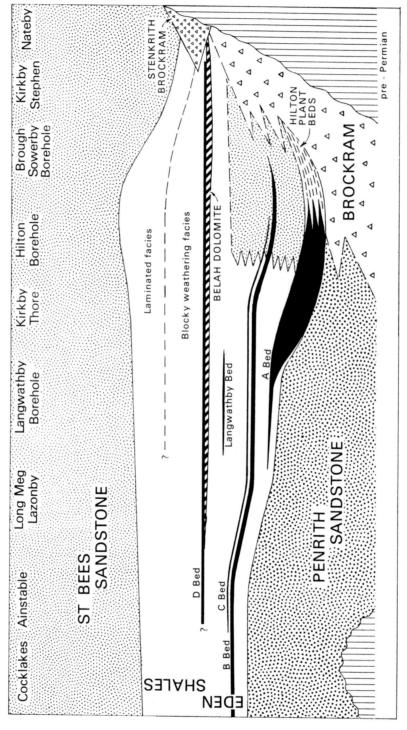

Fig. 65. Vale of Eden: Section showing the interrelationships of named Permian rock divisions. Gypsum-anhydrite shown black. B-Bed taken as datum. Length of section about 55 km; approximate vertical scale 1 cm : 40 m.

Hilton, gradually thinning northwards towards Knot Hill near Cocklakes, and more abruptly southwards, towards Kirkby Stephen, reflecting the thickness variations of the underlying Penrith Sandstone.

The oldest Eden Shales are confined to what seems to have been a contemporaneous topographical hollow on the Penrith Sandstone in the Kirkby Thore area. Here they accumulated in a desert lake, probably bounded to the north-west by dune-sands (Penrith Sandstone) and on the south-east by a gently rising alluvial plain periodically inundated by flood waters from mountains marginal to the basin. The lake deposits comprise finely laminated, grey, carbonate-rich siltstones and mudstones, together with the nodular and crudely laminated A-Bed gypsum of the Kirkby Thore mines. These lacustrine deposits suggest a rise in the regional water table following the deposition of the Penrith Sandstone, and this rise is thought to correlate with the first (Saltom) of the Upper Permian marine transgressions of west and south Cumbria (Burgess and Holliday 1974). However, the existence of a direct marine connexion to the north of the Lake District at this time remains unproved. Contemporaneous vegetation around the A-Bed lake is indicated by the abundance of derived carbonaceous plant debris, present both in the A-Bed itself and in the grey siltstones and sandstones (Hilton Plant Beds) of the adjoining flood-plain.

A period of increased aridity, probably related to a phase of marine regression, followed, marked in the Hilton district by red-brown sandstone and farther north by red-brown siltstone. With the infilling of the Kirkby Thore hollow, an almost flat depositional plain extended far to the north, the deposits overlapping a shelf of silicified Penrith Sandstone, which until then had extended from Cliburn to the northern end of the basin. At the top of this sequence, a rise in the contemporary water table is indicated by the presence of a few metres of plant-bearing grey siltstone with gypsum nodules in the Hilton and Langwathby bores. This heralded a second period of marine influence (correlated with the Sandwith transgression of west Cumbria), during which time B-Bed was formed.

B-Bed is up to 6.5 m thick, and extends throughout all but the southernmost part of the Vale around Brough and Kirkby Stephen and into at least the southern part of the Carlisle Plain. The presence of algal mat structures in its lower part suggests that true marine conditions may have been briefly established, the direction of marine influx being apparently from the north.

The succeeding C-Bed, separated from B-Bed by some 2 to 4 m of red and plant-bearing grey silty mudstone, and marking a third period of strong marine influence (Fleswick Cycle), is consistently thinner and more restricted laterally. It is present southwards as far as Kirkby Thore, but in the Hilton Borehole it is represented only by scattered gypsum-anhydrite nodules in red and grey siltstone. The sediments between C-Bed and the Belah Dolomite are similar to those below B-Bed; around Hilton they comprise red-brown sandstones with sporadic anhydrite nodules, and apparently accumulated on a flood-plain, while farther north they consist of red-brown mudstones and siltstones. The topmost 15 m include grey plant-bearing layers; anhydrite nodules are common in these beds, and laminated gypsum-anhydrite of presumed detrital origin, as for example in the Langwathby Bed (Arthurton 1971), is also present, apparently deposited in small saline lakes on the surface of a continental sabkha. The lacustrine beds are thought to indicate a rise in regional water-table preceding the next marine transgression into the basin which is marked by the Belah Dolomite (Burgess and Holliday 1974), a marine carbonate member traceable from the extreme south of the basin at Kirkby Stephen northwards only as far as Long Meg, near Lazonby (Meyer 1965). Thus, in contrast to the earlier marine transgressions into the Vale of Eden basin, this influx appears to have entered by way of the Stainmore Depression, directly from the North Sea basin to the east (Burgess 1965).

In the south of the basin, around Brough Sowerby, the Belah Dolomite comprises 5 m of fossiliferous, shallow-water marine dolomites but passes north-westwards into tidal-flat dolomites with algal-mats in the Hilton area (Burgess and Holliday 1974). In the Lounthwaite and Langwathby boreholes grey dolomitic mudstones occur at this level.

The overlying anhydrite, D-Bed, is up to 3 m thick, and has a distribution similar to that of the Belah Dolomite; it is unknown in the northern part of the Vale, and though it is not seen in the Brough Sowerby–Kirkby Stephen area, its former presence there is indicated by collapse breccias and solution residues in the Brough Sowerby Borehole (Ann. Rept. Inst. geol. Sci. for 1967).

The top of D-Bed marks an abrupt change in depositional environment. The succeeding strata are brick-red in colour in contrast to the purple-red and grey colours that predominate below the Belah Dolomite. They consist largely of blocky-weathering muddy siltstones and sandstones, commonly with gypsum-anhydrite nodules, with a varying proportion of large (up to 3 mm) 'millet-seed' quartz grains. Adhesion ripples and deflation surfaces are the commonest sedimentary structures, and most of the deposit appears to be of aeolian origin (Burgess and Holliday 1974).

The highest beds of the Eden Shales consist of alternations of laminated or cross-laminated fine-grained sandstones, siltstones and mudstones, also mainly brick-red in colour. They appear to have been water-laid and show desiccation features such as mud-cracks, sand dykes, injection structures, and mud-flake conglomerates (Burgess and Holliday 1974). In the extreme south, around Kirkby Stephen, a northwards thinning wedge of wadi-breccia, the Stenkrith Brockram, separates the aeolian from the water-laid facies, and indicates the continuing proximity of contemporaneous uplands; east of the town at Hartley, and southwards at Wharton, this breccia overlaps all earlier strata to rest directly on Carboniferous rocks (Burgess 1965).

West and south Cumbria

Upper Permian strata crop out locally on the coastal strip of west and south Cumbria, on the eastern margins of the East Irish Sea Permo-Triassic Basin (Colter and Barr 1975, Fletcher and Ransom, Chapter 16). Their stratigraphy (Fig. 66) is known largely from boreholes, drilled mostly in exploration for anhydrite, coal and iron-ore (Hollingworth 1942, Dunham and Rose 1949, Taylor 1961, Arthurton and Hemingway 1972).

In west Cumbria (Fig. 66) they comprise the partly marine St Bees Evaporites (Arthurton and Hemingway 1972), and the red St Bees Shales. They also include the lower part of the Brockram (Smith 1924), a basin margin deposit similar to the Lower Permian breccia wedges of the Vale of Eden.

The St Bees Evaporites include three sedimentary cycles in their type section at St Bees Head, and three comparable cycles are present at the base of the Upper Permian sequence in the Roosecote Borehole. Marine shells are abundant in the first, and are also recorded at the base of the second.

The transgressive sediments of the first (Saltom) cycle comprise two facies, their distribution indicating an eastward or northward marginal shallowing of the Bakevellia Sea against a land mass broadly coincident with the present Lake District. Both

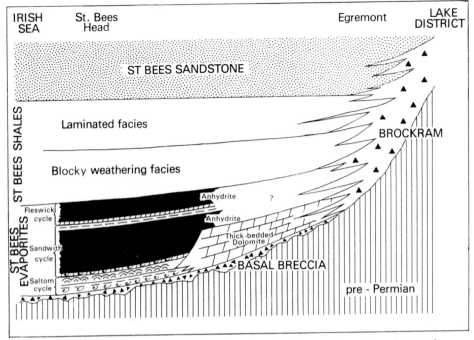

FIG. 66. West Cumbria: Section showing the interrelationships of named Permian rock divisions. Gypsum-anhydrite shown black. Base of St Bees Sandstone taken as datum. Length of section about 10 km; approximate vertical scale 1 cm : 15 m.

facies rest sharply on Basal Breccia, and their bases mark an abrupt and profound change from the earlier desert environment. One includes the Grey Beds of south Cumbria (Dunham and Rose 1949) and the Saltom Siltstone at the base of St Bees Evaporites, and consists of dark grey calcareous or dolomitic siltstone up to 7 m thick. It contains carbonaceous plant fragments, derived presumably from contemporaneous on-shore vegetation.

The other facies comprises thick-bedded and partly shelly limestone or dolomite up to 21 m thick, a unit referred to as the Magnesian Limestone in south Cumbria (Rose and Dunham in press), and previously in west Cumbria also (Smith 1924). It is regarded as having formed in banks or shoals in shallow water at the margin of the marine basin. In west Cumbria the thick-bedded carbonates crop out at Saltom Bay, and they are known, from boreholes, to thin as they approach the Lake District massif to the south-east.

In south Cumbria, thick-bedded oolites of the marginal facies at Old Holbeck, near Barrow (C. M. Jones, Pers. comm.), are represented in the Roosecote bore, about 1 km farther south, by fining upwards and slumped units with shells and ooliths, apparently slurry deposits containing material derived from adjoining shell and oolith banks.

An abrupt fall of sea-level followed the Saltom transgression, and the subsequent sediments of the cycle, mostly laminated dolomites, apparently never extended over the marginal thick-bedded carbonates.

The strata of the second (Sandwith) cycle are locally unconformable on those of the first, which themselves are reddened to a depth of about 2 m below their top. They comprise a basal dolomite member rarely more than 2 m thick, and overlying this, anhydrite up to 20 m thick; the 'bottom seam' of the Sandwith Mine workings. The dolomite is thick-bedded to massive and shelly, and marks a short-lived fully marine episode. Algal lamination is present at the top in some sections, and this heralds the establishment of a coastal flat or sabkha environment in which the overlying and largely nodular anhydrite was formed.

Deposits of the third (Fleswick) cycle, like those of the second, show a local unconformity at their base. The basal member is 2 to 4 m thick and consists of lami-nated red or grey siltstone, and unfossiliferous, laminated, fine textured dolomite. It includes cut-and-fill structures and pellet breccias, and apparently accumulated on a coastal flat which became intertidal. The overlying anhydrite, the 'top seam' of the Sandwith workings, is up to 7 m thick at St Bees Head, and both nodular and laminated varieties are present.

The St Bees Shales of west Cumbria are continental deposits overlying the St Bees Evaporites, although in south Cumbria they are taken to include strata down to the top of the Grey Beds or the Magnesian Limestone (Rose and Dunham in press). They comprise two principal facies; in the lower part, blocky weathering siltstone and silty mudstone, commonly gypsiferous and with coarse sandy bands and calcareous concretions, and above, laminated micaceous siltstone and mudstone and subordinate sandstone, with load-casts, desiccation cracks and mud-flake breccias. A comparable change from dominantly wind to water deposition within the upper part of the Eden Shales was noted previously. This would seem to result from a regional climatic change with increased rainfall in the areas of high relief that were the main sources of sedi-ment. Gypsum and anhydrite, both laminated and nodular, are present particularly towards the base of the formation in the blocky facies. The St Bees Shales thickens away from the Lake District, the proved maximum being about 200 m in the Haverigg Haws bore (Dunham and Rose 1949). At the basin margins mudstones and siltstones overlap the carbonate and evaporite rocks and intercalate with the Brockram (Fig. 66), alluvial fan deposits of breccia derived from the contemporaneous uplands of Carboniferous limestones and Lower Palaeozoic rocks in the area of the present Lake District (Smith 1924, Dunham and Rose 1949, Rose and Dunham (in press). The Brockram attains a thickness of 120 m at Calder Bridge, near Egremont (Trotter et al. 1937). At Humphrey Head in south Cumbria, a borehole sunk recently by the Institute of Geological Sciences proved 257 m of Permo-Triassic conglomerates and sandstones resting on Carboniferous strata, and similar conglomerates are exposed nearby at Roughholme Point. The precise age of these deposits is not known and it is even possible that they lie high in the sequence rather than at the base of the Permian (Rose and Dunham in press).

Correlation

The palaeontological evidence for correlation between the Upper Permian rocks of Cumbria and NE England is limited and has been summarized by Smith (1970b), Pattison (1970), Pattison et al. (1973) and Pattison (*in* Smith et al. 1974). The com-moner species in the Saltom and Sandwith Cycles are *Bakevellia binneyi*, *Permophorus costatus* and *Schizodus obscurus*, which are thought to indicate an approximate correlation with the EZ1 or EZ2 cycles of the Zechstein Basin. The fauna of the

Belah Dolomite of the Vale of Eden, *Liebea squamosa*, *Schizodus obscurus*, and *Calcinema permiana* is considered to favour the correlation of this bed with the Zechstein EZ3 carbonates. Though plant remains are locally abundant (Stoneley 1958) they have proved of little value in correlation. Palynological studies may be of value (Clarke 1965, Visscher 1971), but remain unproved. Warrington (*in* Arthurton and Hemingway 1972, p. 573) has examined miospore assemblages from the Saltom Cycle and suggested that they are older than the higher part of EZ2. In the absence of further fossil control, more detailed correlations must depend on inferences made on general stratigraphical and sedimentological grounds. The following discussion assumes that Upper Permian sedimentary cycles are laterally continuous throughout much of north-west England and are broadly synchronous with those of the Zechstein Basin (Table 4).

The Saltom Cycle of the Cumbrian coast is incomplete in that, at least in the known areas, it contains no evaporite member; it is overlain by the more complete Sandwith Cycle which contains a thick anhydrite member. This sequence compares closely with that of the EZ1 cycle of north-east England (Smith 1970b, 1971; Smith et al. 1974). In Durham this cycle comprises the Lower Magnesian Limestone at the base, separated by a disconformity from the overlying Middle Magnesian Limestone and Hartlepool Anhydrite. Similarly in Yorkshire the EZ1 carbonate member, overlain by the Permian Middle Marl and the Hayton Anhydrite is in two parts, separated by beds indicative of shallow water and emergent conditions (Smith et al. 1974). If this correlation is correct, then the Fleswick Cycle is likely to be equivalent to the EZ2 cycle of the north-east i.e. that containing the Concretionary Limestone, the Hartlepool and Roker Dolomites and 'Seaham Residue' Evaporite (Smith 1970b, 1971).

Between the Cumbrian coast and the Vale of Eden, the Sandwith and Fleswick Anhydrites and B- and C-Beds respectively have long been regarded as correlatives (Hollingworth 1942, Meyer 1965) on the basis of similarities of thickness and sequence, and this is accepted here. If correct, it would imply that A-Bed of the Vale formed during the Saltom transgression, and that the continental deposits separating A-Bed from B-Bed accumulated during the subsequent regression (Table 4).

It has previously been inferred that, during the deposition of the Belah Dolomite and D-Bed, the sea entered the Vale of Eden from the east, in the vicinity of the Stainmore Depression. Thus the Belah Dolomite could be expected to correlate with one of the major carbonate beds of the north-eastern sequence and the most likely correlative is the Seaham Beds (EZ3 of Smith 1971 (Table 4)), a view supported, as noted earlier, by palaeontological evidence. Strata of the EZ4 cycle include only a poorly developed carbonate member (Smith 1970b) and are therefore unlikely to correlate with the Belah Dolomite.

TRIASSIC

In this region the base of the Triassic has been arbitrarily fixed at the base of the St Bees Sandstone. In places north of Barrow Triassic rocks overlap the Permian, and rest on Carboniferous and Lower Palaeozoic rocks (Rose and Dunham in press). The Triassic sequence has two main divisions. The lower comprises red sandstones mostly and belongs to the Sherwood Sandstone Group, here probably largely or wholly Scythian in age, and the upper is mainly mudstones belonging to the Mercia Mudstone Group. The general relationships of the named units are shown in figure 67.

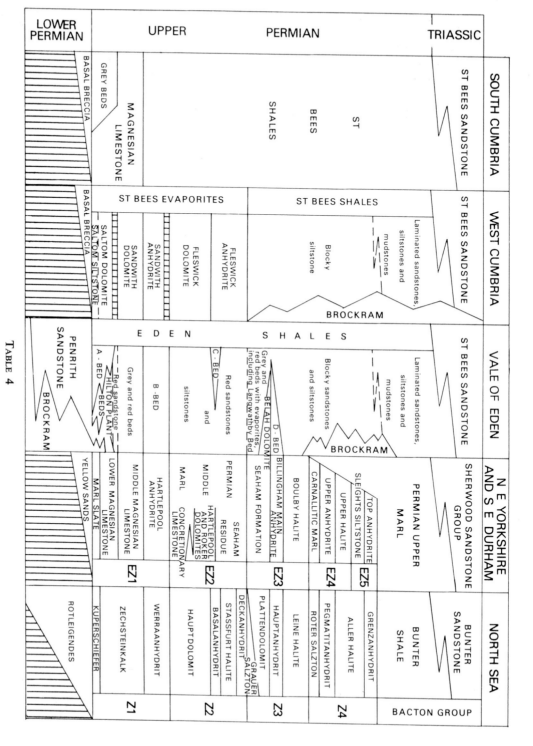

TABLE 4

Proposed correlation of Upper Permian sequences in Cumbria and the Zechstein Basin. Not to scale.

Sherwood Sandstone Group

Rocks of this group crop out continuously to the east and north of the Lake District from the head of the Vale of Eden at Kirkby Stephen to the Solway shore at Maryport, to the west, in the coastal strip from St Bees Head to Kirksanton, and to the south at Barrow in Furness. The thickest proving is 975 m in a boring at Seascale in west Cumbria (Gregory 1915), but neither the top nor the base was penetrated. Two formations have been recognized within the group, the St Bees Sandstone (Murchison and Harkness 1864) and the Kirklinton Sandstone (Holmes 1881). The St Bees Sandstone is widely distributed and forms all or at least the major part of local sequences. The Kirklinton Sandstone is mainly restricted to the Carlisle Plain, though lithologies of Kirklinton type have been recorded both in the Vale of Eden (Goodchild 1893) and in the Gosforth district (Trotter et al. 1937).

St Bees Sandstone

The St Bees Sandstone is generally well exposed, nowhere better than at its type locality, St Bees Head near Whitehaven, where it forms sea-cliffs 90 m high. The formation has been penetrated by numerous boreholes, particularly in west Cumbria, drilled either for mineral exploration in the underlying strata, or for water supply, the formation having considerable importance as an aquifer. Of the many published accounts of the St Bees Sandstone, those given in the memoirs of the Geological Survey are the most comprehensive, and the following description is based largely on those works.

In each of its main outcrops, the St Bees Sandstone ranges up to at least 500 m thick. It is typically dull red, fine to medium grained sandstone, compact and generally well-cemented, and has been widely used as a building stone. Grey sandstone has been encountered at the top of the formation in boreholes in Furness, and in one bore such sandstone contained hydrocarbon traces (Rose and Dunham in press). The sandstone is generally even-bedded, forming posts from a few centimetres to over 2 m thick. Chocolate-coloured, commonly micaceous mudstone, forms partings up to more than a metre thick, and such layers are particularly common in the lowermost 50 m. Pellets of mudstone within the sandstone, and minor disconformities provide evidence of contemporaneous erosion, while records of rain-prints and desiccation cracks point to an environment that was periodically if not wholly emergent.

The sandstone is composed of angular to sub-angular quartz grains of fine to medium grade, together with scattered grains of feldspar and flakes of white mica. Coarser wind-rounded quartz grains are usually sparingly scattered through the rock. A calcareous cement has been recorded in borings in west Cumbria, while ferruginous or more rarely siliceous cements have been noted in the Furness district (Rose and Dunham in press).

In some sections, notably in the River Calder near Calder Bridge, the upper part of the formation includes cross-bedded sandstone in which wind-rounded grains are common, interspersed with the more typical even-bedded sandstones with thin mudstones. These cross-bedded units tend to be poorly cemented and bright brick-red or orange in colour, and were described as being of 'Kirklinton type' by Trotter et al. (1937), though grouped by these authors with the St Bees Sandstone.

In the Whitehaven and Gosforth districts of west Cumbria the St Bees Sandstone intercalates towards the Lake District with layers of breccia, and south of Egremont these have been recorded at intervals up to 80 m above the base of the formation. To

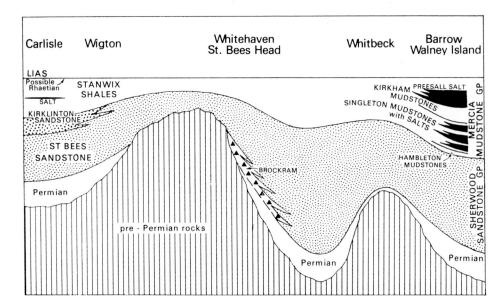

FIG. 67. Triassic sediments: Section showing interrelationships of named Triassic rock divisions from Carlisle Plain through West Cumbria to South Cumbria. Salt shown black. Estimated base of Lias taken as datum. Length of section about 110 km; approximate vertical scale 1 cm : 300 m.

the north of the Lake District, a pebbly layer has been noted in the mid part of the sequence, in the banks of the River Caldew near Dalston, south of Carlisle. The St Bees Sandstone was deposited on an alluvial plain, crossed by a number of braided rivers. The sandstones accumulated in stream courses, while the layers of mudstones were laid down during widespread flooding of the plain. According to Warrington (1970) this area formed part of a depositional tract occupying much of western and central southern England, and fed by a river system with its source perhaps as far removed as Brittany.

Kirklinton Sandstone

In the Carlisle Plain, soft-weathering, brightly coloured sandstones, of the type described above as layers within the St Bees Sandstone of west Cumbria form the Kirklinton Sandstone (Holmes 1881), which is up to 100 m thick. In the Carlisle and Brampton districts it is underlain by 75 m of passage beds, making the base of the formation imprecise. In the Cockermouth district boreholes at Wigton proved only about 10 m of Kirklinton Sandstone, stained black in places, between the top of the St Bees Sandstone and the base of the Stanwix Shales. To account for this marked south-westerly attenuation, Eastwood et al. (1968) suggested the possibility of a non-sequence at the base of the Stanwix Shales.

The strongly cross-bedded nature of the Kirklinton Sandstone, and its lack of mudstone layers are generally accepted as indications of aeolian deposition. It is doubtful, however, that a sand-sea on the scale of the Penrith Sandstone desert was ever established. Indeed, the intercalated St Bees and 'Kirklinton-type' sandstones of west Cumbria and the passage beds of the Carlisle Plain show that aeolian and fluviatile or lacustrine sedimentation were interspersed. There is evidence however for a general increase of aeolian at the expense of fluviatile accumulation towards the top of the Group.

Mercia Mudstone Group

The upper, argillaceous part of the Triassic sequence belongs to the Mercia Mudstone Group, and is preserved in a broad syncline in the Carlisle Plain, and at Walney Island in the Furness district of south Cumbria. In both outcrops these strata form low-lying, drift-covered country and surface exposures are restricted to the vicinity of Carlisle. Knowledge of the stratigraphy comes mainly from boreholes.

Carlisle Plain

In the Carlisle Plain the group comprises only one formation, the Stanwix Shales (Holmes 1881), which is capped by mudstones and argillaceous limestones of proved Liassic age (Hettangian) at Great Orton, just south-west of Carlisle. The existence of strata of Rhaetian age, which are widespread at the top of the Triassic sequence elsewhere in Britain, has been suggested by Dixon (in Dixon et al. 1926) but is as yet unproved.

The thickest proving of Stanwix Shales is near Abbeytown in a boring at Kelsick Moss, which penetrated some 170 m of the mudstones before entering Kirklinton Sandstone. Other boreholes at Wigton, proved a shale breccia at the base of the formation, which was interpreted by Trotter (in Eastwood et al. 1968) as indicating a possible non-sequence on the Kirklinton Sandstone.

The detailed stratigraphy and sedimentology of the Stanwix Shales are not well known, but in general sections include two contrasting argillaceous facies (Trotter, in Dixon et al. 1926). These are red 'marls' composed of poorly sorted mud and silt in which sedimentary lamination is usually absent, and well laminated mudstones and silty mudstones, commonly micaceous and alternating red and grey-green. In the lower part of the sequence the laminated deposits are calcareous and partly sandy, and beds of dolomitic sandstone are present. Higher up, there are thin beds of marly dolomite or impure limestone. Small cavities, probably after halite, are present particularly in the sandier lithologies, while in the borehole at Kelsick Moss salt is recorded in a 1.5 m layer of red shale some 150 m above the top of the St Bees Sandstone. Gypsum is common throughout.

Trotter (in Dixon et al. 1926) drew attention to the alternations of the two facies, the red marl forming layers about 1 to 1.5 m thick and separated by laminated deposits, and was inclined to the view that they were a result of seasonal changes.

Walney Island

In Furness the stratigraphy of the Mercia Mudstone Group is much better known, though it is probably that the sequence is incomplete, the youngest strata being perhaps Ladinian in age. Following the work of Dunham and Rose (1949), Evans and Dunham (*in* Rose and Dunham in press) have shown that the sequence is very similar to that established recently by Evans and Wilson (1975) 20 km to the south-east near Blackpool.

Three principal divisions are recognized within the group (Fig. 67). The basal Hambleton Mudstones comprise grey mudstones or sandy mudstones some 20 m thick. These are overlain first by the Singleton Mudstones, red mudstones with impersistent layers of salt, and then by the Kirkham Mudstones, well laminated and alternately red and green. These two units together form a sequence 300 to 400 m thick. The Kirkham Mudstones include a salt unit at least 100 m thick at Biggar, where it was formerly pumped as brine and fed to a saltworks at the southern end of the island. This salt is almost certainly the lateral equivalent of the Preesall Salt of Blackpool and possibly is of the same age as the Lower Keuper Saliferous Beds of the Cheshire Plain.

By analogy with the Blackpool sequence the Hambleton and Singleton Mudstones of Walney are of Scythian age, and the Kirkham Mudstones are Anisian and Ladinian. Examination of the cores from a recent borehole on the coast west of Biggar by Dr A. A. Wilson and R. S. Arthurton has shown that the Singleton Mudstones are largely composed of apparently structureless, poorly sorted units, while the generally well-laminated Kirkham Mudstones include pseudomorphs after halite and desiccation cracks at many levels, with cross-lamination and current ripple lamination in the siltier beds. In both formations gypsum and anhydrite are present as nodules and laminae, and veins of both gypsum and halite are present. Where syn-sedimentary halite is preserved, it occurs as the 'haselgebirge' facies, formed apparently by displacive growth within soft sediment. The upper part of this borehole penetrated, immediately beneath the drift cover, a thick collapse breccia/solution residue, formed by the removal of the Preesall Salt by circulating groundwater.

The deposits of the Mercia Mudstone Group are believed to have accumulated in prominently arid conditions on extensive alluvial flats or 'playas', subject to periodic floods. In the basins around the Lake District these floods may have been derived locally, from the dwindling marginal uplands of the Lake District, the Southern Uplands, and the northern Pennines, or from farther afield to the south, via the remains of the once major river system responsible for the supply of detritus in the Sherwood Sandstone Group, and still, in Anisian/Ladinian times at least, supplying arenaceous detritus to central southern England (Warrington 1970). Whether the playas were wholly continental or were coastal flats is not clear. Glennie (1970) and Glennie and Evans (1976) have suggested that the sedimentation of the Keuper Marl (Mercia Mudstone Group) may have been akin to the modern sedimentation on the Rann of Kutch in India, a coastal alluvial flat more than 100 km wide with a surface lying very near sea level, inundated not only by flood-waters from the hinterland but also by marine surges. Certainly the existence of a 100 m thick halite deposit within the group lends support to the 'coastal plain' model, the sea being by far the most convenient source of such large quantities of halite.

Acknowledgement

This chapter is published by permission of the Director, Institute of Geological Sciences.

REFERENCES

ARTHURTON, R. S. 1971. The Permian evaporites of the Langwathby Borehole, Cumberland. *Rep. No. 71/17, Inst. Geol. Sci.* 18 pp.

—— and HEMINGWAY, J. E. 1972. The St Bees Evaporites — a carbonate-evaporite formation of Upper Permian age in West Cumberland, England. *Proc. Yorks. geol. Soc.* **38**, 565–92.

BINNEY, E. W. 1855. On the Permian beds of the north-west of England. *Mem. Proc. Manchr. lit. phil. Soc.* (Ser. 2), **12**, 209–69.

—— 1857. Additional observations on the Permian beds of the north-west of England. *Mem. Proc. Manchr. lit. phil. Soc.* (Ser. 2), **14**, 101–20.

BURGESS, I. C. 1965. The Permo-Triassic rocks around Kirkby Stephen, Westmorland. *Proc. Yorks. geol. Soc.* **35**, 91–101.

—— and HOLLIDAY, D. W. 1974. The Permo-Triassic rocks of the Hilton Borehole, Westmorland. *Bull. geol. Surv. Gt. Br.* **46**, 1–34.

—— and WADGE, A. J. 1964. *The geology of the Cross Fell area.* vii + 91 pp. H.M.S.O. London.

CLARKE, R. F. A. 1965. British Permian saccate and monosulcate miospores. *Palaeontology.* **8**, 322–54.

COLTER, V. S. and BARR, K. W. 1975. Recent developments in the geology of the Irish Sea and Cheshire Basins. *In* WOODLAND, A. W., Ed. Petroleum and the Continental Shelf of NW Europe. *Applied Science Publishers, London.* **1**, 61–73.

DIXON, E. E. L., MADEN, J., TROTTER, F. M., HOLLINGWORTH, S. E. and TONKS, L. H. 1926. The geology of the Carlisle, Longtown and Silloth district. *Mem. geol. Surv. U.K.* xiii + 113 pp.

DUNHAM, K. C. 1932. Quartz-dolerite pebbles (Whin Sill type) in the Upper Brockram. *Geol. Mag.* **69**, 425–27.

—— and ROSE, W. C. C. 1949. Permo-Triassic geology of South Cumberland and Furness. *Proc. geol. Ass.* **60**, 11–40.

EASTWOOD, T., DIXON, E. E. L., HOLLINGWORTH, S. E. and SMITH, B. 1931. The geology of the Whitehaven and Workington district. *Mem. geol. Surv. U.K.* xvii + 304 pp.

—— HOLLINGWORTH, S. E., ROSE, W. C. C. and TROTTER, F. M. 1968. Geology of the country around Cockermouth and Caldbeck. *Mem. geol. Surv. U.K.* x + 298 pp.

EVANS, W. B. and WILSON, A. A. 1975. 1:50 000 New Series Geological Sheet 66 (Blackpool). Institute of Geological Sciences.

GLENNIE, K. W. 1970. *Desert Sedimentary Environments.* Elsevier, Amsterdam.

—— and EVANS, G. 1976. A reconnaissance of the Recent sediments of the Ranns of Kutch, India. Sedimentology. **23**, 625–47.

GOODCHILD, J. G. 1893. Observations on the New Red Series of Cumberland and Westmorland, with especial reference to classification. *Trans. Cumb. Westrm. Assoc.* **17**, 1–24.

GREGORY, J. W. 1915. A deep bore at Seascale in Cumberland. *Geol. Mag.* **52**, 146–49.

HARKNESS, R. 1862. On the sandstones and their associated deposits in the Vale of Eden, the Cumberland Plain and the south-east of Dumfriesshire. *Q. Jl geol. Soc. Lond.* **18**, 205–18.

HICKLING, G. A. 1909. British Permian footprints. *Mem. Proc. Manchr. lit. phil. Soc.* **53**, 1–30.

HOLLINGWORTH, S. E. 1942. Correlation of gypsum-anhydrite deposits in the north of England. *Proc. geol. Ass.* **53**, 141–51.

HOLMES, A. and HARWOOD, H. F. 1928. The age and composition of the Whin Sill and the related dykes of the north of England. *Mineralog. Mag.* **21**, 493–542.

HOLMES, T. V. 1881. The Permian, Triassic and Liassic rocks of the Carlisle Basin. *Q. Jl geol. Soc. Lond.* **37**, 286–98.

KENDALL, P. F. 1902. On the brockrams of the Vale of Eden, and the evidence they afford of an inter-Permian movement of the Pennine Faults. *Geol. Mag.* **39**, 510–13.

MEYER, H. O. A. 1965. Revision of the stratigraphy of the Permian evaporites and associated strata in north-western England. *Proc. Yorks. geol. Soc.* **35**, 71–89.

MURCHISON, R. I. and HARKNESS, R. 1864. On the Permian rocks of the north-west of England, and their extension into Scotland. *Q. Jl geol. Soc. Lond.*, **20**, 144–65.

MYKURA, W. 1965. The age of the lower part of the New Red Sandstone in south-west Scotland. *Scott. J. Geol.* **1**, 9–18.

PATTISON, J. 1970. A review of the marine fossils from the Upper Permian rocks of northern Ireland and north-west England. *Bull. geol. Surv. Gt. Br.* **32**, 123–65.

—— SMITH, D. B. and WARRINGTON, G. 1973. A review of late Permian biostratigraphy in the British Isles. *In* LOGAN, A. V. and MILLS, L. V., eds. *Mem. Can. Soc. Petrol. Geol.* **2**, 220–60.

RHYS, G. H. (Compiler) 1974. A proposed standard lithostratigraphic nomenclature for the southern North Sea and an outline structural nomenclature for the whole of the (UK) North Sea. *Rep. No. 74/8, Inst. Geol. Sci.* 14 pp.

ROSE, W. C. C. and DUNHAM, K. C. In press. Geology and hematite deposits of South Cumbria. *Mem. Geol. Surv. U.K.*

SEDGWICK, A. 1832. On the deposits overlying the Carboniferous series in the valley of the Eden, and on the north-western coasts of Cumberland and Lancashire. *Proc. geol. Soc. Lond.* **1**, 343–45.

—— 1836. On the New Red Sandstone Series in the basin of the Eden, and north-western coasts of Cumberland and Lancashire. *Trans. geol. Soc. Lond.* **4**, 383–407.

SHERLOCK, R. L. and HOLLINGWORTH, S. E. 1938. Gypsum and anhydrite; celestine and strontianite, 3rd Ed. *Mem. geol. Surv. U.K.* v + 98 pp.

SMITH, B. 1924. On the west Cumberland brockram and associated rocks. *Geol. Mag.* **61,** 289–308.

SMITH, D. B. 1970a. The Palaeogeography of the British Zechstein. *In* RAU, J. L. and DELLWIG, L. F. eds. Proc. 3rd Symp. on Salt. *Northern Ohio Geological Society, Cleveland, Ohio.* **1,** 20–23.

—— 1970b. Permian and Trias. *In* HICKLING, G. ed. Geology of Durham County. *Trans. Nat. Hist. Soc. Northumb.* **41,** 66–91.

—— 1971. The stratigraphy of the Upper Magnesian Limestone in Durham: a revision based on the Institute's Seaham Borehole. *Rep. No. 71/3, Inst. geol. Sci.* 12 pp.

—— 1972. The Lower Permian in the British Isles. *In* FALKE, H., ed. Rotliegende essays in European Lower Permian. *Brill, Leiden.* 1–33.

—— BRUNSTROM, R. G. W., MANNING, P. E., SIMPSON, S., and SHOTTON, F. W. 1974. A correlation of Permian rocks in the British Isles. *Special Reports of Geological Society of London,* No. 5, 45 pp.

SMITH, G. V. 1884. On further discoveries of the footprints of vertebrate animals in the Lower New Red Sandstone of Penrith. *Q. Jl geol. Soc. Lond.* **40,** 479–81.

STONELEY, H. M. 1958. The Upper Permian flora of England. *Bull. Br. Mus. nat. Hist.; Geol.* **3,** 295–337.

TAYLOR, B. J. 1961. The stratigraphy of exploratory boreholes in the west Cumberland coalfield. *Bull. geol. Surv. Gt Br.* **17,** 1–74.

TROTTER, F. M. 1939. Reddened Carboniferous beds in the Carlisle Basin and Edenside. *Geol. Mag.* **76,** 408–16.

—— 1953. Reddened beds of Carboniferous age in north-west England and their origin. *Proc. Yorks. geol. Soc.* **29,** 1–20.

—— and HOLLINGWORTH, S. E. 1932. The geology of the Brampton district. *Mem. geol. Surv. U.K.* xviii + 223 pp.

—— —— EASTWOOD, T. and ROSE, W. C. C. 1937. Gosforth District. *Mem. geol. Surv. U.K.* xii + 136 pp.

VERSEY, H. C. 1939. The Petrography of the Permian rocks of the southern part of the Vale of Eden. *Q. Jl geol. Soc. Lond.* **95,** 275–94.

VISSCHER, H. 1971. The Permian and Triassic of the Kingscourt outlier, Ireland. *Spec. Pap. geol. Surv. Irel.* **1,** 114 pp.

WALKER, T. R. 1976. Diagenetic origin of continental red beds. *In* FALKE, H., ed. The continental Permian in Central, west and south Europe, *D. Reidel Publishing Co., Dordrecht.* 240–82.

WARRINGTON, G. 1970. The stratigraphy and palaeontology of the 'Keuper' Series of the central Midlands of England. *Q. Jl geol. Soc. Lond.* **126,** 183–223.

WAUGH, B. 1970a. Petrology, provenance and silica diagenesis of the Penrith Sandstone (Lower Permian) of north-west England. *J. Sedim Petrol.* **40,** 1226–40.

—— 1970b. Formation of quartz overgrowths in the Penrith Sandstone (Lower Permian) of north-west England as revealed by scanning electron microscopy. *Sedimentology,* **14,** 309–20.

—— 1973. The distribution and formation of Permian-Triassic Red Beds. *In* LOGAN, A. V. and MILLS, L. V., eds. The Permian and Triassic Systems and their mutual boundary. *Mem. Can. Soc. Petrol. Geol.* **2,** 678–93.

R. S. ARTHINGTON, B.Sc., I. C. BURGESS, B.Sc., and D. W. HOLLIDAY, Ph.D.
Institute of Geological Sciences, Ring Road, Halton, Leeds LS15 8TQ.

14

Quaternary Geology

WINIFRED PENNINGTON (Mrs T. G. TUTIN)

In the mountains and valleys of the Lake District proper, the most striking changes brought about during the 2 Ma of Quaternary time have been caused by erosion. The dome of Palaeozoic rocks raised during Tertiary time appears to have developed a pattern of radial drainage partly independent of the structure of the present bedrock — a pattern which was then emphasized by the effects of the outward flow of ice during repeated episodes of glaciation in the later part of the Pleistocene. The present major lakes occupy rock basins produced by glaciation in the valleys which radiate from the central mountains. On the surrounding lowlands glacial deposition has been the strongest influence in formation of the present land surface. The superficial Quaternary deposits consist almost entirely of the products of the most recent (Devensian) glacial episode, and most probably date from the period of maximum advance of Devensian ice, c. 25,000–15,000 BP.

The considerable thicknesses of Devensian till present all around the Cumbrian mountains, and the thin sheets of drift on the mountainsides, appear from both geomorphological and lithological evidence to represent deposition from an independent and outwardly moving Lake District ice cap. The map showing directions of movement of the ice given by Marr (1916, Fig. 29) was based on the direction of glacial striae and the relation to parent sources of glacial erratics, as observed by himself and the older workers including Clifton Ward (1873, 1874). The conclusions illustrated by this map have not been seriously challenged by later work. There is abundant geomorphological evidence as to the relationship between Lake District ice and the flood of ice which was moving southwards from Scotland, filling the Irish Sea basin and diverting the outward flow of Lake District ice (e.g. Hollingworth 1931).

The land forms of the central mountains include most of those characteristic of highly glaciated uplands; well developed corries, occasional arêtes, overdeepened valleys with rock basins, flat-floored glacial troughs with tributary valleys hanging up to 250 m above the floor, striated surfaces of glacially modified bedrock and roches moutonnées, rock-cut channels unrelated to the present drainage pattern, and an abundance of steep crags formed by frost-shattering of hard and well-jointed bedrock (Marr 1916). At the same time, the existence, particularly on west-facing slopes, of smooth hillslopes carrying a continuous mantle of vegetation shows that the most intense forms of glacial erosion must have been somewhat localized during the Devensian glaciation (Hollingworth 1951). Mitchell (1931) drew attention to the contrast in the eastern part of the Lake District between the severely glaciated valleys of Mardale, Swindale, Kentmere and Troutbeck and valleys such as Mosedale (the upper continuation of Swindale) which are scarcely modified from their pre-glacial form.

The land forms of glacial deposition in the lowlands include both those formed sub-glacially, such as drumlins and eskers, and those indicative of ablation *in situ*. At a few places there are exposures showing a gradual upward passage from bedrock (usually Triassic sandstone) into boulder clay of similar composition which is interpreted as a compact lodgement till formed at the base of an ice-sheet. Most of the boulder clay is of more assorted composition and is interpreted as the residue of melting, detritus-charged ice. The freshest morainic topography, found in the innermost heads of the main dales and in the corries (Manley 1959) has been firmly dated by investigation of associated lake sediments to the period of Younger Dryas time and the Loch Lomond Readvance in Scotland (Sissons 1974); that is to *c.* 11,000–10,000 BP (Pennington 1947, 1970, 1977).

Interglacial deposits

No trace has been found in Cumbria of Pleistocene deposits older than the later Pleistocene glaciations, and it may be supposed that any sediments of Lower or Middle Pleistocene age are likely to have been removed by severe erosion during later glaciations. It seems probable that northern England suffered, during Lower Pleistocene time, climatic vicissitudes contemporaneous with those revealed by biostratigraphical studies in East Anglia (West 1968), where deciduous forest was repeatedly replaced by a more cold-tolerant vegetation of dwarf-shrub heath or tundra, and that subsequent glaciations of northern England were separated by temperate episodes of sufficient length to compass the complete interglacial cycles known from Hoxnian and Ipswichian times. No organic deposits of possible Hoxnian type are known from Cumbria, but three instances of possible Ipswichian deposits are known from boreholes.

i. Dalton-in-Furness [247 746]. More than a century ago, organic deposits beneath till were found in Low Furness during mining operations to extract haematite from Carboniferous limestone. Kendall (1881) commenting on earlier descriptions of the plant remains recovered from beneath nearly 30 m of boulder clay (Bolton 1862, Hodgson 1862) pointed out their probable interglacial origin. Unfortunately the techniques of investigation then available did not produce any critical plant identifications from this 'black muck' discovered beneath the till in boreholes near Lindal and Crossgates. The question therefore remains as to whether the black muck represents an Early or Middle Devensian interstadial, dating from before the ice advance of the Late Devensian, or an interglacial. Subsequent decay of the disused mines has left the organic deposits inaccessible, and no samples. If the record of Beech (*Fagus*) 'leaves and fruit receptacle' made by Bolton (1862) from the black muck is accepted,

facing page

PLATE 12. A. The U shaped glacial valley of Boredale, east of Ullswater. [425 182]. Place Fell is on the right.

B. In the foreground, Esthwaite Water and drumlins [36 96]. In the background, the Coniston and Langdale Fells. A Late-Devensian late-glacial sequence similar to that described from Low Wray Bay, Windermere (Table 5) is present in the bay behind the drumlin (Strickland Ees) which projects from the far side of the lake.

Photographs: A. *F. Moseley*
B. *C. H. Wood, Bradford*

A

B

PLATE 13. Nethermost Pike and the Helvellyn range forms the background with Eagle Crag in the foreground [335 140]. The Eagle Crag lead vein can be traced from the Pike to spoil heaps at the base of Eagle Crag. It traverses gently dipping Borrowdale volcanic rocks (mostly tuff), with the bedding difficult to see, but just visible in the snow covered areas. Nethermost Cove, to the right, shows immature corrie development, with a hanging valley.   *Photograph: F. Moseley*

then the deposit is clearly interglacial. In Mitchell et al. (1973) Evans and Arthurton tentatively assign it to the Ipswichian interglacial, and the patchy thin till reported from below it to the Wolstonian.

    ii. Wigton [253 487]. A similar age is suggested by the same authors for the marine clay containing *Turritella communis*, foraminifera and ostracods which was found at the base of the drift in a borehole near Wigton.

    iii. Near Appleby a tooth of ox, indicative of non-glacial climate, was found in sands beneath boulder clay of the Main Glaciation (Dakyns, Tiddeman and Goodchild 1897).

GLACIATION

Chronology

Wolstonian

    Since the most recent Main Glaciation of the Lake District is firmly correlated by radiocarbon dating with the Devensian glacial stage, equivalent to the Weichselian of northern Europe (Mitchell et al. 1973), older glacial deposits are likely to be of Wolstonian age. Trotter and Hollingworth (1932) described from the Irthing valley [622 661] a basal Boulder Clay which, though impossible to date with certainty, appears from its position and lithology to have been deposited by an earlier ice sheet than that of the most recent Main Glaciation of the Lake District. If this belonged to a previous glacial stage it would now be referred to the Wolstonian.

Devensian

    Shotton (in Mitchell et al. 1973) suggested the date of 26,000 BP for the lower boundary of the Upper (or Late) Devensian stage, and the upper boundary falls at 10,000 BP which is the beginning of the Flandrian (post-glacial). Evans and Arthurton (op. cit.) placed the whole of the glacial deposits of Cumbria, except for the above-mentioned basal boulder clay, within the Late Devensian. Glacial erosion, however, must have been cumulative from the beginning of Pleistocene glaciations, and there is no way of establishing with certainty how much of the rock erosion of the Cumbrian mountains was the work of glaciations earlier than the Devensian.

Late Devensian, late-glacial sub-stage

    The lowest organic sediments, both in a kettlehole in the surface of Devensian drift and in the lake deposits of Windermere, have been dated to *c.* 14,500 years BP (Pennington 1977). The period 15,000–10,000 BP is commonly taken to constitute the late-glacial sub-stage of the Late Devensian glacial stage (Pennington 1977 and Table 5).

Glacial erosion

    The classical effects of glaciation in forming the mountain scenery of the Lake District were described and illustrated by Marr (1916). He explained how the bold jointing of the Borrowdale Volcanics responded to glacial erosion in such a way as to produce the most rugged mountain landscapes, those characteristic of the central fells round the heads of Ennerdale, Wasdale, Dunnerdale, Langdale and Borrowdale which are so well known and widely illustrated.

Corries

Corries are found at altitudes from 300 to 850 m (Fig. 68) and many contain corrie tarns, ranging in depth from 7 m [Blind Tarn, Coniston; 263 967] to 63 m [Blea Water; 448 108]. Manley (1959) discussed corrie morphology, and the importance of snow drifting leading to accumulation of ice in pre-formed corries during subsequent episodes of glaciation. As in other parts of Highland Britain (Seddon 1957, Sissons 1967) there are many more corries facing north or north-east than south or west. The importance of this orientation with respect to snow drift and prevailing wind, and to protection from snow melt, has been discussed by many authors (e.g. Temple 1956). In attempts to reconstruct palaeoclimate from corrie morphology, however, it has to be remembered that 'the cirques that we now see are mostly the products of not one but several glaciations of differing magnitude . . . a cirque may be composite in the sense that it may have been reoccupied and deepened in successive glacial episodes' (Embleton and King 1968). The development of corries in the Lake District may be locally very irregular, as seen in the contrast between the classical corrie of Blea Water (Lewis 1960) and the head of the adjacent and similarly NE facing valley of Riggindale which has practically no corrie development.

The absence of well-developed corries from most of the Skiddaw Slate mountains was related by Marr to the lithology. Scales Tarn [392 282] and Bowscale Tarn [336 313], both facing east to north-east on the north-east slopes of the Saddleback massif, are the only true corries in the Skiddaw Slate. In the Silurian fells the absence of corries is probably related to the generally low altitudes rather than to lithology. In the Howgill Fells, where there appears to have been a small local ice-cap of low erosional power, Cautley Crags enclose a corrie-like hollow so orientated that its floor is unusually shaded [685 973].

The final episode in the history of the corries and the deposition of corrie moraines will be discussed later.

Valley erosion: lakes

The classically flat-floored and steep-sided valleys of glacially eroded mountain regions are highly characteristic of the Lake District, and the inner parts of all the major dales and of their larger tributaries are of this pattern. The 13 major lakes occupy rock basins in the valley floors, and the depth of water is increased by the presence of morainic material at their lower ends. In Windermere (altitude 40 m) Echo-sounding of the sediments has suggested a total thickness of *c.* 21 m in the North Basin (maximum water depth 66.8 m) and *c.* 40 m in the South Basin (maximum water depth 42 m) (Howell 1971). The highly irregular rock floor of this lake is therefore 34–37 m below present sea level. Intermediate in size between the major lakes of the valleys and the corrie tarns is a group of upland lakes, e.g. Burnmoor Tarn [184 004], which occupy comparatively shallow hollows which have been scooped by the ice in saddles and plateaux, and increased in depth by morainic deposits on the lip.

Channel erosion

Both within the mountains and on the surrounding lower ground there are deep channels incised into the rock which do not relate to the present drainage pattern. Until recently these have been interpreted as cut by surface (subaerial) meltwater —

drainage channels either marginal to the ice-sheet, or spilling over from temporary ice-dammed lakes (Trotter 1929, Hollingworth 1931, B. Smith 1932). There are many channels, now practically dry, incised into the bedrock of S.W. Cumbria and of the western Pennines on the flanks of the Vale of Eden [e.g. Great Comb on Renwick Fell, 595 455]. Recent re-interpretations have identified most of these channels as more probably the results of sub-glacial drainage, during periods of rapid flow beneath the ice, of streams whose flow was laterally confined along alignments imposed by the ice-sheet (Arthurton and Wadge in preparation, R. A. Smith 1967).

The dendritic pattern of some systems of linked channels in north-eastern and south-western Cumbria implies a system of sub-glacial drains of considerable lateral and vertical extent. Some of the channels from the Pennine escarpment fall c. 300 m to the Vale of Eden, implying that the thickness of ice was at least of this order (Arthurton and Wadge in preparation). Huddart (1967) re-interpreted the Nannycatch system of channels as formed by sub-glacial streams during a period when the ice was thick, and R. A. Smith (1967) has shown how channels cut in the Eskdale Granite of Muncaster [115 985] and Waberthwaite [140 935] Fells can be interpreted as subglacial, rather than as the results of the escape of surface waters from glacier lakes impounded in Wasdale and Eskdale. Within these channel systems are examples of the type of channel which includes sections within which the gradient is reversed; these are now interpreted as the consequence of the movement of water within an ice-sheet under hydrostatic pressure. There are some good examples of such 'humped' channels between the rivers Petteril and Eden (Arthurton and Wadge in preparation).

Glacial deposition

Deposits of the Late Devensian glaciation mantle much of the Cumbrian lowland in till, interstratified in many places with laminated clay, silts and sands. The sequence has been divided by many workers into Lower Boulder Clay at the base, overlain by respectively Middle Sands, Upper Boulder Clay and Upper Sands (e.g. Eastwood et al. 1931), but no type sites have been designated. Evans and Arthurton (in Mitchell et al. 1973) take the view that 'current work suggests that this classification is over-simplified.'

Erratics

These include the high-level erratic boulders now interpreted as the products of the higher and cleaner parts of the Late Devensian ice-sheet. There are boulders of Ennerdale Granophyre on the Skiddaw Slate summit of Starling Dodd [142 158] at just over 650 m, and boulders of Borrowdale Volcanic rocks on the Skiddaw Slate summit of Maiden Moor [240 182] at c. 620 m (Clifton Ward 1876). The absence from the Borrowdale Volcanic mountains of any erratics of either Skiddaw Slate or Silurian rocks was regarded as 'sufficient to prove a general outward movement of ice from the central part of the district' (Marr 1916).

Drumlins

Drumlin fields are found south, east, north and north-west of the Cumbrian mountains, where outflowing ice impinged on flatter ground, and there are drumlins in the outer parts of certain of the radiating valleys, including Borrowdale, Windermere and Esthwaite. The orientation of the drumlins agrees with other evidence for the direction of ice movement: for drumlin maps see King 1976, figures 4.6 and 7.1. In the Vale of Eden, drumlins elongated from SSE (stoss ends) to NNW (tail ends) represent the deposits of a period when ice was moving NNW down the Vale. If it is assumed that the ice was continuous from the Lake District mountains into the Vale, the drumlin fields near Penrith must have formed beneath about 1,000 feet of ice. To the north-east of the mountains, both drumlin orientation and the evidence from erratics indicate how ice pouring out of the Lake District was eventually able to push anticlockwise, round the mountains from the Vale of Eden and over the lowland of north Cumberland to the Solway (Hollingworth 1931).

In the Windermere area, the drumlins beside the lake and in the neighbouring valleys represent the deposits of a period when thick ice, confined and directed by the valley sides, was moving southwards; near the head of the present lake a branch stream of drumlins indicates the diversion of some ice through the cross valley which now contains Blelham Tarn, into the Esthwaite valley near Hawkshead [352 980] (Gresswell 1952). South of this point the Esthwaite drumlins follow the orientation of this valley and form conspicuous promontories in Esthwaite Water (Plate 12).

Boulder Clay

Much of the till, laminated clays, silts and sands, the 'Boulder clay/sand complex' of Evans and Arthurton (in Mitchell et al. 1973), represents englacial or supraglacial debris, constituting the accumulated residues from melting, detritus-charged ice. The stratified deposits, commonly present in complex association with the boulder clay, were probably formed subglacially or englacially, though local bodies of surface water may have been present periodically. Contortions and faulting which are recorded in stratified deposits in some sections may have resulted either from subsequent collapse due to ice melting, or from subsequent overriding of the deposits by ice (Arthurton and Wadge in preparation).

In the lowlands of north and west Cumberland there is some evidence that the 'Upper Boulder Clay' is of different composition from lower deposits, and contains more material of Scottish origin. Evans and Arthurton (op. cit.) found insufficient evidence to support the earlier views of Trotter (1929) and Trotter and Hollingworth (1932) that the Upper Boulder Clay is everywhere, in north and north-west Cumberland, the product of a later and distinct readvance of Scottish ice. The deposits at St Bees described by Eastwood et al. (1931) with supposedly interglacial organic deposits between a Lower and an Upper Boulder Clay have disappeared and the organic deposit now exposed in the cliff south of St Bees is undoubtedly of Late-Glacial age. Huddart (1972) analysed drift sections in the low cliffs of the coast west of Black Combe, and described from Annaside Banks [090 855] and Gutterby [010 843] a complex in which two tills are separated by up to 8.5 m of sandy clay including laminated deposits. The upper till contains fragments of marine shells and its lithological assemblage includes a percentage of Scottish erratics. In Huddart's view the lower till represents the deposit of the Main Glaciation, and the overlying deposits the products of a later oscillation of the ice in the northern Irish Sea.

Sand and gravel; eskers, kames and sediments of glacial lakes

Eskers are often found associated with systems of sub-glacial drainage channels in the marginal lowlands of west Cumberland and the Vale of Eden, and are now interpreted as the infilled portions of such systems. Other eskers, particularly on the plain of north Cumberland, have been interpreted as deltaic deposits formed at the margin of a retreating ice-sheet (e.g. Dixon et al. 1926). Kames are abundant in the areas where the temporary existence of glacial lakes has been demonstrated. There are many in west Cumberland in the area where glacier-lakes Loweswater, Pardshaw, Ennerdale, Wasdale and Eskdale were formed during stages in deglaciation (B. Smith 1932, Eastwood et al. 1931). The delta deposits of streams which entered Glacier-lake Ennerdale are conspicuous on the slopes of Herdus north of the present lake [105 165], and many others have been described. Huddart (1967) and R. A. Smith (1967) have suggested that the earlier workers were in error when they believed the ice to have formed an impenetrable barrier so that the glacier lakes overflowed through a series of cols in the solid rock, thereby incising deep channels. It is now thought more likely that the glacier lakes drained into or beneath the ice, and that the series of deltas on the fellsides represent falling lake levels as the ice melted downwards. Sometimes, in kame deposits accumulated in partially ice-confined lakes, the level top of the kame terraces can be related to the existence of pre-existing drainage channels through which the lake waters flowed out, for example in the valley of Croglin Water, tributary to the Eden (Arthurton and Wadge in preparation).

Deep-water sediments of glacial lakes were described from the sites of Glacier-lakes Loweswater to Eskdale in west Cumberland; they consist of non-fossiliferous laminated clays and silts which reach considerable depths in lower Ennerdale and Wasdale. Many of the laminated clays, both in west Cumberland and in the Vale of Eden, do not show the graded bedding of true waterlain varves.

Moraines

During the 60 years since publication of Marr's book there has been much discussion of the geomorphology of the abundant ridges and hummocks of drift. Other authors (e.g. Hay 1944, Gresswell 1952) followed Marr in attempts to interpret the ridges as terminal, lateral or medial moraines of a postulated stage of valley glaciers. More recent interpretations have followed the lead of Hollingworth (1951) in questioning whether the deglaciation of the Lake District did in fact involve significant periods during which mountain snowfields were in equilibrium with valley glaciers. Huddart (1967, 1971) and others have recently emphasized the change in interpretation which must result from acceptance of the concept that deglaciation was rapid, involving mainly ablation and downmelting *in situ*. Evans and Arthurton (in Mitchell et al. 1973) have raised doubts as to the reality of a distinct 'Scottish Readvance' of a retreating ice-front, and this removes the need to postulate any corresponding standstill phase, with frontal moraines, of the Lake District ice. Recent work has demonstrated the comparatively short time spanned by the advance and retreat of the Late Devensian ice (Mitchell et al. 1973), and this has supported the new concept of a rapid climatic amelioration and decay *in situ* of the ice, rather than the long period of orderly retreat of an ice front which was visualized by the earlier workers.

The sites of the major lakes are now known to have been free of ice since about 14,500 BP (Pennington 1977). There is no biostratigraphical evidence from Cumbria for any oscillation involving readvance or recrudescence of glaciation except that of Younger Dryas time (11,000–10,000 BP) and this agrees with recent interpretations from the fauna of ocean cores from the North Atlantic (Ruddiman, Sancetta and McIntyre 1977). The absence of frontal moraines from the floor of the main valleys, except for the controversial ridge at Rosthwaite [259 148] discussed by Hay (1944), agrees with the concept of rapid decay *in situ* of the residual ice of the Main Glaciation.

The hummocky moraines found in the heads of the main valleys, both on valley floors and in the corries, and the arcuate moraines which dam the corrie tarns, have now been firmly correlated with Younger Dryas time and so with the Loch Lomond Readvance in Scotland (Sissons 1974).

THE LATE-GLACIAL PERIOD, c. 15,000 – 10,000 BP

The Windermere section; the Windermere Interstadial

Table 5 represents the sedimentary sequence from the littoral region of Windermere, as established from numerous cores taken by the Freshwater Biological Association (Pennington 1943, 1947, 1977 and unpublished). Fifteen ^{14}C dates have now been obtained from a core of five inch diameter taken from Low Wray Bay [376 013] (Pennington 1977). The section illustrated has been related by biostratigraphy and chronostratigraphy to 'pollen zones I, II and III' of the final part of the Late Devensian (Mitchell et al. 1973) and to Late Weichselian chronostratigraphy (Mangerud et al. 1974). This constitutes the type site for the Windermere Interstadial of the Late Devensian (Coope and Pennington 1977) and the other late-glacial deposits of Cumbria will be described with reference to this chronology.

The Lower Laminated Clay

The oldest deposits obtained from Windermere are cobble gravels and finer gravels in clay; these pass upwards through sandy layers into the Lower Laminated Clay, in which alternating laminations of silt and clay decrease upwards in thickness from about 2 cm to about 0.3 cm. By comparison with the sediment sequence in Nigardsvatn, S. Norway (Østrem 1975), (a lake which came into existence in 1937 by the retreat of the snout of Nigardsbre, a valley glacier originating in the Jostedalsbre plateau ice-cap) the transition from gravel to a regularly laminated sediment in Windermere is interpreted as marking the retreat of active ice from the lake basin. The paired laminations of silt and clay are interpreted, again by comparison with the record in Nigardsvatn, as the deposits of the period during which valley glaciers were maintained by the declining ice-cap on the Lake District mountains. In littoral areas such as Low Wray Bay where the full sequence of sediments has been sampled by corers, the paired laminations, if annual, represent the deposits of only a few centuries; however in the parts of the lake where the water is more than 30 m deep, the 6 m cores so far obtained represent only a fraction of the total depth of sediment indicated by echo-sounding (21 to 40 m Howell 1971), and the Lower Laminated Clay is assumed to be much thicker in deep water than in the littoral areas.

Coope and Pennington (1977) interpreted the upper boundary of the Lower Laminated Clay as representative of the termination of the main Devensian glaciation of north-west England and the beginning of interstadial sedimentation (Table 5).

Interstadial sediments

The Lower Laminated Clay passes upwards through a thin unlaminated barren clay into the fine grey to grey-brown organic (fossiliferous) silts which represent the biostratigraphical interstadial. These were originally correlated with pollen zone II

(Pennington 1947) but have been shown subsequently, by more detailed pollen analysis and ^{14}C dating, to represent the deposits of a much longer period than Alleröd time, and to include pollen zones which do not correlate with the I, II & III sequence used in south-eastern Britain (Pennington 1975a, 1977). The horizon at which microfossils first appear is believed to fall at about 14,000 BP, from backward projection of the time-scale based on younger dates (Pennington 1977, Fig. 5), since the δ ^{13}C values for some of the lowest ^{14}C dates suggest some hardwater error. It is however not possible as yet to quantify this error.

In the deposits of the period 14,000–13,000, the pollen spectra resemble those found in surface samples (lake sediments and moss polsters) from among plant communities of snowbed and pioneer vegetation near the glaciers of Jostedalsbre and Jotunheim, S. Norway, but the insect fauna is made up of species now found only in temperate climates (Pennington 1977, Coope 1977). The interpretation is that at this time summer temperatures round Windermere must have been appreciably higher than in arctic environments today, but on the plant evidence some feature of the climate delayed processes of soil maturation and kept even the lowland vegetation in a comparatively pioneer stage. On palaeoentomological grounds the interstadial begins almost immediately after the end of deposition of the Lower Laminated Clay, but on the palaeobotanical definition of 'interstadial', which has hitherto required a 'woodland biozone' with good representation of the pollen of trees and shrubs, the interstadial begins at $c.$ 13,000 with the sudden expansion of juniper pollen, which is followed a few centuries later by the expansion of tree birch pollen (Table 5).

The Windermere Interstadial, beginning at $c.$ 14,000 or $c.$ 13,000 (according to the definition adopted) and lasting until 11,000 BP, is divided at a horizon dated by radiocarbon to $c.$ 12,000, by a small oscillation in lithostratigraphy and biostratigraphy which corresponds in time with the Bölling-Alleröd stade (zone Ic) of the Late Weichselian. Some environmental change at this date, seemingly a climatic deterioration, brought about both sufficient increase in soil movements to introduce small angular rock fragments into the littoral sediments of Windermere, and a change to a more cold-demanding insect fauna (Coope 1977, p. 329). However the effect on vegetation appears to have been small.

The deposits of Alleröd time (11,800–11,000 (Mangerud et al. 1974)) contain both insect species indicative of lower temperatures than during the period $c.$ 13,000–12,000, and chemical evidence for increasing soil erosion. The representation of tree birches, particularly with respect to fruits and catkin-scales, remains high though fluctuating. Comparison of evidence from many sites in western Britain has led to the conclusion that representation of tree birches can no longer be used as an index to late-glacial temperatures in this region (Pennington 1977). Figure 1 in Pennington 1975a compares this sequence with pollen zones I and II of the Late Weichselian sequence. The early onset of interstadial conditions in western Britain correlates with the faunal evidence of warming in the eastern North Atlantic dated to $c.$ 13,500 by Ruddiman and McIntyre (1976) and Ruddiman et al. (1977), and with the faunal warming shown by the *Zirphaea* beds in N. Denmark which is now used to define the base of the Bölling chronozone at 13,000 BP (Mangerud et al. 1974).

The Upper Laminated Clay

This deposit overlies conformably the interstadial sediments in the Windermere section and contains about 400 paired varves, within some 50 cm thickness of clay.

LITHOLOGY	FOSSILS: PLANTS	FOSSILS: ANIMALS: INSECTS	FOSSILS: ANIMALS: OTHERS	WINDERMERE C-14 DATES (ALL BP)
Rock fragments increase. ↑ UPPER LAMINATED CLAY	*Artemisia* PAZ Pollen very sparse	Sparse fauna + characteristic arctic alpine species		(11000)
Becomes less organic ↑ Organic detritus silt	Tree birch macroscopics; Cyperaceae; *Betula*-*Juniperus* PAZ	Woodland species present; Fauna less rich without the southern spp	Cladocera (Decline to zero) ↑; *Cristatella* statoblasts	11344 ± 90
Paler silt Rock fragments	*Betula*–*Rumex* PAZ	*Helophorus glacialis*		12213 ± 150
Organic detritus silt	*Betula* PAZ; *Juniperus* PAZ	Sparse fauna at base, more diverse & abundant towards top. Many species absent from modern arctic Europe.	Cladocera (Sudden increase) ↓	12112 ± 125; 12517 ± 150; 12441 ± 82; 12913 ± 120; (13000)
Somewhat organic silt; ↑ Carbon increasing	*Rumex*-Gramineae PAZ (Tree birches present)			13185 ± 170; 13938 ± 210*; 13863±270*
Slightly organic silt	*Salix herbacea*-Cyperaceae-*Lycopodium selago* PAZ	Very sparse fauna, no alpine sp.	*Cristatella* statoblasts ↓	14557±280; 14623±360
Clay	Moss stems	Chironomid larvae and Trichoptera		
LOWER LAMINATED CLAY				

TABLE 5

The Windermere Interstadial; correlation table for the Low Wray Bay, Windermere section. (Modified from Coope and Pennington 1977; *Phil. Trans. R. Soc.*) Double lines in column 2 represent the limits of the interstadial on palaeobotanical criteria. Asterisks against radiocarbon dates indicate those for which

[*continued opposite*

INTERPRETATION					
ENVIRONMENT		CLIMATE		DURATION OF WINDERMERE INTERSTADIAL	
Seasonal melt from mountain glaciers		Very cold			
Increased snowfall Declining temperatures Increased soil erosion — Re-establishment of *Betula - Juniperus* woods		Cool temperate becoming colder towards the top		Palaeobotanical	Palaeo entomological
Increased soil erosion Woodland reduced		Sudden cooling			
		Temperate throughout			
Pioneering flora & fauna Herbs common including non-arctic species Tree birches present but no woodland developed		Plants: Inconclusive on temperature Unfavourable for trees	Insects: Temperate throughout		
Plants: Compare Mid-alpine - zone of N.W. Europe	Insects: Lack of alpine species suggests early pioneer fauna	Plants: Alpine conditions	Insects: Inconclusive		
Seasonal melt from mountain glacier		Very cold			

the $\delta^{13}C$ values show hardwater error. PAZ = pollen assemblage zone. In the column headed Environment, the blank space represents the first establishment of a woodland biozone, which is not accompanied by any change in the insect fauna.

The varves are therefore very narrow, but show clearly graded bedding, indicating that between *c.* 11,000 and 10,500 the lake must have been turbid with clay and silt produced by ice melt. This indicates recrudescence of glaciation in the high corries of the Windermere catchment during Younger Dryas time, which was confirmed by demonstration that interstadial sediments are absent from two high basins in the catchment Red Tarn, Pike o'Blisco [267 037] at 515 m (Pennington 1964) and Langdale Combe [263 057] at 455 m (Walker 1965). There must have been active ice in these basins in Younger Dryas time. The very low rates of pollen deposition in this varved sediment have been explained by comparison with contemporary low annual rates in a Norwegian lake, where the sediment is similarly varved, where there is very rapid throughput of water from seasonal melt of mountain glaciers (Pennington 1977). The insect fauna of the Upper Laminated Clay includes exclusively arctic-alpine species (Coope 1977). The Upper Laminated Clay passes conformably upwards into post-glacial (Flandrian) organic muds.

Late-glacial deposits of Cumbria

St Bees

G. R. Coope writes: 'The late-glacial deposits exposed on the coast south-west of St Bees, have for long been the focus of geological and palaeontological attention (see Walker 1956 for a summary of earlier investigations). At the moment only one exposure of the late-glacial sequence is available [965 114] and this is almost without doubt Walker's Site 6, the same site from which Pearson (1962) obtained an assemblage of fossil Coleoptera. The fact that the modern exposure is the same as that worked upon by these earlier workers can be established from an unpublished photograph in Pearson's thesis (1960). Walker's pollen analysis showed a typical zone I, II and III sequence. Unfortunately it is not possible to allocate Pearson's records of Coleoptera to the stratigraphy, and a reinvestigation of the beetles from the site was commenced in 1972 by Coope. The results of this new analysis have not yet been published but they may be summarised as follows. In the centre of the exposed depression, grey sand underlies the thick layer of brown, felted detritus mud. This grey sand and the lowest 10 cm of the detritus mud contains an entirely temperate insect fauna including the following species: *Asaphidion cyanicorni, Bembidion octomaculatum, Cymindis angularis, Cleonus piger, Baris peramoena* and *Smicronyx coecus*. No exclusively northern species occur at this level. Sticks washed out of the lowest 5 cm of the detritus mud gave a ^{14}C date of 12,650 $\pm$ 170 BP (Birm 378) thus providing an age free of any imputation of hard water error for the maximum concentration of temperate insect species at this site. All the relatively southern species of Coleoptera are replaced by species of more northern distribution between 10 and 15 cm above the base of the detritus mud, that is, at about 12,200 BP (Williams and Johnson 1976). There is no sign of any significant increase in the temperate species of Coleoptera during the period allocated to the Alleröd, and the climate apparently remained consistently cool throughout. At about 11,000 BP the insect fauna becomes greatly enriched in arctic-alpine species, and these form an important component of the insect fauna of the largely minerogenic silts and clays that overlie the detritus mud.'

The sediments of existing lakes

In each of the radiating valley lakes whose catchments include mountain corries there is the same sedimentary sequence as in Windermere; an Upper and a Lower

Laminated Clay separated by somewhat organic interstadial silts. The thickness of individual varves of the Upper Laminated Clay varies with the degree of development of corries in the catchment, reaching a maximum in Haweswater (where the Upper Laminated Clay is up to 150 cm thick) but the total number of varves is constant at 400–450 pairs. The existence of unlaminated organic interstadial sediments in Haweswater is interpreted as showing that even in the largest corries, which produced the maximum output of sediment in the subsequent Younger Dryas corrie glaciation, no active ice remained during the Late Devensian (Windermere) Interstadial. The interstadial deposits of Ullswater prove the same point with respect to the eastern corries of the Helvellyn ridge.

In lakes whose catchments do not include corries, the sedimentary sequence differs, in that the post-interstadial deposits are not varved, but consist of unlaminated clays, interpreted as deposited by solifluction during the cold period (c. 11,000–10,400) when the corries were re-occupied by ice. The lithological unit of solifluction clay in these profiles always coincides with a pollen zone characterized by *Artemisia*, which equates with pollen zone III, and indicates that an incomplete vegetation cover of steppe-tundra plants, tolerant of continuous soil disturbance, was present at all altitudes throughout Cumbria during the time when ice re-occupied the mountain corries.

Complete late-glacial profiles, constituting proof that the site was not again occupied by ice after the end of the Main Glaciation, have been found in non-corrie lakes up to altitudes of c. 476 m. At these high sites the vegetation contemporaneous with the woodland of the Windermere Interstadial included no trees or shrubs, but produced pollen spectra similar to those now found among arctic-alpine grass-sedge heaths. Some of the sites occupied by ice (corries) are at lower altitudes than 476 m for example Easedale Tarn [308 087] at 277 m, and Levers Water at 410 m. This emphasizes the importance of drifting of snow into pre-formed corries in determining the sites of glaciation in Younger Dryas time, cf. Manley (1959).

From pollen and geochemical evidence in interstadial silts, at altitudes up to 476 m, it is clear that during the period 14,000–11,000 BP a vegetation cover developed over the whole of Cumbria, up to at least this altitude (1,570 feet), and processes of maturation of the skeletal soils left on deglaciation became well advanced during the interstadial (Pennington 1970).

The corrie glaciation of Younger Dryas time (correlated with the Loch Lomond Readvance) and its moraines

Manley (1959) discussed in detail the distribution in the Lake District of the fresh moraine which Marr (1916) had supposed to represent the deposits of a period of glacial recrudescence. The map (Fig. 68) shows some results of recent re-survey of these moraines by members of the Brathay Exploration Group. It distinguishes between hummocky moraine present both on the innermost part of the floors of main valleys and within corries, and the arcuate moraines which dam corrie tarns. Table 6 presents the data from 17 corrie basins of which the sediments have been investigated; from Langdale Combe (Walker 1965) and from the 16 sites (two infilled basins and 14 corrie tarns) from which sediment cores have been obtained by the Freshwater Biological Association.

At all these sites cores taken in the deepest part of the tarn or former tarn found impenetrable stony clay or gravel at the base of post-glacial mud, in contrast to the

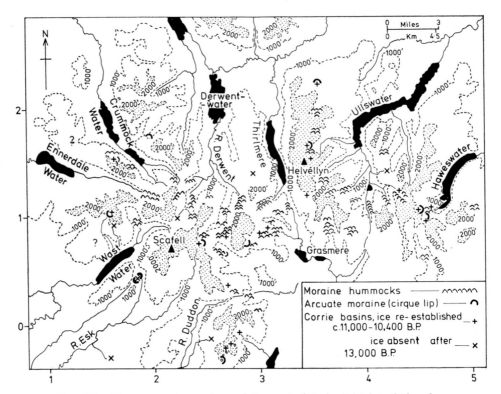

FIG. 68. The youngest moraines of the central Lake District, dating from 11,000–10,000 BP. Map prepared from field maps of the Brathay Exploration Group by A. B. Ware (Principal of Brathay) and G. de Boer; with acknowledgements to G. Manley (Manley 1959; Fig. 2). The corrie and other upland tarns indicated were sampled by the Freshwater Biological Association's coring team.

readily penetrable soft clay (above interstadial sediments) found in non-corrie lakes. From most of the tarns several cores were taken and found to correspond, and from Blea Water seven mutually consistent cores were obtained from different positions in the tarn. Corrie lakes have not yielded any late-glacial sediment except fine gravel in clay, and the major lakes into which they drain accumulated varved sediments during Younger Dryas time, whereas lakes in shallow open basins which lack fresh moraine (even at higher altitudes than some of the corries) contain the full sequence of late-glacial sediments and their Younger Dryas deposits are unlaminated solifluction clays. This correlation together with the radiocarbon dated section from Windermere dates the formation of the most recent moraines to Younger Dryas time.

The basal sediments reached in the 14 corrie basins cored by the Freshwater Biological Association could not be penetrated by a Mackereth corer, which is driven into the sediment with considerable force by compressed air. In many instances the coring tube sustained damage at its lower end indicative of having met an immovable obstacle, either boulders or solid rock. This coarse minerogenic sediment is interpreted

TABLE 6

Artemisia	Gramineae-*Rumex*-*Empetrum*	*Juniperus*			*Betula*
		base	middle	top	
Bowscale Tarn	Blea Water Moss	Angle Tarn			Blea Water
Goatswater	Grisedale Tarn	Bowfell			
Langdale Combe	Levers Water	Easedale Tarn			
Low Water	Red Tarn, Helvellyn		Keppelcove Tarn		
Red Tarn, Pike o'Blisco	Scoat Tarn Wolf Crags Moss			Hayeswater	
				Small Water	

Sites of Younger Dryas glaciation (*c.* 11,000–10,400); pollen zones to which the lowest (earliest) polleniferous sediments belong.

as the detritus from the Younger Dryas ice; at present there seems to be no way to find out whether this coarse sediment in, for instance, Blea Water, overlies interstadial deposits, or whether the Younger Dryas glacier removed all then existing sediments.

Manley (1959) pointed out, and figure 68 illustrates, that whilst in some localities for example Red Tarn, Helvellyn [349 152] there is no fresh moraine below the arcuate ridge, which therefore seems to mark the limit of Younger Dryas ice in that valley, in most of the main valleys there is fresh hummocky moraine on the floor, which is most easily interpreted as the product of decay of glacier tongues which had decended from higher basins at the peak of the Younger Dryas glaciation, and then became cut off from the ice source at an early stage in deglaciation. The size of each glacier at its maximum, and the altitude of the snout, as measured by the areal extent of fresh drift, show no correlation with altitude, but were related by Manley to the amount of high ground available as a source from which snow could be driven by prevailing winds into the corrie basins.

An estimate of the size of each Younger Dryas glacier is provided by the age of the first post-glacial sediments deposited in each basin, on the assumption that the largest glaciers would be the last to melt. The age of the earliest organic sediment cannot be critically determined by ^{14}C because of the low carbon content and inaccessibility to wide-diameter samplers, but a provisional date for the deglaciation of each corrie basin is possible by correlation of pollen zones, assuming that the expansion of *Juniperus* at the base of the post-glacial profile represents a synchronous horizon, i.e. the response of juniper to the major climatic warming (cf. Iversen 1960, Pennington 1977). At sites outside the limits of glaciation the solifluction clay of Younger Dryas age corresponds with an *Artemisia* pollen zone, and at a few sites of Younger Dryas glaciation the lowest recoverable deposits include a few samples which fall within the top of this pollen zone. Their presence shows that the ice melted comparatively early in these basins, before the end of the period of general periglacial soil disturbance indicated by pollen spectra of the *Artemisia* zone. Sites in this group are either peripheral; Goatswater [266 977] and Low Water [275 983] in the Coniston

Fells and Bowscale Tarn in the north-east, or are shallow basins rather than true corries; Langdale Combe and Red Tarn, Pike o'Blisco. In a second group, of six sites, the lowest polleniferous deposits in the cores post-date the *Artemisia* zone and fall within the following Gramineae-*Rumex-Empetrum* zone, cf. Table 6.

A significantly later melt is indicated by the group of five sites, all in deep basins, where polleniferous sediments did not begin to accumulate until after the major climatic warming correlated with the base of the *Juniperus* zone (Table 6) — and by the sediments of Blea Water, where in seven cores no sediments earlier than the forested (Flandrian) period which followed the time of the *Juniperus* zone could be found. This evidence for late melt of the ice in Blea Water is supported by the consistent sequence of sites in Mardale. i Blea Water Moss, the lowest basin of accumulation inside the hummocky moraine left by the Mardale glacier, where organic mud began to accumulate in the Gramineae-*Rumex-Empetrum* pollen zone, ii Small Water, at 452 m, where organic mud did not begin to accumulate until near the top of the *Juniperus* zone, and iii Blea Water in its large corrie at 484 m, in which the beginning of accumulation of sediments was delayed until after the end of the time represented by the *Juniperus* zone.

Table 6 shows the consistent pattern provided by the lowest pollen zone found in each basin; clearly the last to melt were the corrie glaciers on the east sides of the High Street, Helvellyn and Bowfell–Langdales ranges.

The rapid melt at the end of the Younger Dryas glaciation must have led to much redistribution of the products of periglacial erosion, and it is possible that the deposition of valley infill, giving flat stretches at the head of most of the major lakes, and separation of the twin lakes by deltaic deposits, date from this period. It would be expected that the formation of screes must have been active during the period of glacial recrudescence, and much of the existing scree at the foot of crags probably dates from this period.

POST-GLACIAL (FLANDRIAN) DEPOSITS

Head

Evans and Arthurton (in Mitchell et al. 1973) take the view that head deposits may date from any time from Late Devensian to recent. Though the periglacial conditions of Younger Dryas time must have been particularly conducive to the formation of head, some of these deposits must be included in those of post-glacial time.

Peat

The only type of peat known from a late-glacial environment is the lacustrine peat or detritus mud formed in shallow depressions, such as that now sectioned by coastal erosion at St Bees. Formation of this kind of peat in shallow basins continued during early post-glacial time; in many of them an upward progression from open-water mud through reed-swamp (fen) peat to *Sphagnum* peat can be found (for example Scaleby Moss [431 635], Walker 1966). Many lowland basins now contain *Sphagnum* peat of this kind.

From the time of the expansion of alder (*Alnus*) at *c*. 7,500 BP onwards, there has been more widespread formation of non-basin (ombrogenous) peat, both as large raised bogs in the lowlands and as blanket peat in the uplands. Some of the raised bogs have developed on top of lacustrine and fen peats, but the largest raised bogs, those of the Solway plain and the Morecambe Bay estuaries developed on the surface of the marine clay which represents deposition from the marine transgression of the mid-post-glacial period. Great domes of *Sphagnum* peat were built up, from late-Atlantic time onwards, succeeding an initial growth of reed-swamp on the ill-drained clay flats exposed by the falling sea level. Recurrence surfaces with the *Sphagnum* peat which appear to be contemporaneous between, for instance, Foulshaw Moss and Helsington Moss on the Kent estuary (A. G. Smith 1959) have been interpreted as possibly indicative of the sensitivity of the growing peat surface to post-glacial climatic change.

The high relief and consequent good drainage of the central Lake District mountains has limited the development of upland blanket peat, and extensive peat deposits are found only in the marginal areas of gentler slopes, NE, north and west of the central mountains. Some of this upland blanket peat represents lateral extension, during the second half of post-glacial time, of the peat deposits of shallow basins, and some of it has developed on the top of old woodland soils, both mineral soils and organic soils (*mor* humus layers) (Pennington 1975 b).

Lake sediments

In the organic muds of the large and small lakes of the Lake District is recorded the sequence of environmental changes, through the last ten thousand years, which have simultaneously affected soils and vegetation. Parallel investigations of the sediments using pollen (Pennington 1964, 1970) and geochemical analyses (Mackereth 1966) have related the changing composition of the sediments to soil dynamics on the catchments (Pennington and Lishman 1971).

LAND-SEA LEVEL CHANGES

Off the present coast of Cumbria the eustatic rise from low late-glacial sea levels took place early in post-glacial time, laying down marine deposits over peaty layers, containing wood fragments, which had formed before the general expansion of alder (*Alnus*) in pollen zone VIc. The 'Submerged Forests' of the intertidal zone of the Cumbrian coast include some dating from this period, for example that between Mowbray and Allonby on the southern shore of the Solway described by Dixon et al. 1926 (Tooley 1972). Peaty layers found in boreholes in Barrow and Heysham Harbours, and in boreholes in the Leven estuary, have yielded ^{14}C dates between *c.* 9,000 and 8,000 BP; these layers are overlain by marine deposits of the mid-post-glacial marine transgression (Tooley 1974). The culmination of the rise in sea level has been dated by pollen analysis and ^{14}C dating of organic deposits below and above marine deposits, to the early part of zone VIIa (Atlantic time); suggested dates range from 7,000–6,600 BP in the Solway, at *c.* 7 m OD (Walker 1966) to just before 6,000 BP at Silverdale Moss on the southern coast of Morecambe Bay, at *c.* 5 m OD (Oldfield 1960). These differences were interpreted by these authors as indicative of isostatic uplift of northern Cumbria relative to the Morecambe Bay area. At Drigg [046 985], King (1976) obtained ^{14}C dates of 6,720 $\pm$ 120 and 6,200 $\pm$ 140 from a peat immediately below the raised beach shingle which marks the culmination of the marine transgression, at *c.* 7 m OD at this place. At its maximum the sea laid down marine clay, up to 6 m thick, in the inner parts of the estuaries. The top of this clay is now at *c.* 4.5 m in the estuarine plains of Morecambe Bay, and on it have developed the great raised bogs of the estuaries of the Kent, Leven, Duddon and Solway. Round parts of Morecambe Bay a wave-cut notch in the base of cliffs of Carboniferous Limestone, at *c.* 5 m OD, is associated with the maximum of the marine transgression (Oldfield 1960). A later marine transgression has been deduced by Tooley (1972) from borings in the Kirkby Pool area of the Duddon estuary.

REFERENCES

BOLTON, J. 1862. On a deposit with insects, leaves, etc. near Ulverston. *Proc. geol. Soc. Lond.* **18**, 274–77.

COOPE, G. R. 1977. Fossil coleopteran assemblages as sensitive indicators of climatic changes during the Devensian (Last) cold stage. *Phil. Trans. R. Soc.* B **280**, 313–40.

—— G. R. and PENNINGTON, W. 1977. The Windermere Interstadial of the Late-Devensian. *Phil. Trans. R. Soc.* B. **280**, 337–9.

DAKYNS, J. R. TIDDEMAN, R. H. and GOODCHILD, J. G. 1897. The geology of the country between Appleby, Ullswater and Haweswater. *Mem. Geol. Surv. U.K.* Sheet 102 SW.

DIXON, E. E. L., MADEN, J., TROTTER, F. M., HOLLINGWORTH, S. E. and TONKS, L. H. 1926. The geology of the Carlisle, Longtown and Silloth District. *Mem. Geol. Surv. U.K.* Sheets 11, 16 and 17.

EASTWOOD, T., DIXON, E. E. L., HOLLINGWORTH, S. E. and SMITH, B. 1931. Geology of the White-haven and Workington District. *Mem. Geol. Surv. England and Wales.*

EMBLETON, C. and KING, C. A. M. 1968. *Glacial and periglacial geomorphology.* Arnold, London. 608 pp.

EVANS and ARTHURTON. 1973. In Mitchell et al. 1973. A correlation of Quaternary deposits in the British Isles. *Geol. Soc. Lond. Spec. Rep.* **4**, 99.

GRESSWELL, R. K. 1952. The glacial geomorphology of the southeastern part of the Lake District. *L'pool. Manchr. geol. J.* **1**, 57–70.

HAY, T. 1934. The glaciology of the Ullswater area. *Geogrl. J.* **84**, 2, 136–48.

—— 1944. Rosthwaite moraines and other Lakeland notes. *Geogrl. J.* **103**, 3, 119–24.

HODGSON, E. 1862. On a deposit containing Diatomaceae, leaves, etc., in the iron-ore mines near Ulverston. *Q. Jl geol. Lond.* **19**, 19–31.

HOLLINGWORTH, S. E. 1931. The glaciation of western Edenside and adjoining areas and the drumlins of Edenside and the Solway basin. *Q. Jl geol. Soc. Lond.* **87**, 281–359.

—— 1951. The influence of glaciation on the topography of the Lake District. *J. Inst. Wat. Engrs.* **5**, 486–96.

HOWELL, F. T. 1971. A continuous seismic profile survey of Windermere. *Geol. J.* **7**, 2, 329–34.

HUDDART, D. 1967. Deglaciation in the Ennerdale area, a re-interpretation. *Proc. Cumb. geol. Soc.* **3**, 63–75.

—— 1971. A relative glacial chronology from the tills of the Cumberland lowland. *Proc. Cumb. geol. Soc.* **3**, 21–32.

—— 1972. In Huddart, D. and Tooley, M. J. (Editors), *The Cumberland lowland.* Q.R.A. handbook.

IVERSEN, J. 1960. Problems of the early Post-glacial forest development in Denmark. *Danm. geol. Unders.* **4**, 1–32.

KENDALL, J. D. 1881. Interglacial deposits of West Cumberland and North Lancashire. *Q. Jl geol. Soc. Lond.* **37**, 29–39.

KING, C. A. M. 1976. *The geomorphology of the British Isles: Northern England.* Methuen and Co. Ltd. 213 pp.

LEWIS, W. V. 1960. *Norwegian cirque glaciers.* R.G.S. Research Series 4.

MACKERETH, F. J. H. 1966. Some chemical observations on Post-glacial lake sediments. *Phil. Trans. R. Soc. Lond.* B **250**, 165–213.

MANGERUD, J., ANDERSEN, S. T., BERGLUND, B. E. and DONNER, J. 1974. Quaternary stratigraphy of Norden, a proposal for terminology and classification. *Boreas* **3**, 109–28.

MANLEY, G. 1959. The late-glacial climate of north-west England. *L'pool. Manchr. geol. J.* **2**, 188–215.

MARR, J. E. 1916. *Geology of the Lake District.* Cambridge. 220 pp.

MITCHELL, G. H. 1931. The geomorphology of the eastern part of the Lake District. *Proc. L'pool. geol. Soc.* **15**, 322–38.

MITCHELL, G. F., PENNY, L. F., SHOTTON, F. W. and WEST, R. G. 1973. A correlation of Quaternary deposits in the British Isles. *Geol. Soc. Lond. Spec. Rep.* **4**, 99.

OLDFIELD, F. 1960. Late Quaternary changes in climate and vegetation and sea-level in Lowland Lonsdale. *Inst. Brit. Geogr. Trans. and Pap.* Pub. No. 28.

ØSTREM, G. 1975. Sediment transport in glacial meltwater streams. *Glaciofluvial and glaciolacustrine sedimentation.* Sp. Pub. No. 23, 101–22.

PEARSALL, W. H. 1971. *Mountains and Moorlands*. Collins. Revised edition. 312 pp.

—— and PENNINGTON, W. 1973. *The Lake District*. Collins. 320 pp.

PEARSON, R. G. 1960. The Coleoptera from some Late-Quaternary deposits. and their significance for zoogeography. Unpublished Ph.D. thesis, University of Cambridge.

—— 1962. The Coleoptera from a Late-glacial deposit at St Bees, Cumberland. *J. Anim. Ecol.* **31**, 129–50.

PENNINGTON, W. 1943. Lake sediments: The bottom deposits of the North Basin of Windermere, with special reference to the Diatom Succession. *New Phytol.* **42**, 1–27.

—— 1947. Studies of the Post-glacial History of British Vegetation. VII. Lake sediments: Pollen diagrams from the bottom deposits of the North Basin of Windermere. *Phil. Trans. R. Soc.* B **233**, 137–75.

—— 1964. Pollen analyses from the deposits of six upland tarns in the Lake District. *Phil. Trans. R. Soc.* B **248**, 205–44.

—— 1970. Vegetation history in the north-west of England: a regional synthesis. In Walker, D. and West, R. G. (editors), *Studies in the vegetational history of the British Isles*. Cambridge University Press. 41–79.

—— 1975a. A chronostratigraphic comparison of Late-Weichselian and Late-Devensian subdivisions, illustrated by two radiocarbon-dated profiles from western Britain. *Boreas.* **4**, 157–71.

—— 1975b. The effect of Neolithic man on the environment in north-west England: the use of absolute pollen diagrams. In Evans, J. G., Limbrey, S. and Cleere, H. (editors), *The effect of man on the landscape: The Highland Zone. Council for Brit. Archaeol. Res. Rep.* 11, 74–86.

—— 1977. The late-glacial flora and vegetation of Britain. *Phil. Trans. R. Soc.* B **280**, 247–71.

—— and LISHMAN, J. P. 1971. Iodine in lake sediments in northern England and Scotland. *Biol. Rev.* **46**, 279–313.

RUDDIMAN, W. F. and MCINTYRE, A. 1976. Northeast Atlantic paleoclimatic changes over the last 600,000 years. *Geol. Soc. Amer. Mem.* **145**, 111–46.

—— SANCETTA, C. D. and MCINTYRE, A. 1977. Glacial/Interglacial response rate of subpolar North Atlantic waters to climatic change: the record in oceanic sediments. *Phil. Trans. R. Soc.* B **280**, 119–42.

SEDDON, B. 1957. Late-glacial cwm glaciers in Wales. *J. Glaciol.* **3**, 94–8.

SHOTTON, F. W., BLUNDELL, D. J. and WILLIAMS, R. E. G. 1970. Birmingham University Radiocarbon dates IV. *Radiocarbon.* **12**, 385–9.

SISSONS, J. B. 1967. *The evolution of Scotland's scenery*. Oliver and Boyd, Edinburgh.

—— 1974. The Quaternary in Scotland: a Review. *Scott. J. Geol.* **10**, 311–37.

SMITH, A. G. 1959. The Mires of South-Western Westmorland: Stratigraphy and Pollen Analysis. *New Phytol.* **58**, 105–27.

SMITH, A. J. 1959. Structures in the stratified late-glacial clays of Windermere, England. *J. sedim. Petrol.* **29**, 447–53.

SMITH, B. 1932. The glacial lakes of Eskdale, Miterdale and Wasdale, Cumberland, and the retreat of the ice during the main glaciation. *Q. Jl geol. Soc. Lond.* **88**, 57–83.

SMITH, R. A. 1967. The deglaciation of south-west Cumberland. *Proc. Cumb. geol. Soc.* **2**, 76–83.

TEMPLE, P. H. 1965. Some aspects of cirque distribution in the west central Lake District, northern England. *Geogr. Ann.* **47**, A, 185–93.

TOOLEY, M. J. 1972. In Huddart, D. and Tooley, M. J. (Editors), *The Cumberland Lowland*. Q.R.A. Handbook.

—— 1974. Sea-level changes during the last 9,000 years in north-west England. *Geogr. J.* **140**, 18–42.

TROTTER, F. M. 1929. The glaciation of eastern Edenside, the Alston block and the Carlisle plain. *Q. Jl geol. Soc.* **85**, 549–612.

—— and HOLLINGWORTH, S. E. 1932. The glacial sequence in the north of England. *Geol. Mag.* **69**, 374–80.

WALKER, D. 1956. A late-glacial deposit at St Bees Cumberland. *Q. Jl geol. Soc. Lond.* **112**, 93–101.

—— 1965. The post-glacial period in the Langdale fells, English Lake District. *New Phytol.* **64**, 488–510.

—— 1966. The Late Quaternary history of the Cumberland lowland. *Phil. Trans. R. Soc.* B **251**, 1–210.

WARD, C. 1873. The glaciation of the northern part of the Lake District. *Q. Jl geol. Soc.* **29**, 422–41.

—— 1874. The origin of some of the lake basins of Cumberland. *Q. Jl geol. Soc.* **30**, 174.

—— 1876. The geology of the Northern part of the English Lake District. *Geol. Surv. Mem. S.o.* London.

WEST, R. G. 1968. *Pleistocene geology and biology*. Longmans. 377 pp.

WILLIAMS, R. E. G. and JOHNSON, A. S. 1976. Birmingham University radiocarbon dates X. *Radiocarbon.* **18**, 249–67.

WINIFRED PENNINGTON (MRS T. G. TUTIN) PH.D.
Freshwater Biological Association, Windermere, and Botanical Laboratories, University of Leicester

16

15

Epigenetic Mineralization

R. J. FIRMAN

HISTORY

The Lake District was Britain's leading copper producer in the 16th century and the chief source of haematite during the 19th. From the mid 16th to the mid 19th century graphite was obtained from the unique deposit in Borrowdale. Carrock is the only mine outside Devon and Cornwall to have yielded commercial quantities of wolfram. Argentiferous lead, zinc and lesser amounts of manganese, antimony, arsenic, cobalt and nickel ores as well as barytes, umber and china clay have also been produced from various localities.

Early references mention an iron mine at Egremont in the late 12th century, and the Goldscope copper mine in the early 13th century (Postlethwaite 1913). Silver-lead mines in the Caldbeck Fells and copper mines near Keswick were working during the reign of Edward III and in 1474 a commission investigated mines at Alston and Keswick. The Elizabethan Company of the Mines Royal (Donald 1955), with its German expertise, rapidly made Keswick Britain's most important centre for mining and smelting copper-lead-silver ores. Existing mines, especially in the Newlands, Caldbeck and Coniston areas were revitalized, new areas prospected and new mines established.

Base metal mining declined during the 17th century whilst graphite and iron ore mining flourished. Ore was obtained from veins in the Borrowdale Volcanics of the Central Fells, from Tongue Gill near Grasmere, Low Furness and West Cumberland. Throughout the 18th and 19th century new haematite mines were developed, at least eighteen of which were in the Carboniferous Limestone (Smith 1924). Mines in the Skiddaw Slates were opened at Knockmurton in 1853 and veins in the Eskdale Granite became important by 1870. Throughout this period output rose to a peak of almost 3 million tons in the 1880s. Output from Furness rapidly declined but 20th century developments at Hodbarrow, Florence and Beckermet ensured continued haematite mining though now only small reserves remain.

Copper mining was less successful: no new sources were discovered and by the end of the 17th century mines in both the Newlands and Caldbeck areas were virtually exhausted. Mines continued to be worked near Coniston, the richest and deepest (Bonsor) closing in 1895 due to a slump in copper prices. Attempts to reopen these mines have failed. Nineteenth-century Lake District copper production was insignificant but lead mining was more successful. A major deposit was discovered, or rediscovered, at Greenside. Between 1822 and its closure in 1962 this produced over 200,000 tons of lead concentrates far exceeding the total for the rest of the Lake District. Between 1845 and 1860 eleven mines reopened, nine for lead and two for copper. Many also worked blende which, due to metallurgical advances, had recently become a source of zinc. The revival was short lived and many closed and reopened several times, some reopening to work other minerals such as barytes, umber, manganese or tungsten ores (Postlethwaite 1913). A few non-ferrous metal mines and barytes mines have worked spasmodically this century (Shaw 1970).

LOCATION, STRATIGRAPHICAL AND STRUCTURAL CONTROL

Base metal veins are concentrated in the more mountainous areas, and haematite occurs in both high and low ground (Fig. 69). Most veins are underlain by Bott's (1974) batholith (Fig. 70). With few exceptions they are confined to Skiddaw Slates,

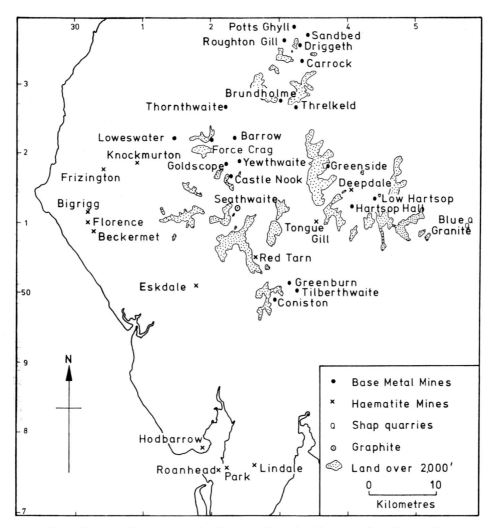

FIG. 69. Locality map showing the most important base metal and haematite mines mentioned in the text. Note that the base metal mines tend to be in the hilliest districts.

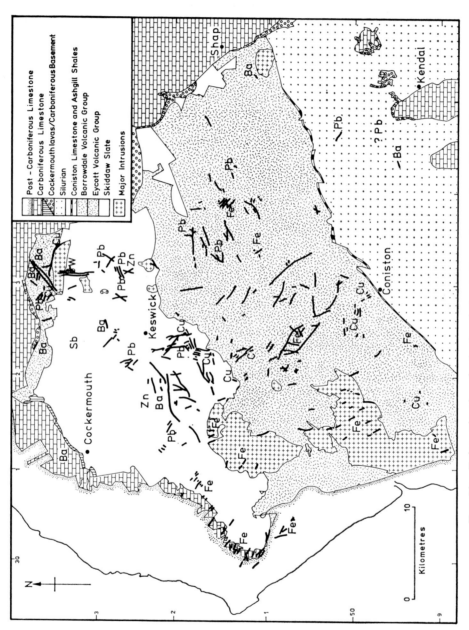

Fig. 70. Epigenetic mineral veins in relation to the surface geology. Chemical symbols indicating the principal ore produced are inserted where the evidence is unambiguous. Data derived principally from Geological Survey 6 inch maps, Postlethwaite (1913) and Shaw (1970).

Ordovician volcanics, intrusives or Carboniferous Limestone. Other formations, notably the Bala and Silurian, are virtually barren. Possibly the argillaceous strata within the Bala and Llandovery prevented mineralizing fluids penetrating upwards except where, as at Shap, these strata are either absent or thermally metamorphosed. In the Skiddaw Slates and Borrowdales only the areas underlain by the batholith show any significant mineralization. Of these, the competent formations are the most intensively mineralized. There is no preferential concentration in anticlines, the shape of the buried granite and the host rock lithology probably controlling the location of the veins.

There is no uniformity of vein orientation, sequence or type of mineralization. Copper veins may precede or succeed lead-zinc veins but Eastwood notes (1921) 'there is a general rule (subject to many exceptions) that veins carrying ores of lead and zinc range within 45° of a north-and-south direction, and hade to the east, while copper-veins trend more nearly east-and-west, hading south'. Similarly the majority of haematite veins trend between 310° and 25°. Possibly pre-Bala and more certainly main Caledonian earth movements initiated the fracture pattern and isostatic movements of the granite basement controlled the sequence of opening to receive mineralizing fluids.

CLASSIFICATION

Because of the apparent correlation between 'mining fields' (cf. Postlethwaite 1913, Eastwood 1959) and basement structures the writer proposes the following classification.

1. Veins in, over and around the three principal granite masses:
   a) Eskdale   b) Shap   c) Skiddaw
2. Veins above the roof region of the granite batholith in:
   a) Skiddaw Slates   b) Borrowdales
3. Mineral deposits above the walls and boundary faults of the batholith:
   a) Coniston copper deposits   b) West Cumbria haematites
4. Mineral deposits apparently unconnected with the batholith:
   a) Millom and Furness haematites   b) W and SW of Kendal
   c) North of the batholith.

Eskdale Granite

Gravity data suggests that the Eskdale Granite extends northward beneath the Ennerdale Granophyre and ENE under the Scafell Range (Bott 1974 and Chapter 3). The five milligal contour marks a change in gradient of the Bouguer anomaly and delimits the eastern boundary of solely haematite mineralization. To the north, the five milligal line roughly corresponds with the margin of the Ennerdale Granophyre, north of which, haematite veins in the Skiddaw Slates and in the Carboniferous Limestone seem to be more closely associated with the north wall than with the main mass of the batholith.

The veins in the granite, granophyre and volcanics consist of haematite with traces of manganese oxides and variable, though usually small, amounts of carbonates such as ankerite, dolomite or calcite, whilst some chalcopyrite and malachite occurs in Eskdale. In the granite the haematite veins are thought to occupy early complementary wrench faults opened and filled during later uplift (Firman 1960). Kendall (1893) and

Smith (1924) suggest that the haematite replaces pre-existing carbonate veins but cite little supporting evidence. Though a post-Carboniferous age is indicated by their similarity to haematite veins in the Carboniferous Limestone, it is possible that iron rich waters may have been re-cycled several times.

Shap Granite

In the aureole barytes was worked on a small scale in Sherry Gill [124 102], small mine dumps exist between the A6 road and Shap Wells and a baryte-calcite vein 0.3 m wide was encountered in the Birkbeck tunnel [560 031]. On the fringe of the gravity anomaly, lead was mined unsatisfactorily at Stavely Head (Kendall 1884) and two veins were explored for copper near Haweswater (Dakyns et al. 1897). This paucity of economic deposits contrasts with the abundance and variety of uneconomic metasomatic alteration and mineralization.

In the granite, hydrothermal minerals include very rare scheelite (Davidson 1957) which is succeeded by molybdenite; quartz veins; joint coatings of quartz-calcite with bismuthinite, chalcopyrite or zinc-blende and veins with either quartz-calcite-haematite-fluorite or baryte (Grantham 1928). Pyrite is the most abundant sulphide occurring both with quartz and as a joint filling unaccompanied by other minerals. Galena occurs as scattered traces in zinc-blende.

In the aureole abundant fissure metasomatism (Firman 1957) is accompanied by numerous hydrothermal minerals. Garnet bearing veins were fractured and infiltrated by fluids which precipitated mostly quartz and calcite with pyrite and more rarely, marcasite, haematite, molybdenite, magnetite and rare pyrrhotite. Later, shearing was accompanied by chlorite, nacrite, hydrous mica, haematite and pyrite. Then tension joints were filled with further haematite, goethite, pyrite, chalcopyrite, galena, zinc-blende and occasionally molybdenite, malachite, psilomelane and ery-thrite. Laumontite, very rarely prehnite, pectolite and saponite represent a later hydrous phase of mineralization followed by pink barytes (sometimes with galena). As in the granite, quartz or calcite or both occur as gangue minerals at almost any stage and baryte is the last mineral to form. A similar, though less complex suite, occurs in Silurian rocks in the Birkbeck tunnel (P. Redfern personal communication).

Potash metasomatism which accompanies some of the pyrite mineralization in the Blue Quarry (Firman 1957) is unusual. Most of the late hydrothermal mineral-ization is only rarely accompanied by wall rock alteration. Illitic clay formed in fault zones may precede or succeed mineralization but does not appear to be contempora-neous with any phase other than pyrite. K-Ar dating is thus unlikely to directly indicate the age of base metal or barytes veins. The lack of mineralization in the nearby Carbo-niferous; the apparent close connexion between metamorphism, metasomatism and mineralization; and the correlation between mineralization within the granite with that in the aureole all favour a Caledonian age.

Skiddaw Granite

Although the veins form a distinct structural group over and around the Skiddaw gravity anomaly, the Carrock Mine, Caldbeck Fells and the Blencathra–Skiddaw areas have sufficiently different mineralogies to justify separate descriptions.

Carrock Mine area

Mineralization is most intense in, and to the north of, the Grainsgill Greisen where N–S tensional fractures, apparently parallel to the long axis of the cupola, have opened several times. According to Hitchen (1934) many early fractures in the greisen were filled with barren quartz, quartz-microcline and quartz-apatite. Later movements led to the formation of three N–S wolfram veins which have a complex paragenetic history. Precipitation of wolfram and quartz was followed by stringers of quartz with bismuthinite, rare native bismuth and bismuth sulpho-tellurides. At about the same time veinlets of molybdenite formed both in the wolfram and earlier veins. Arsenopyrite and pyrite locally replaced by ferriferous zinc blende are accompanied by a little chalcopyrite. Scheelite, replacing wolfram, probably formed at this stage. Finally a pervasive carbonitization filled fractures less than 2 mm wide with calcite, dolomite and a little arsenopyrite.

Shepherd et al. (1976) claim that the final stage consisted of barren quartz with minor carbonates, whereas Hitchen believed that barren quartz with small pockets of white mica occurred before wolfram mineralization. If Hitchen is correct the K-Ar dates of 386 and 388 $\pm$ 6 Ma obtained by Shepherd et al. (1976) from muscovite in the barren quartz dates an event which preceded wolfram mineralization and thus gives a maximum age for these deposits. Ewart (1962) has shown that wall rock alteration adjacent to these wolfram veins includes the introduction of pyrite and arsenopyrite so the 282 Ma date (Ineson and Mitchell 1974) may correspond to this arsenopyrite phase and hence give a minimum age for wolfram mineralization.

At least three periods of fracturing and mineralization occurred after the final carbonitization of the N–S wolfram bearing veins. With so many periods of mineralization it is inevitable that isotopic 'overprinting' should have occurred. Thus wallrock alteration cannot be dated confidently. Accepting the dates quoted by Shepherd et al. (1976); the suggestions of Ineson and Mitchell (1974) and Hitchen's (1934) paragenetic sequence, the events may have been as follows:

1   intrusion of granite (minimum age 392 $\pm$ 4 Ma).

2   greisenization and crystallization of barren quartz veins (mean age 385 $\pm$ 4 Ma).

3   undated wolfram bismuth and molybdenite mineralization.

4   arsenopyrite mineralization (minimum age 282 Ma).

5   the lead-antimony sulphosalts (Kingsbury and Hartley 1956) and the E–W lead veins are undated but Ineson and Mitchell have recognized isotopic events at 233 Ma and 197 Ma which they tentatively assigned to E–W cross-courses.

6   undated ankerite veins in faults cutting all other structures.

Mineralogical studies (Hitchen 1934, Ewart 1962) and oxygen isotope and fluid inclusion data (Shepherd et al. 1976) suggest a genetic connexion between tungsten mineralization and Grainsgill Greisen. Shepherd et al. (1976) have shown that, 'the mineralising fluids were moderately saline NaCl solutions, periodically charged with CO_2 and enriched in tungsten. Isotopically they were depleted in ^{18}O relative to magmatic fluids (δ ^{18}O > 5%), and considered therefore to have contained a major non-magmatic component'. Fluid inclusion studies suggest temperatures of 240–95°C. for greisenization and 265–95°C. for wolfram mineralization, at 800 bars pressure. The barren quartz appears to have formed at slightly cooler temperature which, if Hitchen's sequence is correct, indicates a period of cooling after greisenization and before wolfram mineralization. The source of tungsten is problematical. Shepherd et al., seem to favour a source beyond the Skiddaw Granite. They postulate that moderately saline, non-magmatic, fluids in the country rocks were drawn into the northern part of the Skiddaw Granite during cooling causing greisenization in the Grainsgill area and chloritization further south, enriching the fluids in silica. The convective system then expelled the fluids into N–S fractures where they met and mixed with further quantities of unreacted non-magmatic fluids which, on cooling, deposited quartz with small amounts of wolfram and scheelite. This fails to explain the genesis of the other mineral episodes in the tungsten veins and differs from Hitchen (1934) in regarding scheelite as a co-precipitate of wolfram and in attributing the barren quartz to a late stage.

Caldbeck Fells

Carrock Mine and the Caldbeck Fells have long been famous for rare minerals. Many lists have been published (e.g. Goodchild 1885, Davidson and Thomson 1951 and Davidson 1957) and many discoveries made during the last 30 years. Earlier descriptions of the principal mines are reviewed by Eastwood et al. (1968) and Thimmaiah (1956) has written an unpublished account of the mineralization.

The abundance and variety of secondary minerals make difficulties in establishing the nature and distribution of the primary ores. Chalcopyrite predominates at Haygill, Potts Gill and Carrock End Mines; galena, zinc blende and chalcopyrite occur in varying proportions in most of the other veins. Quartz is the usual gangue mineral, although calcite up to 9 m wide has been recorded at Roughton Gill. Barytes is more restricted, occurring at Roughton Gill, east of Drygill, and in commercial quantities at Potts Gill and Sandbeds mines. Arsenopyrite has been noted in a cross-cut between the latter two mines (Davidson and Thomson 1951).

The wolfram veins appear to pass northward into manganese bearing quartz veins and Hollingworth (in discussion, Hitchen 1934) suggests that a belt of manganese surrounds the wolfram 'zone'. Possibly some manganese was derived from the metasomatic replacement of wolfram by scheelite. Similarly the replacement of arsenopyrite by ferriferous zinc blende (Hitchen 1934) could account for the various arsenates found, both north and south of Carrock Mine, in a broad N–S belt. There is no comparable metasomatic source of the phosphorous in the pyromorphite minerals which abound in the Caldbeck Fells. Possible sources include residual fluids from the early mineralization which deposited the quartz-apatite veins at Carrock; re-mobilization of early phosphate minerals, and phosphate rich waters draining from overlying strata.

The later stages are characterized by oxides and hydroxides of manganese, lead chlorophosphates, and a bewildering variety of chloroarsenates, arsenates, vanadates, carbonates, sulphates and hydrous silicates of copper, lead, zinc or iron or combinations of these metals. Lead molybdates and tungstates occur near the wolfram 'zone' and in copper rich areas secondary minerals are plentiful. Rastall (1942) ascribes this assemblage to normal copper-lead-zinc mineralization 'modified only by reaction in the zone of oxidation with materials provided by the earlier Skiddaw–Grainsgill group' of minerals. Alteration products of the secondary minerals occur (Kinsbury and Hartley 1957a) but other minerals such as the well crystallized pyromorphites and campyllites may be primary. Such minerals could be formed from an ore fluid from which H_2S had been removed by precipitation of sulphides, evaporation and oxidation (Strens 1964).

The veins exhibit many stages of mineralization, at least one of which, as shown by the barytes at Ruthwaite, must be post-Carboniferous. K-Ar dates from altered wall rocks show a wide range of ages (Ineson and Mitchell 1974) from 272 to 178 Ma suggesting an episodic mineralization beginning in the Upper Carboniferous and ending in Lower Jurassic times.

Skiddaw and Blencathra ranges

South and west of the Caldbeck Fells the mineralization is entirely in the Skiddaw Slates. Mineral assemblages are simpler and less oxidized though some secondary minerals occur. Cerussite was worked at Threlkeld and Glenderaterra mines (Postlethwaite 1913) and malachite and the rare zinc arsenate, adamite, also occur at Threlkeld. Arsenopyrite has not been recorded although it is plentiful further south at Wanthwaite, in St John's Vale, where the southernmost vein consists 'mostly of arsenopyrite and pyrite with some blende, galena, stibnite, and sulpho-salts, in a quartz-carbonate gangue' (Kingsbury and Hartley 1958, p. 814). Stibnite was also found in drift near Troutbeck station and was worked at Robin Hood, Bassenthwaite [229 327], (Postlethwaite 1893).

Marginal veins tend to be either at right angles or tangential to the gravity contours. They were chiefly worked for lead; some zinc was produced at Threlkeld, barytes at Carlside and a little copper ore at Glenderaterra. Quartz is ubiquitous and pyrites and baryte are common. Cores of chalcopyrite surrounded by galena and baryte at Glenderaterra (Dewey and Eastwood 1925) suggest two periods of mineralization and at Threlkeld a lead vein is displaced by a rich vein in zinc blende (Postlethwaite 1913). K-Ar dates on illitic spoil heap material at Threlkeld and the nearby Brundholme Mine gave ages ranging from 340 to 272 Ma (Ineson and Mitchell 1974). Two distinct ages, late Devonian and Namurian/Westphalian were recognized at Brundholme but not at Thelkeld where ages of 316, 329 and 331 Ma were reported.

Veins in Skiddaw Slates above the roof region of the batholith

West and SW of Skiddaw, base metal veins are most prolific in the Vale of Newlands and in the Buttermere–Crummock Valley but further west at Knockmurton haematite mineralization predominates. A few haematite veins outcrop amongst the copper and lead-zinc veins suggesting 'Eskdale' type mineralization. Haematite also occurs as a minor constituent in a few other veins including the Cobalt vein on Causey Pike. This is probably the oldest vein in the area. It trends N–S, contains mostly quartz with apatite, muscovite, arsenopyrite, löllingite and probably scheelite (Kingsbury and Hartley 1957, p. 498), and is emplaced in thermally metamorphosed Skiddaw Slates which may be underlain by a granite cupola (Rose 1954).

In the Vale of Newlands E–W copper veins are cut and appear to be shifted sinistrally by N–S lead-zinc veins. The copper veins contain chalcopyrite with lesser amounts of galena, zinc blende, pyrite and arsenopyrite. The lead-zinc veins lack arsenopyrite, commonly contain pyromorphite, and galena is the most abundant ore mineral. Psilomelane and pyrolusite are not infrequent constituents. Quartz is ubiquitous, baryte occurs in some of the lead-zinc veins, and fluorite in abundance in the upper levels in the Old Brandlehow lead vein. Pseudomorphs of quartz after baryte and fluorite in the middle levels at Old Brandlehow and quartz after baryte at Yewthwaite (Strens 1962) suggest two periods of mineralization in the N–S lead-zinc veins. The lead-zinc mineralization is not confined to N–S veins and at Force Crag there is extensive lead-zinc type mineralization in E–W fractures which contain dolomite, psilomelane and rare fluorite. Eastwood (1959) suggested that lead-zinc minerals were followed by baryte-dolomite and then psilomelane-pyrite.

In the Newlands area, therefore, early apatite-scheelite-löllingite is followed by chalcopyrite-arsenopyrite and two or more periods of galena-blende-pyromorphite mineralization. Ineson and Mitchell (1974) report ages of 395 to 345 Ma on thirteen samples collected from five localities and tentatively suggest a 'copper' mineralizing epoch at about 390 Ma and a lead mineralization about 360 Ma. Further north at Thornthwaite lead mines they confirmed the 360 Ma period and noted a late resurgence close to 315 Ma. No ages comparable to those from the Caldbeck Fells were reported.

Veins in Borrowdale Volcanics above the roof region of the batholith

Workable haematite veins tend to be most frequent over the postulated granite ridge joining Eskdale with Shap (Fig. 71). E–W copper veins outcrop near Wythburn and Thirlspot (Dewey and Eastwood 1925). Copper minerals are otherwise rare, the main metal produced being lead with a little zinc. Barytes in the upper levels of the Greenside mine and a barytes vein east of Blowick, Ullswater, are the only records of workable barium minerals (Wilson et al. 1922). Among the more unusual gangue minerals, witherite, fluorescent calcite, dolomite and ankerite occur at Greensides and fluorite at Hartsop Hall mine. Phosphates and arsenates seem to be absent and arsenopyrite has been reported from the Seathwaite (Borrowdale) area only. Tetrahedrite and bournonite at Greensides and stibnite at St Sunday Crag mines, suggest that the antimony mineralization may have spread southward from Skiddaw.

Greensides was the third richest lead mine in Britain yielding 200,000 of lead concentrates in an otherwise impoverished area. This N–S vein from its surface outcrop to a depth of 790 m maintained an average grade of more than 7 % Pb. Baryte and silver values decreased and zinc blende and chalcopyrite increased with depth (Gough 1965). Chalcopyrite appears to be an early phase and late baryte is replaced by quartz in a similar manner to that observed at Newlands and in the Caldbeck Fells. Ore grades increase in the more steeply dipping parts of the vein and tend to be poorest where the strike is west of north and where a quartz porphyry dyke forms both the foot and hanging walls. Vein widths vary from a few millimetres to 9 m averaging about 2 m. Branches sub-parallel to the main vein occur adjacent to the dyke. The ore body is deeper than it is wide, tending to spread furthest north at the deepest levels. All structural features are consistent with a simple fissure filling of a normal fault with variable easterly hade. The porphyry dyke may have acted as a channelway through the relatively impermeable Skiddaw Slates and the N–S elongated dome proved at the base of the Borrowdale Volcanics might indicate a buried cupola which channelled fluids into this area but the full explanation of this unusual concentration of galena is unknown. K-Ar ages of 324, 325 and 372 Ma, the latter possibly due to incomplete isotopic resetting, suggest Namurian/Westphalian mineralization (Ineson and Mitchell 1974).

The origins of the Seathwaite Graphite deposit are more enigmatic. No comparable deposit has been found elsewhere. Graphite occurs in stringers and nodular masses ranging in size from a few millimetres to 1 m or more in diameter. Eight 'veins' and highly inclined pipes up to 1 m × 3 m in cross sections and 2 to over 100 m deep, are shown on mine plans and sections (Ward 1876). Both veins and pipes seem to have been composed of nodular graphite, which were not fissure fillings. Strens' (1965) reinterpretation showed that the graphite occurs as joint coatings and replacements in an intrusive diabase and in surrounding volcanic rocks. These deposits are the only example known to the writer, not either associated with high grade regional metamorphism or due to thermal metamorphism of carbonaceous material.

A unique mineral deposit demands a unique explanation. Strens (1965) favours the reaction $2CO \lessgtr CO_2 + C$ catalysed by iron oxides, pyrite, iron silicates and quartz. The production of sufficiently large volumes of carbon monoxide is a severe problem. Strens (1965) suggests that iron-rich pyroxenes might react with CO_2, but, if his proposed reaction ($3FeSiO_3 + CO_2 \lessgtr Fe_3O_4 + CO$) were generally applicable to the decomposition of pyroxenes in the presence of CO_2, large volumes of CO should be evolved whenever limestone is incorporated into basic igneous magmas and graphite deposits of the 'Seathwaite-type' should be very common.

Late Caledonian K-Ar ages of 382 and 376 Ma for the diabase and graphite veins respectively were reported by Ineson and Mitchell (1975) providing further evidence of the probable genetic connexion between the diabase and the graphite.

Coniston Copper deposits

A recent discussion of Copper deposits at Coniston is contained in Dagger (1977) and Firman (1977). This district has produced virtually no ores other than copper. In the late 19th century more than 50,000 tons of dressed ore, running at 5–13% Cu, was produced with twelve tons of lead and three tons of cobalt and nickel ore (Eastwood 1959). The productive veins mostly trend E–W but south of Levers Water they veer SE. Many are cut and shifted by barren crosscourses or by lead veins trending N–S or NE–SW. Fault breccias filled with ore are richer than thick quartz veins. The richest veins are SW of Levers Water and appear to be associated with the Paddy End Rhyolites and Ignimbrites (Mitchell 1940). Most veins hade northward but the deepest and richest, the Bonsor Vein, hades north at outcrop, steepens with depth and hades south from about 240 m to the deepest working 360 m below the surface. All veins cut obliquely across the strike of the Borrowdales.

The typical mineral assemblage is quartz, chalcopyrite, arsenopyrite, and pyrites, other minerals being rare. Copper minerals include chalcocite at Seathwaite and tenorite, malachite, azurite and chysocolla in veins near Levers Water. The cobalt and nickel arsenides, smaltite and niccolite, together with erythrite and tetrahedrite were indentified by Russell (1925). Shaw (1959) reported that at depth the Bonsor vein contained 'an ever-increasing amount of magnetite, unfortunately at the expense of copper ore'. Stanley (personal communication 1977) shows that this magnetite is associated with native bismuth and bismuthinite (cf. Wheatley 1971). Among the non-metallic minerals a little dolomite and calcite occurs and Kendall (1884) recorded saponite coating calcite and chalcopyrite.

Five wall rock samples (Ineson and Mitchell 1974) yielded a mean K-Ar age of 388 Ma comparable to the ages obtained from copper veins in the Newlands area.

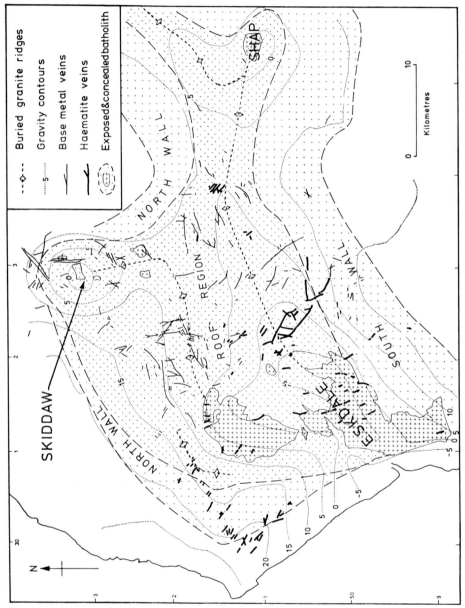

FIG. 71. Epigenetic mineral veins in relation to the inferred subsurface batholith.
Gravity contours are in milligals and all subsurface interpretation is based on
Bott (1974).

West Cumbrian haematites

Above the north wall of the batholith as defined by Bott (1974) and near the Western Boundary Fault, haematite occurred in commercial quantities in veins in the Skiddaw Slates and in replacements in Carboniferous Limestone a few of which are still worked. Smaller amounts have also been noted in the overlying Coal Measures, Brockram and St Bee's Sandstone (Trotter 1945). Most of the iron ore veins and replacements are predominately haematite. Recently discovered fluorite-haematite-dolomite ores reported by Goldring and Greenwood (in Evans 1977) to contain up to 48 % CaF_2 suggest that fluorite may be more abundant in the Carboniferous Limestone than was previously thought. Manganite, chalcopyrite, pyrite, siderite, baryte, quartz, dolomite and calcite also occur, usually in small amounts. Paragenesis is uncertain: apparently early sulphides commonly occur in lenses, late non-metallic minerals line cavities (Trotter 1945) and haematite pseudomorphs after baryte occur (Rudler 1905, p. 47). In the Skiddaw Slates most veins are confined to the Loweswater Flags, trend NW–SE and dip east at 80–45°. The large ore-bodies in the Carboniferous Limestone are metasomatic, with faults acting as conduits and the flat shape of many replacements is controlled by the stratification (Kendall 1884).

Millom and Furness haematite deposits (Fig. 72)

Apart from a replacement of Coniston Limestone (Kendall 1876) these ores are in Carboniferous Limestone. At Hodbarrow ore was won from a large, thick flat deposit containing unaltered blocks of dolomitized limestone. Further south ore occurs in downward tapering 'sops' up to 210 m deep and 130 m diameter, often incorporating blocks of Trias sandstone and water worn limestone. These 'sops' show evidence of both collapse and replacement. The ore was usually soft with few cavities and associated minerals. South of Lindale and Dalton 'sops' are rarer and most ores were worked from NW–SE veins (Smith 1924). Pseudomorphs of haematite after fluorite or pyrite at Hodbarrow, and the chalcocite-pyrite ore mixed with haematite at Anty Cross near Dalton (Smith 1924) may represent an early phase of mineralization.

West and south-west of Kendal

This area south of the batholith overlies positive magnetic anomalies (Fig. 72) which Evans and Maroof (1976) suggested might be due to buried granodiorite or diorite intrusions containing magnetite. They recommended that the Carboniferous west and south-west of Kendal might be worth exploring. In fact both the Carboniferous and the Silurian west of Kendal have been prospected and a full description will be found in Rose and Dunham (1978). Two prospects probably, in the Carboniferous Limestone, are described in the 1602 Mines Royal report (Donald 1955) and two veins were worked during the 19th century (Aveline et al. 1888). A more pronounced anomaly further south-west has no known base metal deposits associated with it (Fig. 72). Perhaps a modern survey is overdue.

North of the batholith

Apart from the poor N–S vein in Skiddaw Slate and diorite near Embleton (Eastwood 1921) no workable veins are known in the Lower Palaeozoics. In the Carboniferous Limestone a chalcopyrite vein was tried near Papcastle (J. A. D. Dickson,

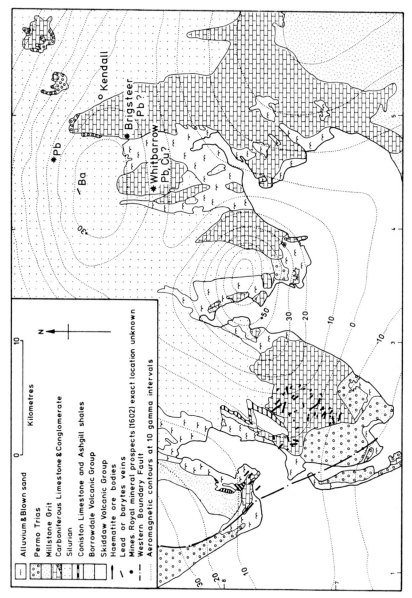

FIG. 72.   Known mineral veins and mineral prospects west and south-west of Kendal in relation to the total force magnetic anomalies (I.G.S. 1st ed. 1965).

personal communication) and limestones near Cockermouth frequently have joint coatings of baryte. Barytes was mined near Greengill, 6 km north of Cockermouth, and further east at Ruthwaite [240 367]. These Carboniferous localities are 5 to 10 km north of the buried batholith so lateral migration must have occurred if the mineralizing fluids were derived from the batholith. These mineral occurrences are also associated with a positive magnetic anomaly (cf. Evans and Maroof 1976).

SPECULATIONS AND CONCLUSIONS

The loci of mineralization appear to vary in time and place. Copper mineralization is associated with early or contemporaneous vulcanicity (Wheatley 1971) and bismuth, antimony and arsenic are characteristic of volcanic exhalations (White and Waring 1963). Thus the chalcopyrite-arsenopyrite-pyrite veins and associated minerals might have originally had a volcanic source. Possibly pre-Bala volcanic exhalations contributed to formation waters which were later recirculated when the Lower Devonian granites were intruded. Convective systems proposed by Shepherd et al. (1976) for the Grainsgill Greisen probably operated in other areas including Coniston and Newlands. This early (c. 390 Ma) mineralization was probably concentrated around a few granite cupolas where veins were emplaced under a thick cover of Ordovician and Silurian strata, in dominately E–W fissures opened by the intruding batholith, or in fractures released as Caledonian stresses relaxed.

In contrast, lead-zinc mineralization began (c. 360 Ma) when the Lake District was being rapidly uplifted and the Mell Fell and Carboniferous conglomerates were forming. This uplift seems to have opened relatively shallow fissures in and above the central and eastern parts of the batholith forming channels for mineralizing formation waters, now depleted in Cu, As etc, which were then convectively recycled. The dominately N–S trend of the most productive lead veins suggests an overall E–W tension. Continued intermittent uplift and accompanying mineralization throughout the Carboniferous, Trias and Jurassic is implied by the K-Ar ages, sulphates and carbonates forming when rising formation waters met ground-waters percolating from overlying Carboniferous and Triassic strata.

The genesis of haematite is still controversial (Trotter 1945, Dunham 1952, Shepherd 1974). The recognition of the importance of convection cycles makes possible reconciliation of views of the 'ascentionists' and 'descentionists' and Shepherd's (1974) demonstration of the geochemical homogeneity and spacial correlation with the concealed Eskdale granite favours a granitic source. But if alteration of the Eskdale granite resulted in the formation of an iron rich brine (Shepherd 1974) why are the vast bulk of haematite ores post-Carboniferous and not Caledonian? Possibly conditions were not favourable for much haematite formation until the (?) Kimmerian uplift allowed oxidizing waters draining a lateritic subsoil (cf. Dunham 1952) to mix with iron-rich brines rising from the concealed Eskdale Granite and its boundary faults. In contrast to the earlier lead-zinc mineralization haematite depositing fluids were channeled preferentially through NW-trending faults suggesting an overall NE–SW tension at that period. Sir Kingsley Dunham (Rose and Dunham 1978) has recently added a new element to this controversy by suggesting that the ferric iron was derived from the deeper parts of the Irish Sea Basin, where it was leached from the New Red Sandstone by warm hypasaline brines which were later driven up-dip by tectonic pressures.

Recent predictions based on the statistical analysis of stream sediment data (Cruzat and Meyer 1974) suggest that the Lake District contains reserves of lead, copper and zinc equal respectively to almost twice, nine times and twenty-seven times the previous production. Their astonishing forecast of 20 copper deposits in the Silurian and 77 zinc deposits concentrated over and around the Eskdale granite gravity low contrasts with the more conservative views of Eastwood (1959) and Shaw (1970) and with the ideas presented in this chapter. A re-investigation (Nottingham University unpublished research) suggests that they greatly exaggerated the mining potential of the Lake District.

REFERENCES

AVELINE, W. T. and HUGHES, T. McK. 1872. The geology of the country around Kendal, Sedbergh, Bowness and Tebay. 2nd Ed. revised and enlarged by A. Strahan. Mem. Geol. Surv. vii + 944 pp.

BOTT, M. H. P. 1974. The geological interpretation of a gravity survey of the English Lake District and the Vale of Eden. Jl geol. Soc. Lond. 130, 309–31.

CRUZAT, A. C. E. and MEYER, W. T. 1974. Predicted base-metal resources of north-west England. Trans. Instn. Min. Metall. 83, B131–34.

DAGGER, G. W. 1977. Controls of copper mineralisation at Coniston, English Lake District. Geol. Mag. 114, 195-202.

DAKYNS, J. R., TIDDEMAN, R. J. and GOODCHILD, J. G. 1897. The geology of the country between Appleby, Ullswater and Haweswater. Mem. Geol. Surv. vi + 110 pp.

DAVIDSON, W. F. and THOMSON, N. 1951. Some notes on the minerals of Westmorland and Cumberland. N.W. Naturalist, 23, 136–54.

—— 1957. Mines and Minerals in 'The Changing Scene'. Joint Trans., Eden Field Club, Penrith and District Natural Hist. Soc. and Kendal Natural Hist. Soc. No. 1, pp. 61–63.

DEWEY, H. and EASTWOOD, T. 1925. Copper ores of the Midlands, Wales, the Lake District and the Isle of Man. Mem. Geol. Surv. Spec. Rep. Mineral Resources Gr. Br. 30, ix + 87 pp.

DONALD, M. B. 1955. Elizabethan Copper: the history of the Company of the Mines Royal, 1568–1605. London.

DUNHAM, K. C. 1952. Age relations of the epigenetic mineral deposits of Britain. Trans. geol. Soc. Glasgow. 21, 395–429.

EASTWOOD, T. 1921. The lead zinc ores of the Lake District. Mem. Geol. Surv. Spec. Rep. Mineral Resources. Gr. Br. 22.

—— 1959. The Lake District mining field: pp. 149–174 in 'The future of non-ferrous mining in Great Britain and Ireland' xxvi + 614 pp. Instn. Min. Metall. London.

—— HOLLINGWORTH, S. E., ROSE, W. C. C. and TROTTER, F. M. 1968. Geology of the country around Cockermouth and Caldbeck. Mem. Geol. Surv. Gt. Br. x + 298 pp.

EVANS, A. M. and MAROOF, S. I. 1976. Basement controls on mineralisation in the British Isles. Mining Mag. 134, 401–11.

EVANS, T. 1977. The nature and genesis of mineral deposits. A conference report. The British Geologist. 3, No. 1, 8–9.

EWART, A. 1962. Hydrothermal alteration in the Carrock Fell area, Cumberland, England. Geol. Mag. 99, 1–8.

FIRMAN, R. J. 1957. Fissure Metasomatism in volcanic rocks adjacent to the Shap Granite, Westmorland. Q. Jl geol. Soc. Lond. 113, 205–22.

—— 1960. The Relationship between the joints and fault patterns in the Eskdale Granite (Cumberland) and the adjacent Borrowdale Volcanic Series. Q. Jl geol. Soc. Lond. 116, 317–47.

—— 1977. Lake District copper (correspondence with a reply by G. W. Dagger) Geol. Mag. 114, 483.

GOODCHILD, J. G. 1881–2. Contributions towards a list of minerals occurring in Cumberland and Westmorland. Trans. Cumb. & West. Assoc. Adv. Lit & Sci. No. 7, 101–26.

GOUGH, D. 1965. Structural analysis of the ore shoots at Greenside Lead Mine, Cumberland, England. Econ. Geol. 60, 1459–77.

GRANTHAM, D. R. 1928. The Petrology of the Shap Granite. Proc. Geol. Ass. 39, 299–331.

HITCHEN, C. S. 1934. The Skiddaw Granite and its residual products. Q. Jl geol. Soc. Lond. 90, 158–200.

INESON, P. R. and MITCHELL, J. G. 1974. K-Ar isotopic age determinations from some Lake District localities. Geol. Mag. 111, 521–37.

—— —— 1975. Potassium-argon ages from the graphite deposits and related rocks of Seathwaite, Cumbria. Proc. Yorks. Geol. Soc. 40, 413–18.

KENDALL, J. D. 1876. The Haematite in the Silurians. *Q. Jl geol. Soc. Lond.* **32**, 180–83.
—— 1884. The minerals veins of the Lake District. *Trans. Manchester Geol. Soc.* **17**, 292–341.
—— 1893. The Iron Ores of Great Britain and Ireland. *Crosby Lockwood and Son. London.*
KINGSBURY, A. W. G. and HARTLEY, J. 1956. Cosalite and other lead Sulpho-salts at Grainsgill, Carrock Fell. *Min. Mag.*, **31**, 298–300.
—— —— 1957. Childrenite from the Lake District. *Min. Mag.* **31**, 498.
—— —— 1957a. Beaverite from the Lake District. *Min. Mag.* **31**, 701–702.
—— —— 1958. Jarosite and natrogarosite from the Lake District. *Min. Mag.* vol. 31, pp. 813–15.
MITCHELL, G. H. 1940. The Borrowdale Volcanic Series of Coniston, Lancashire. *Q. Jl geol. Soc. Lond.* **96**. 301–19.
—— 1956. The Borrowdale Volcanic Series of the Dunnerdale Fells, Lancashire. *Liverpool & Manchester geol. Journ.* **1**, 428–49.
MOORBATH, S. 1962. Lead isotope abundance studies on mineral occurrences in the British Isles and their geological significance. *Phil. Trans. R. Soc.* **A254**, 295–360.
POSTLETHWAITE, J. 1913. Mines and Mining in the (English) Lake District. 3rd Edit., *W. H. Moss: Whitehaven.* 164 pp.
RASTALL, R. H. 1942. The Ore Deposits of the Skiddaw District. *Proc. Yorks. geol. Soc.* **24**, 329–43.
ROSE, W. C. C. 1954. The sequence and structure of the Skiddaw Slates in the Keswick–Buttermere area. *Proc. geol. Soc.* **65**, 403–406.
—— and DUNHAM, K. C. 1978 (in press). Geology and hematite deposits of South Cumbria. *Mem. Geol. Surv. G.B.*
RUDLER, F. W. 1905. A Handbook to a collection of minerals in the British Isles. *Mem. Geol. Surv.*
RUSSELL, A. 1925. A notice of the occurrence of native arsenic in Cornwall; of Bismuthinite at Shap Westmorland and of smaltite and niccolite at Coniston, Lancashire. *Min. Mag.* **20**, 299–304.
SHAW, W. T. 1959. Joint discussion on the Lake District Mining Field: 219–24 *in* 'The future of non-ferrous mining in Great Britain and Ireland'. *Instn. Min. Metall.* London.
—— 1970. Mining in the Lake Counties. *Dalesman Publishing Co. Ltd., Clapham,* Yorks. 128 pp.
SHEPHERD, T. J. 1974. Geochemical evidence for basement control of the West Cumberland haematite mineralisation (Ph.D. thesis abstract). *Trans. Instn. Min. Metall.* **83**, B47–48.
—— BECKINSALE, R. D., RUNDLE, C. C. and DURHAM, J. 1976. Genesis of Carrock Fell tungsten deposits: fluid inclusion and isotopic study. *Trans. Instn. Min. Metall.* **85**, B63–73.
SMITH, B. 1924. The Haematites of West Cumberland, Lancashire and the Lake District. *Mem. Geol. Surv.*, 2nd Edit. vi + 236 pp.
STRENS, R. G. J. 1962. The geology of the Borrowdale–Honister district (Cumberland) with special reference to the mineralisation. *Unpub. Ph.D. thesis, University of Nottingham.*
—— 1964. Pyromorphite as a possible primary phase. *Min. Mag.* **33**, 722–23.
—— 1965. The graphite deposit of Seathwaite in Borrowdale, Cumberland. *Geol. Mag.* **102**, 393–406.
THIMMAIAH, T. 1956. Mineralisation of the Caldbeck Fells area, Cumberland. *Unpub. Ph.D. thesis. University of London.*
TROTTER, F. M. 1945. The origin of the West Cumberland Haematites. *Geol. Mag.* **82**, 67–80.
WARD, J. C. 1876. The Geology of the Northern part of the Lake District. *Mem. Geol. Surv.* 12 + 132 pp.
WHEATLEY, C. J. V. 1971. Aspects of metallogenesis within the Southern Caledonides of Great Britain and Ireland. *Trans. Instn. Min. Metall.* **80**, 211–23.
WHITE, D. E. and WARING, G. A. 1963. Volcanic emanations. In *Data of Geochemistry, 6th Edn. Prof. Pap. U.S. geol. surv.* 440K, 23 pp.
WILSON, G. W., EASTWOOD, T., POCOCK, R. W., WRAY, D. A. and ROBERTSON, T. 1922. Barytes and Witherite. *3rd Ed. Mem. Geol. Surv. Spec. Rep. Miner. Resour. G.B.* **2**.

R. J. FIRMAN, PH.D.
Department of Geology, The University, Nottingham

16

Offshore Cumbria

B. N. FLETCHER and C. R. RANSOME

The great diversity of relief and rock type in Cumbria is not continued offshore where the sea bed shows only subdued topography (Fig. 73). The solid rocks are predominantly Permo-Triassic in age and are largely covered by thick Quaternary deposits.

Before the systematic exploration of the continental shelf began, the geology of offshore Cumbria could only be guessed at by the extension seawards of onshore outcrops. The Coal Measures of the Cumberland coalfield had been worked several kilometres offshore and Permian rocks had been proved in exploratory boreholes in Risehow Colliery about 5 km west of Maryport. It seemed likely that the Permo-Triassic rocks thickened westwards to underlie much of the northern Irish Sea. Similarly boreholes drilled for salt and coal in the drift-covered northern part of the Isle of Man proved Permo-Triassic rocks and led Boyd-Dawkins (1902) to suggest that they might extend beneath the Irish Sea between the island and the Lake District. Glacial erratics of Permo-Triassic rocks on Anglesey also pointed to the occurrence of these rocks in Liverpool Bay (Greenly 1919).

The Geological Survey produced an aeromagnetic map of Great Britain in 1964 which covered the Irish Sea area and in the same year the results of a gravity survey of the north-east Irish Sea were published (Bott 1964). This survey showed two major gravity anomalies, one between the Solway Firth and the Point of Ayre (IOM) and the other between the Duddon Estuary and Douglas (IOM).

The systematic survey of the offshore geology began in 1967 with shallow seismic investigations and sonar, magnetometer and gravimetric surveys by the Institute of Geological Sciences. This geophysical programme was followed by sampling with vibracorer, gravity corer and shipek grab. In addition, fourteen shallow boreholes were sunk to prove the solid rocks beneath the extensive superficial deposits.

In the search for hydrocarbons, oil companies have conducted extensive geophysical surveys and drilled several deep exploratory boreholes which have confirmed that the Permo-Triassic succession in the Irish Sea is similar to that in Cheshire (Colter and Barr 1975) with the implication that the two areas were inter-connected during sedimentation.

Carboniferous rocks

Carboniferous beds crop out around the coasts of the north-eastern part of the Irish Sea, in the Isle of Man, Cumbria and parts of the Scottish coast bordering the Solway Firth. Sub-drift outcrops of Carboniferous rocks extend offshore to the west of the Whitehaven–Workington coalfield and to the south-east of Ramsey Bay

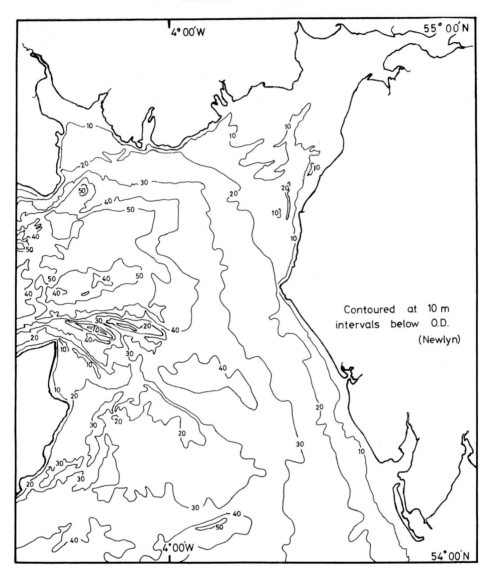

FIG. 73. Offshore Cumbria: bathymetry, depth in metres.

(Fig. 75). Elsewhere in the area the Carboniferous rocks are thought to extend uninterruptedly beneath Permo-Triassic cover. Only two boreholes have proved them however. These were drilled by the Gulf/NCB group, 34 and 43 kms respectively to the south-south-west of Walney Island. In these boreholes the Permian rests upon several metres of barren reddened beds overlying shales and sandstones whose probable age ranges from Namurian to Westphalian B (Colter and Barr 1975).

Throughout the Irish Sea the Carboniferous rocks were uplifted, blockfaulted and eroded before the lowest formation of the Permo-Triassic, the Collyhurst Sandstone,

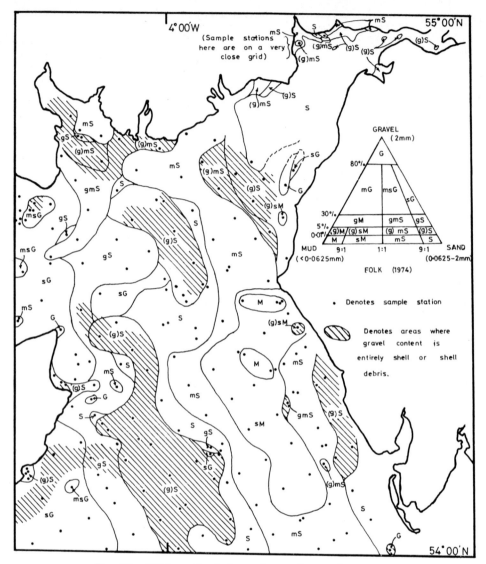

FIG. 74.   Offshore Cumbria: surface sediment distribution.

was laid down. On land the Sandstone rests upon Carboniferous beds ranging in age from Namurian to Westphalian.

Permo-Triassic rocks

These accumulated to great thicknesses in two major basins. In the north the Solway Firth Basin is the seaward extension of the Carlisle Basin. Farther south the North-east Irish Sea Basin is much larger. It includes at least two subsidiary basins (the Manx–Furness and Liverpool Bay basins), and extends landwards into the

Cheshire Basin. Separating the Solway Firth and Manx–Furness basins is the Ramsey–Whitehaven Ridge which can be clearly seen on the gravity maps as a 'high'. Early interpretations showed this as a ridge of Carboniferous rocks but three shallow boreholes on it all penetrated Permo-Triassic rocks which extend over the ridge from one basin to the other.

The precise nature of these offshore basins is still largely unknown, although it is thought that they are original sedimentary basins rather than downfolds delimited by subsequent faulting and folding. The structural framework of the area was determined by the Caledonian Orogeny and persists, with some modification, to the present day. The intrusion of large granites within the Southern Uplands, Lake District and Isle of Man resulted in the formation of upstanding blocks separated by areas of subsidence containing the basins. The Ramsey–Whitehaven Ridge dates from Devonian times and separated the depositional basins to the north and south during both Carboniferous and Permo-Triassic sedimentation. Onshore most of the basins, such as those near Carlisle and in Cheshire, are thought to approximate to the original basins of deposition. Offshore the situation seems analogous with general subsidence over the whole area giving a similar stratigraphy but with centres of more rapid subsidence. Strata within each basin thicken and sometimes change in facies towards the basin centre. The Hercynian Orogeny interrupted deposition at the end of the Carboniferous period and rejuvenated some of the Caledonian structures. The subsequent erosion removed much of the Upper Carboniferous infill. Subsidence and deposition were renewed during the Permian and continued into Mesozoic times, although remnants of the Jurassic and younger rocks only survive onshore as isolated outliers and none have so far been encountered offshore.

North-east Irish Sea Basin

Onshore, typically at St Bees Head, the basal beds of the Permian are a basin-marginal facies consisting of coarse, red and purple Continental breccias or 'brock-rams'. Within the North-east Irish Sea Basin brockrams are absent and except in the centre of the basin, the Collyhurst Sandstone overlies the Carboniferous strata. It consists of an alternating sequence of red-brown to white, medium-grained dune-bedded sandstone, containing well-rounded grains with frosted surfaces, and red-brown micaceous shales and siltstones. Towards the centre of the basin, the shale content increases and the sandstones become subordinate (Colter and Barr 1975). The overlying Manchester Marl near the edge of the basin comprises red-brown marls with evaporites, mainly anhydrite, and thin sandstones. Towards the centre of the basin a salt bed more than 213 m thick lies in the middle of the formation with shale and anhydrite above and below.

The base of the Trias is conventionally taken in Cumbria at the base of the St Bees Sandstone. This remains the most convenient mappable line although the base of the Trias may lie a little higher in the succession, in the lower part of the Sherwood Sandstone Group (Table 7). This red-brown to grey and white sandstone exceeds 1,432 m in thickness near the basin centre. Several shallow boreholes have entered these sandstones, notably those on the Ramsey–Whitehaven Ridge. Borehole 71/61 (Fig. 75) proved red-brown sandstones with grey-green reduction patches beneath 18.5 m of drift and borehole 71/64 proved red-brown friable sandstone with well-rounded grains beneath 75 m of drift. Farther south, borehole 71/41 penetrated red medium-grained sandstone under 27.2 m of drift. None of these sandstones yielded

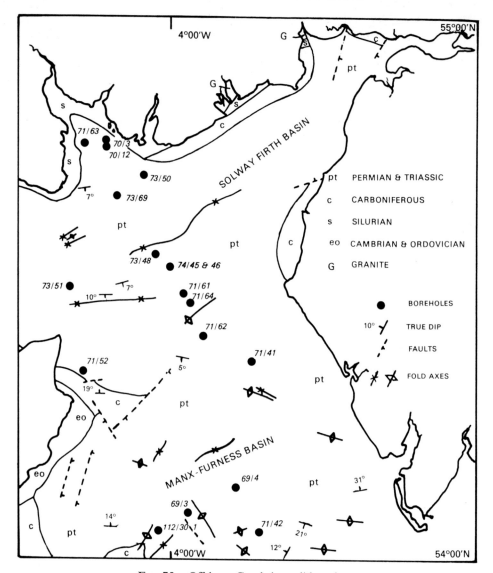

FIG. 75. Offshore Cumbria: solid geology.

fossils, not even spores, but the lithologies are typical of the Sherwood Sandstone Group.

The overlying beds are the Mercia Mudstone Group, a sequence of mudstones and evaporites, the latter thickening towards the basin centre. Towards the top of the group evaporites become less common and the pale grey to medium brown mudstones are interbedded with white to buff, medium-grained friable sandstones with gypsum in veins and nodules. Borehole 71/62 penetrated red-brown and green-grey mudstones of Anisian (middle Triassic) age with traces of gypsum beneath 73.5 m of drift. Boreholes 69/3 and 69/4 proved red-brown calcareous mudstones, with green bands and gypsum, which are considered on lithological grounds to be of Upper Triassic age, beneath 32 and 23 m of drift respectively.

Solway Firth Basin

No deep boreholes have been drilled offshore in the Solway Firth Basin but those on the adjacent land suggest that the infill of the basin resembles that of the North-east Irish Sea Basin. On St Bees Head the brockrams are overlain by the St Bees Evaporites and the St Bees Shale (Arthurton and Hemingway 1972). Correlatives of the St Bees Evaporites do not occur in the Isle of Man but it seems likely that there is a Permian evaporite sequence in the basin centre. Borehole 71/52 east of Ramsey Bay (Fig. 75) drilled a hard, grey, recrystallized dolomite thought to be of Permian age beneath 58 m of drift.

The Sherwood Sandstone Group was proved in two shallow boreholes in Wig-town Bay. Borehole 71/63 drilled a red-brown, fine-grained sandstone with mudstone intraclasts and borehole 73/50, 11 km to the south-east, drilled a red-brown, medium-grained sandstone with whitish-grey bands.

Nearer the centre of the basin the Mercia Mudstone Group crops out beneath boulder clay in boreholes 73/48, 73/51 and 73/69. Borehole 73/69 has been dated palynologically as being Anisian (middle Triassic) in age. On the Isle of Man the Mercia Mudstone group contains an evaporite sequence which is thought to thicken offshore.

The Permo-Triassic rocks of offshore Cumbria are gently folded with fold axes aligned both NE–SW and NW–SE. Dips are generally low, although they may reach 20° or more. Numerous faults are seen on shallow seismic records but are difficult to correlate from one seismic line to another.

Jurassic rocks

Lower Liassic dark shales and argillaceous limestones occur in outliers at Great Orton, west of Carlisle, and at Prees and Wem in the Cheshire Basin but no Jurassic rocks have been found offshore. It is possible however that Rhaetic or Jurassic beds may be preserved in local outliers within the North-east Irish Sea Basin.

Quaternary deposits

Rockhead over the whole area is covered by Quaternary deposits, which in general upward succession are boulder clay, proglacial water-laid sediments and marine sediments (Pantin 1977). The reddish-brown, stiff boulder clay was deposited by the Irish Sea Devensian glaciation and locally occurs very close to the sea bed where it is covered by only a thin veneer of marine sediments. The proglacial beds are well bedded characteristically muddy, with sand and coarse silt laminae which contain no macrofauna. These beds are thought to have been deposited during the late Devensian (Pantin 1977), whilst the overlying sediments date from late Devensian to the present day. Mud, sandy mud and muddy sand is accumulating at the present time off the Cumbria coast in a belt extending from the Solway Firth to Morecambe Bay (Fig. 74). These muddy sediments contain large numbers of shells of the gastropod *Turritella communis* Risso. To the west the mud belt passes into sand and shelly gravel and to the east into dominantly sandy sediments near the Cumbria coast. Off the

TABLE 7

Generalized classification and correlation of the Permo-Triassic

Age		Lithostratigraphic Units		North-East Irish Sea Basin	St Bees Head	Vale of Eden
Triassic	Keuper	Mercia Mudstone Group		Keuper Marl Lower Saliferous Beds Keuper Waterstones	Stanwix Shale	
	Bunter	Sherwood Sandstone Group		Lower Keuper Sandstones St Bees Sandstones	Kirklinton Sandstone St Bees Sandstone	St Bees Sandstone
Permian	Upper			Manchester Marl	St Bees Shale St Bees Evaporites	Eden Shales and Evaporites Penrith Sandstone and Brockram
	Lower			Collyhurst Sandstone	Brockram	

eastern coast of the Isle of Man the shell content of the recent sediments is high and consists mainly of bivalve debris, confined to the coarse sand to fine gravel range. The thickness of the Quaternary deposits is generally between 20 and 40 m, with a maximum of 75 m recorded in shallow borehole 71/64.

Hydrocarbons

The Carboniferous beds which underlie the Permo-Triassic sediments throughout the offshore area are important as hydrocarbon source rocks, whilst the Permo-Triassic sequence contains potential reservoir rocks and cap rocks. In 1969 the two deep boreholes south of Walney mentioned above, 110/8–1 and 110/8–2, were drilled by the Gulf/NCB group to depths of 1307 and 3115 m respectively. Since that time

Cluff Oil have drilled in block 112/30 (Fig. 75) and a number of boreholes have been drilled by Hydrocarbons GB Ltd, a subsidiary of the British Gas Corporation, in block 110 in the Liverpool Bay area. A gas field has been discovered in this latter area and at the present time further appraisal wells are being drilled to determine whether the structure is of sufficient size to justify economic exploitation.

Acknowledgements

This chapter is published by permission of the Director, Institute of Geological Sciences. The writers would like to thank the following colleagues from the Institute of Geological Sciences for their help and criticism, H. M. Pantin, G. H. Rhys, A. J. Wadge and G. Warrington.

REFERENCES

ARTHURTON, R. S. and HEMINGWAY, J. E. 1972. The St Bees Evaporites — a carbonate-evaporite formation of Upper Permian age in West Cumberland, England. *Proc. Yorks. geol. Soc.* **38**, 565–92.

BOTT, M. H. P. 1964. Gravity measurements in the north-eastern part of the Irish Sea. *Q. Jl geol. Soc. Lond.* **120**, 369–96.

COLTER, V. S. and BARR, K. W. 1975. Recent Developments in the Geology of the Irish Sea and Cheshire Basins. *In* WOODLAND, A. W., ed. *Petroleum and the Continental Shelf of North-West Europe.* Applied Science Publishers, London. 61–73.

BOYD-DAWKINS, W. 1902. The Carboniferous, Permian and Triassic rocks under the glacial drift in the north of the Isle of Man. *Q. Jl geol. Soc. Lond.* **58**, 647–61.

GREENLY, E. 1919. The Geology of Anglesey. *Mem. geol. Surv.* H.M.S.O. London.

PANTIN, H. M. 1977. Quaternary sediments of the northern Irish Sea. *In* KIDSON, C. and TOOLEY M. J., eds. *The Quaternary History of the Irish Sea. Geol. Jl Spec. Iss.* **7**, 27–54.

B. N. FLETCHER, PH.D., and C. R. RANSOME, M.SC.
Institute of Geological Sciences, Ring Road Halton, Leeds LS15 8TQ.

17

Hydrogeology

C. K. PATRICK

The Lake District and its surrounding area (Plate I) can be divided into two distinct regions on the basis of their hydrological and hydrogeological characteristics. The central area of impermeable Lower Palaeozoic rocks forms an important surface water gathering ground, and contrasts sharply with the surrounding younger rocks which contain several major aquifers.

LOWER PALAEOZOIC

All Lower Palaeozoic rocks, with the exception of the most calcareous parts of the Coniston Limestone, have very low bulk permeabilities. Water balance calculations suggest that the annual infiltration to groundwater in these rocks is less than 40 mm (Wadge 1966). Intergranular permeabilities are negligible but fissures give the rock mass a low bulk permeability. Variations in fissure width, extent, interconnexion and recharge, directly or through the Drift, produce wide variations in groundwater conditions. As a result hydraulic conditions in, and yields from, these rocks are unpredictable. Boreholes generally yield less than 0.005 Ml/d although yields as high as 0.05 Ml/d may be obtained in areas of extensive shattering (Anon 1964). Springs with persistent discharges in excess of about 0.2 Ml/d may be drawing water from the zone of permafrost shattering.

No chemical analyses of normal groundwater have been published. Data derived from analyses of river water during baseflow conditions (Fig. 80) indicate that water derived from granites, Skiddaw Slates and Borrowdale Volcanics, and Silurian rocks form three chemically distinct groups. Mineral and chalybeate springs have been recorded in the area around Derwentwater and Keswick (Ransome 1848, Ward 1876, Eastwood 1921, Kipling 1961) and at Shap (Aveline and Hughes 1872). A warm spring occurs in Force Crag Mine [193 215] (Eastwood 1959, Shaw 1970). All except the Shap spring, which issues from the Coniston Limestone, rise from the Skiddaw Slates. Chemical analyses (Fig. 80) show that three of the springs are similar, suggesting a common origin associated with slow circulation of peaty water along calcite-bearing mineral veins.

LOWER CARBONIFEROUS

Lower Carboniferous limestones, sandstones and the Basal Conglomerate act as aquifers, separated by argillaceous rocks forming aquicludes. Bulk permeability is due almost entirely to fissures except in some of the calcareous sandstones where a secondary intergranular permeability has developed after leaching of the cement.

Intergranular and bulk porosities are always low. Low storage capacity in, and rapid drainage from fissures makes yields from boreholes unpredictable so that the aquifers have been developed only locally (Wilson and Frost 1964, Wadge 1966, Newson 1973). They can be considered as underdeveloped but future developments will probably concentrate on their roles in controlling river baseflow rather than on borehole construction, except in the confined parts of the aquifer (Newson 1973).

Water storage and transmission in the Carboniferous Limestone depend entirely on fissure size, extent and interconnexion. Differences in fissure development determine whether flow is dispersed through the rock mass (diffuse flow), giving a water table, relatively slow drainage and limited storage, or restricted to well-defined passages (conduit flow) with no water table, rapid drainage and negligible storage. Faults

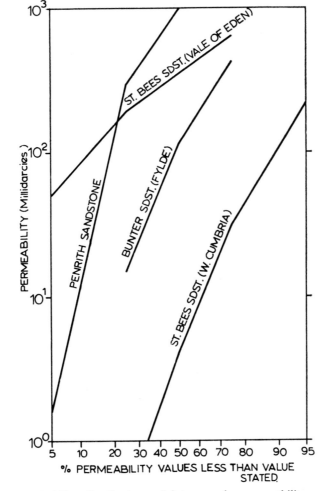

FIG. 76. Probability distributions of intergranular permeability values for samples from the Penrith Sandstone and St Bees Sandstone in West Cumbria and the Vale of Eden, and the Bunter Sandstone in the Fylde (based on Lovelock 1972).

may act as major conduits or impermeable barriers and old mine workings may provide extra storage (Smith 1924). Fissure characteristics and modes of formation have
been studied in the vadose (unsaturated) zone by Firman and Dickson (1968), and in
the epiphreatic and phreatic (saturated) zones by Ashmead (1969) and Patrick (in
prepn.). Earlier periods of weathering have left a range of fossil passages and enlarged
fissures which may still be open, occupied by water or abandoned, or plugged by
sediment (Smith 1924, Ashmead 1974, C. M. Bristow pers. comm.).

The Carboniferous Limestone is recharged at outcrop or through the Drift at
rates depending on Drift type and thickness (Smith 1924, Wadge 1966). Wadge gives
estimates of 420 mm/yr for Drift-free and 95 mm/yr for Drift-covered areas. Recharge
from surface watercourses occurs only in the Carnforth, Furness and Cross Fell areas
where impermeable rocks occur over or adjacent to the limestone. Active cave systems
are restricted to these areas (Ashmead 1974). On the West Cumbrian coast and in the
Vale of Eden the aquifer is confined by the Brockram and St Bees Shales. Limited
recharge, or discharge, takes place through the aquiclude when it is breached by faulting
(Smith 1924, Trotter et al. 1936). Some of the confined areas under the Brockram appear
to be dry (Smith 1924). Natural discharge takes place at springs (Anon. 1964, 1966) and
by lateral recharge into the Permo-Triassic sandstones in Furness (Fleet 1974, Anon.
1964). Spring discharges can be large, up to 1.5 Ml/d, but may fluctuate rapidly in
response to precipitation (cf. Newson 1973). Artificial discharges up to 10 Ml/d and
1 Ml/d are taken from mines (C. F. Johnston pers. comm.) and boreholes respectively
(Anon. 1966).

Groundwater in and draining from (Fig. 80) the Carboniferous Limestone is of
the normal calcium-bicarbonate type (Waltham 1974). The well-developed fissure
systems in the limestone which make it an apparently attractive source of water
resources also interfere with its exploitation (Newson 1973). Water storage is limited
by the rapid fissure drainage and boreholes will yield water only if they intersect
saturated fissures. Abstraction of water may alter hydraulic conditions by unplugging
or plugging fissures with sediment. Pollution of groundwater occurs readily as the
fissures give ready access to surface water and are very attractive to effluent disposers.

UPPER CARBONIFEROUS

In the Upper Carboniferous sandstones form aquifers separated by aquicludes of
mudstone and clay shale, which may be considerably thicker (Eastwood 1930,
Eastwood et al. 1931, 1968, Moseley 1954). The main aquifers occur in the Hensingham
Grit Group and Upper Coal Measures in north-west Cumbria (Wadge 1966) and in
the sandstones of the Pendle Grit and overlying sandstones in the Lancaster area.
Individually the aquifers are less productive than the Permo-Triassic sandstones but
are capable of giving yields as high as 0.5 Ml/d from a single borehole (Anon. 1964,
1966). Recharge averages 60 mm/yr and water balance calculations indicate that the
aquifer would be underdeveloped if mine drainage did not take place (Wilson and
Frost 1964, Wadge 1966).

Intergranular permeability occurs in coarse and conglomeratic sandstones (Earp
1955, Wilson and Frost 1964, Wadge 1966) but fissures are the main cause of the
rock's water-bearing properties. These are closely related to fissure size, extent, and
interconnexion, controlling bulk porosity, permeability and anisotropy, and are
related to lithology, bedding characteristics and tectonic setting. Major fissures occur
along fault planes. Minor fissures are present as joints and bedding plane partings

(Earp 1955). Most fissures close with depth so that permeability decreases downwards within a single aquifer and lower aquifers are generally less permeable than higher ones. Below depths of 250 m the majority of fissures are closed and the sandstones behave as aquifuges (Earp 1955, Wadge 1966). At greater depths fissures can give initially high discharges when intercepted but decrease rapidly to zero, showing that any recharge at depth takes place only very slowly. Faults may yield persistent discharges at depths below 300 m.

The alternations of aquifer and aquiclude give only restricted areas with unconfined conditions so that Upper Carboniferous aquifers are typically confined. The relationships between outcrop, dip and topography frequently give rise to artesian conditions. Recharge takes place at outcrop, below the Drift, from the sea or by upward leakage through aquicludes. In coal mining areas zones of secondary fissure permeability may provide recharge routes for the aquifers, and replenishment of storage in old workings. Natural discharges from the aquifer are to rivers and the sea, and by upward leakage through aquicludes. Borehole pumping has been undertaken on only a very limited scale in the Lancaster area (Anon. 1964) but massive mine pumping in the Cumbrian coalfield has been undertaken for many years.

Chemical analyses of mine drainage water are given by Wadge (1966). These are strongly influenced by infiltrated seawater but suggest that the normal groundwater is probably a sodium-sulphate water with subordinate amounts of chloride water. River baseflow analyses (Fig. 80) indicate that the aquifers discharge sodium-sulphate waters with much lower total dissolved solids contents than the mine drainage waters. This and the similarity to precipitation reflect the limited circulation in the aquifers and the scarcity of soluble minerals.

The Coal Measure rocks have been considered as a source of groundwater on a number of occasions. In the absence of mine drainage and old workings this would be realistic but at present mine drainage abstracts about 20% of the reliable yield, dominates the flow pattern and draws in seawater to pollute the aquifer. Mine workings appear to be attractive for water storage but will yield hard, acid waters subject to pollution from the surface or from other workings.

PERMO-TRIASSIC

These two systems contain both major productive aquifers in the area and two major aquicludes. The Penrith, and St Bees/Kirklinton Sandstones form the aquifers. Aquicludes are formed by the Brockram/St Bees Shales/Eden Shales rock unit, and the Stanwix Shales. Hydrogeological research in the Lake District has concentrated on these aquifers. Detailed investigations have been carried out by Lovelock (1971, 1972), Fleet (1974), Lovelock et al. (1975), Reeves (1975), and Monkhouse and Reeves (1977), in addition to general consideration in Geological Survey Memoirs, by the Surface Water Survey (Wilson and Frost 1964, Wadge 1966), River Authorities, and the North West Water Authority. The aquifers are considered to be the best potential groundwater resource in the area and will probably be exploited by direct abstraction to supply or as sources of water for regulating rivers.

Brockram, St Bees and Eden Shales and Stanwix Shales

The Brockram and its associated sandstones are highly cemented, have only erratically developed bedding and jointing and show a very low to zero bulk permeability. It is generally considered to be an excellent aquiclude or aquifuge (C. F. Johnston pers. comm.). The St Bees and Eden Shales act as an aquiclude containing minor aquifuges formed by evaporite horizons. The Magnesian Limestone has no hydrogeological significance. The lateral passage between the Brockram and the St Bees and Eden Shales provides an almost continuous aquiclude below the St Bees Sandstone except where it is broken by faulting in West Cumbria and in the Pennine Fault Belt. The Stanwix Shales are similar to the St Bees and Eden Shales and form the confining aquiclude over the Kirklinton Sandstone. Limited water movement will probably occur within them, as in the St Bees Shales, but faulting is never sufficiently intense to provide a breach.

Penrith Sandstone

This formation is laterally equivalent to the Brockram and parts of the Eden Shales and its transitional relationships with both play an important part in determining its lithological, structural and hydrogeological characteristics. It has been investigated in detail by Lovelock (1972) and Lovelock et al. (1975). Intergranular and fissure permeability are both present, and play significant roles in determining water storage and transmission. The highest porosity and intergranular permeability values are found in the uncemented and poorly cemented (iron oxide) coarse to very fine sand grade millet seed quartz sandstones south of Cliburn [359 527]. Silica cementation produces sharp reductions in intergranular porosity and permeability (Waugh 1970) and similar reductions are associated with the development of gypsum and calcite cements near the contacts with the Eden Shales and Brockram. Reductions in porosity may be offset locally by resolution of cement at and near outcrop. The relationships between intergranular and fissure flow were investigated by Lovelock et al. (1975) at Cliburn. They showed that aquifer transmissivity is dominated by fissure flow whereas the storativity is controlled by intergranular porosity. Fissures are 1 mm to 3 mm wide and contribute less than 1 % of the total porosity. In contrast it was shown that fissures contribute 93 % of the total transmissivity.

The Penrith Sandstone is confined by the Eden Shales to the east of the River Eden and by the Drift over most of the area to the west. A small Drift-free outcrop area occurs around Penrith. Recharge occurs at outcrop (530 mm/yr) and through the Drift (130 mm/yr). Additional recharge may also occur from the Carboniferous rocks either vertically, or laterally at faulted boundaries. The reliable yield of the aquifer is estimated as 189 Ml/d (Monkhouse and Reeves 1977). Natural discharges take place to the River Eden and its tributaries and through the Drift as springs. It is probable that these will cease if extensive groundwater development is undertaken (Monkhouse and Reeves 1977) and induced infiltration from the rivers will take place instead.

Intergranular permeability data (Fig. 76) show that the Penrith Sandstone is one of the best Permo-Triassic aquifers in north-west England. Yield-drawdown curves (Fig. 77) show that the aquifer behaves in a similar manner to other fissure-permeable sandstone aquifers (Ineson 1959), and confirm that it has the best performance of the three Permo-Triassic aquifers in north-west England.

Chemical analyses of river baseflow (Fig. 80) show that the groundwater in the zone of circulation is of the calcium-bicarbonate type with moderate to low total dissolved solids and low hardness. Agricultural pollution appears to be minimal (Monkhouse and Reeves 1977). Water confined below the Eden Shales and near the limit of their outcrop has higher total dissolved solids and sulphate concentrations as a result of leakage through, and surface drainage over, the shales. Close to the

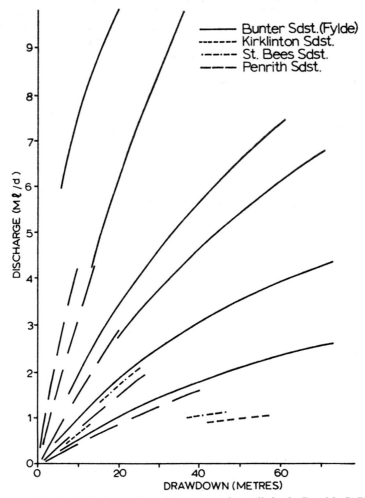

FIG. 77.   Borehole discharge-drawdown curves for wells in the Penrith, St Bees and Kirklinton Sandstones (Monkhouse and Reeves 1977), and the Bunter Sandstone in the Fylde (F. Law, pers. comm.).

Brockram waters with higher hardness develop due to solution of carbonate cement. Water within the St Bees Shales is extremely hard (T. I. Elliott pers. comm.).

Bunter, St Bees and Kirklinton Sandstones

These three formations form a single continuous aquifer up to 1000 m thick. The St Bees Sandstone is a massive to thin bedded medium to fine grained iron oxide cemented quartz sandstone with marl and shale partings which form about half the succession near the base, and decrease in importance upwards. Calcareous cementation occurs locally. Silica cementation does not occur. The Kirklinton Sandstone is a massively bedded cross stratified fine grained calcite cemented quartz sandstone with no marl or shale. The two formations have a transitional relationship. Detailed hydrogeological information is restricted to the St Bees Sandstone (Lovelock 1972, Reeves 1975, Monkhouse and Reeves 1977) but can be applied with caution to the Kirklinton Sandstone. The St Bees Sandstone has the lowest intergranular porosity and permeability values found in the Permo-Triassic sandstones investigated by Lovelock (1972) (Fig. 76). These are mainly due to compaction and cementation and only partly related to texture. Fissures are well developed and contribute 98 % of the total transmissivity in West Cumbria. In the Kirklinton Sandstone the porosity, permeability and specific yield values will probably be higher than those in the St Bees Sandstone and transmissivity will be more closely related to texture.

The aquifer is confined by Drift over most of the outcrop (Wilson and Frost 1964, Wadge 1966) and by Keuper Marl and Stanwix Shales when followed downdip. Recharge rates average about 115 mm/yr giving a yield of about 250 Ml/d (Wadge 1966). Lateral recharge from the Carboniferous Limestone occurs in Furness (Wilson and Frost 1964) and discharges to this aquifer may take place across faults in West Cumbria. Borehole yields of 1 to 2 Ml/d can be expected from all parts of the aquifer except in the Flookburgh area and in the lower parts of the aquifer. Yield-drawdown data (Fig. 77) indicate that wells conform to the type curves for sandstone aquifers (Ineson 1959) but will be less efficient than those in the Fylde Bunter and Penrith Sandstones.

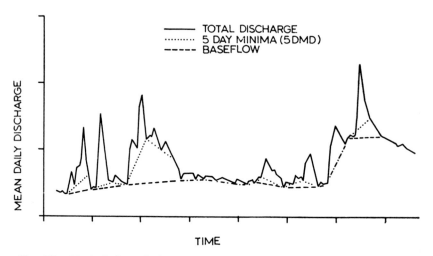

FIG. 78. Typical river discharge hydrograph showing Total Discharge, Five Day Minimum Discharge (5DMD) and Baseflow (data supplied by the Institute of Hydrology).

The groundwater is a hard sodium-bicarbonate water with moderate total dissolved solids concentrations. Water from the Kirklinton Sandstone is harder than that from the St Bees Sandstone. Saline intrusion occurs in areas of heavy pumping near Barrow (Fleet 1974) and probably occurs to some extent all along the coast. Water confined below the Stanwix Shales and close to the Eden and St Bees Shales will have higher hardness, total dissolved solids and sulphate concentrations than elsewhere.

JURASSIC, QUATERNARY AND RECENT

The Lower Lias shales and thin limestones cannot be expected to yield significant amounts of water (Wadge 1966) and have not been exploited, except by shallow wells for farm supplies.

The hydrogeology of the Quaternary and Recent deposits has been considered in all the Geological Survey Memoirs and by Wilson and Frost (1964), Wadge (1966), the Water Resources Board (Anon. 1970) and the North West Water Authority. Aquifers occur within the sand and gravel grade sediments; and the silts, clays and Boulder Clay act as aquitards and aquicludes. All supplies are likely to fluctuate rapidly in response to variations in precipitation. The water is generally hard, due to

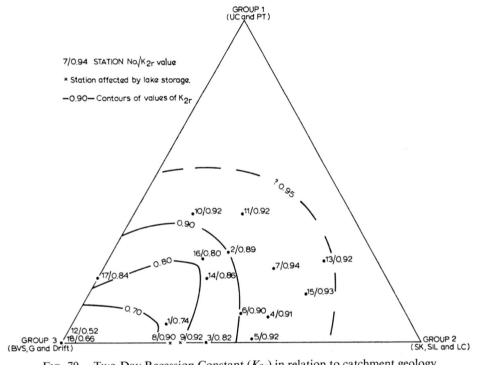

FIG. 79. Two-Day Recession Constant (K_{2r}) in relation to catchment geology (solid and drift). Group 1 — Upper Carboniferous and Permo-Triassic rocks; Group 2 — Skiddaw Slates, Silurian and Lower Carboniferous rocks; Group 3 — Borrowdale Volcanic rocks, granites and Drift. Stream gauging stations:— 1. R. Derwent at Camerton [037 305], 2. R. Eden at Warwick Bridge [471 567], 3. Burns Gill at M6 motorway [608 006], 4. R. Mint at Mint Bridge [524 944], 5. R. Lune at Killington [622 905], 6. R. Rawthey at Brigg Flats [640 912], 7. R. Kent at Sedgwick [508 874], 8. R. Leven at Newby Bridge [371 863], 9. R. Crake at Low Nibthwaite [294 882], 10. R. Calder at Calder Hall [035 045], 11. R. Glenderamackin at Threlkeld [323 248], 12. Inflow to Windermere [372 031], 13. R. Eden at Kirkby Stephen [773 097], 14. Force Beck at M6 motorway [579 134], 15. R. Petterill at Harraby [412 545], 16. R. Keer at High Keer Bridge [522 719], 17. Burrow Beck at [488 495], 18. R. Duddon at Duddon Hall [195 895]. (Data for stations 1, 3–11, 13–15 supplied by the Institute of Hydrology).

bicarbonate or sulphate, and may be ferruginous. Brackish water occurs in the marine deposits along the coast. The aquifers are liable to pollution from agricultural and industrial discharges. The main significance of Quaternary and Recent deposits in the context of regional hydrogeology is that they form the confining bed over the main aquifers and control infiltration and influent water chemistry.

STREAM DISCHARGE AND WATER CHEMISTRY

River low flow discharges and water chemistry are controlled by catchment geology and hydrogeology. Their interrelationships have been investigated (Patrick in prepn.) using the results of an analysis of the relationships between catchment geology and two indices of river discharge made available by the Institute of Hydrology, Wallingford, and supplemented by data from other sources (McClean 1940,

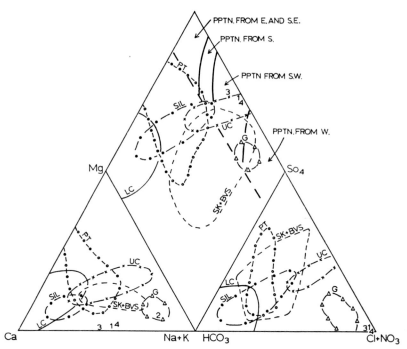

FIG. 80. Representative water analyses of baseflow in rivers draining catchments composed of Skiddaw Slates and Borrowdale Volcanic rocks (SK + BVS), Silurian rocks (SIL), granites (G), Lower Carboniferous (LC), Upper Carboniferous (UC), and Permo-Triassic (PT) rocks. Average values of total dissolved solids (in mg/l) are SK + BVS = 70, SIL = 125, G = 91, LC = 300, UC = 40, PT = 365.

Analyses of waters from mineral springs indicated by numerals; 1 — Manesty, 2 — Holy Well Spa, 3 — Shap Spa, 4 — Brandelhow. Total dissolved solids (mg/l) values are 1 = 23881, 2 = 7390, 3 = 6182, 4 = 23237.

Analyses of precipitation supplied by winds from different directions indicated by areas defined by heavy lines on diamond plot. Mean total dissolved solids concentration in precipitation = 5–10 mg/l.

Sutcliffe and Carrick 1973), chemical analyses of river water during low flow periods (Anon. 1970, Sutcliffe and Carrick 1973, Gascoyne 1974, Laxen 1975) and simplified classifications of catchment geology. River discharge (Fig. 78) between flood events (the baseflow) has been examined as it reflects discharge from groundwater storage. Two indices have been calculated and related to the geology. The Baseflow Index (BFI) indicates the proportion of the total annual discharge derived from long-term storage within rocks, and the storage abilities of the rocks in the catchment. BFI values calculated for individual stream gauging stations have been used to estimate average values for major stratigraphic subdivisions (Table 8). These confirm general conclusions on storage based on purely hydrogeological considerations.

TABLE 8.   Values of Baseflow Index (BFI) for major formations

Formation		BFI
Skiddaw Slates		0.27
Borrowdale Volcanic Series		
overall average		0.51
outcrop west of Scafell		0.32
outcrop east of Scafell		0.68
Silurian		0.34
Granites		0.30
Lower Carboniferous		0.27
Upper Carboniferous	*less than*	0.10
Permo-Triassic		0.69
Glacial gravels	*about*	1.00

The Two-Day Recession Constant (K_{2r}) defines the shape of the baseflow hydrograph through the relationship:

$$Q_2 = Q_0 K_{2r}$$

where Q_0 and Q_2 are discharges at times zero and two days later. The stream Recession Constant depends on the summation of the storage-discharge characteristics of all formations in the catchment. Its relationship to catchment geology has been studied using three groupings of formations designed to reflect their groundwater storage and release behaviour. Fig. 79 shows that a relationship exists between K_{2r} and geology. Similar plots ignoring Drift deposits produce no simple pattern, demonstrating that river recession depends on total catchment geology even though discharge from the Drift takes place rapidly.

River water chemistry is closely related to catchment geology and hydrogeology as these determine the routes by which water enters the streams, the time spent in storage and the minerals available for solution. Under baseflow conditions water chemistry in unpolluted rivers will approximate to the chemistry of the water draining from the contributing aquifers, and can be used in place of borehole sampling and analyses. Analyses of baseflow from 56 sites in the Lake District have been plotted in Fig. 80. This shows that waters from different formations fall within well-defined fields with only limited overlap.

260 C. K. PATRICK

Acknowledgements

I wish to acknowledge the assistance given in the preparation of this chapter by Dr C. M. Bristow, English China Clays; Lovering Pochin and Co. Ltd; Dr T. Cairney; Dr T. I. Elliot, Albright and Wilson; Dr R. J. Firman; O. Gregory, National Coal Board (Western Area); C. F. Johnston, British Steel Corporation, Beckermet Ore Mines; Charlotte Kipling; W. T. Shaw; the Institutes of Geological Sciences and Hydrology; the North West Water Authority; and Water Research Centre (Medmenham Laboratory).

REFERENCES

ANON. 1959. The future of non-ferrous mining in Great Britain and Ireland, a symposium. *Inst. Min. and Metall.*
—— 1964. North Lancashire Rivers Hydrological Survey. *Min. Ho. Loc. Govt.* v + 117 pp. H.M.S.O. London.
—— 1966. Cumbrian Rivers Hydrological Survey. *Min. Ho. Loc. Govt.* v + 87 pp. H.M.S.O. London.
—— 1970. Morecambe Bay Feasibility Study. Water Resources Board. Sir Alexander Gibb and Partners.
ASHMEAD, P. 1969. The origin and development of caves in the Morecambe Bay area. *Trans. Cave Res. Gp. G.B.* 11, 201–08.
—— 1974. The caves and karst of the Morecambe Bay area. in — Waltham, A. C. (Ed.), *Limestones and Caves of North West England.* David and Charles.
AVELINE, W. T., HUGHES, T. McK. 1872. The geology of the country between Kendal, Sedbergh, Bowness and Tebay. *Mem. geol. Surv.* vii + 94 pp. H.M.S.O. London.
EARP, J. R. 1955. The geology of the Bowland Forest Tunnel, Lancashire. *Bull. geol. Surv.* No. 7, 1–12.
EASTWOOD, T. 1921. The lead and zinc ores of the Lake District. *Mem. geol. Surv. spec. Rep. Miner. Resour. Gt. Br.* 22, iv + 56 pp. H.M.S.O. London.
—— 1930. The geology of the Maryport district. *Mem. geol. Surv.* xiii + 137 pp. H.M.S.O. London.
—— 1959. The Lake District Mining Field. 149–74 in Anon. (1959) q.v.
—— DIXON, E. E. L., HOLLINGWORTH, S. E. 1931. The geology of the Workington district. *Mem. geol. Surv.* xvii + 304 pp. H.M.S.O. London.
—— HOLLINGWORTH, S. E., ROSE, W. C. C., TROTTER, F. M. 1968. Geology of the country around Cockermouth and Caldbeck. *Mem. geol. Surv.* x + 298 pp. H.M.S.O. London.
FIRMAN, R. J., DICKSON, J. A. D. 1968. The solution of gypsum and limestone by upward flowing water. *Mercian geol.* 2, 401–08.
FLEET, M. 1974. Report on geophysical logging, Barrow-in-Furness, for North West Water Authority. Unpublished report. Wat. Res. Centre, Medmenham.
GASCOYNE, M. 1974. A study of the chemistry of natural waters of an upland catchment and its application in separation of the flood hydrograph. *Unpub. M.Sc. thesis, University of Lancaster.*
INESON, J. 1959. Yield-depression curves of discharging wells, with particular reference to Chalk wells, and their relationship to variations in transmissibility. *J. Instn. Wat. Engrs.*, 13, 119–63.
KIPLING, C. 1961. A salt spring in Borrowdale. *Trans. Cumb. Westm. Antiq. Archaeol. Soc.* 61 (N.S.), 57–70.
LAXEN, D. P. H. 1975. Water pollution and the highway: a framework for assessment. *Unpub. M.Sc. thesis, University of Lancaster.*
LOVELOCK, P. E. R. 1971. Core analysis results from boreholes in the St Bees Sandstone, West Cumberland. Unpublished report, WD/ST/71/5. *Inst. geol. Sci.* London.
—— 1972. Aquifer properties of the Permo-Triassic sandstone aquifers of the United Kingdom. *Unpublished Ph.D. thesis, University of London.*
—— PRICE, M., TATE, T. K. 1975. Groundwater conditions in the Penrith Sandstone at Cliburn, Westmorland. *J. Instn. Wat. Engrs.* 29, 157–74.
McCLEAN, W. N. 1940. Windermere basin: rainfall, run-off and storage. *Q. Jl Roy. Met. Soc.* 66, 337–62.
MONKHOUSE, R. A., REEVES, M. J. 1977. A preliminary appraisal of the groundwater resources of the Vale of Eden, Cumbria. Tech. Note No. 11, *Cent. Wat. Planning Unit, Reading.*
MOSELEY, F. 1954. The Namurian of the Lancaster Fells. *Q. Jl geol. Soc. London.* 109, 423–54.
NEWSON, M. D. 1973. The Carboniferous Limestone of the U.K. as an aquifer rock. *Geog. J.* 139, 294–305.
RANSOME, T. 1848. Analysis of a saline spring in a lead mine near Keswick. *Mem. Manchr. Lit. and Phil. Soc.* 8, 399–401.

REEVES, M. J. 1975. Proposals for the development of the Triassic sandstone aquifer in the Calder Valley, Cumbria. Tech. Note No. 5. 15 pp. *Cent. Wat. Planning Unit, Reading.*

SHAW, W. T. 1970. Mining in the Lake Counties. *Dalesman Pub. Co. Clapham.*

SMITH, B. 1924. Haematites of West Cumberland, Lancashire and the Lake District. *Mem. geol. Surv. spec. Rep. Miner. Resourc. Gt. Br.* **8,** vi + 236 pp. H.M.S.O. London.

SUTCLIFFE, D. W., CARRICK, T. R. 1973. Studies of mountain streams in the English Lake District. I. pH, Calcium, and the distribution of invertebrates in the River Duddon. *Freshwat. Biol.* **3,** 437–62.

TROTTER, F. M., HOLLINGWORTH, S. E., EASTWOOD, T., ROSE, W. C. C. 1936. The geology of the Gosforth district. *Mem. geol. Surv.* xii + 136 pp. H.M.S.O. London.

WADGE, A. J. 1966. Hydrogeology. 23–29 in Anon. (1966) q.v.

WALTHAM, A. C. (editor). 1974. Limestones and caves of North West England. *David and Charles.* 477 pp.

WARD, J. C. 1876. The geology of the northern part of the Lake District. *Mem. geol. Surv.* xii + 132 pp. H.M.S.O. London.

WAUGH, B. 1970. Petrology, provenance and silica diagenesis of the Penrith Sandstone (Lower Permian) of Northwest England. *J. Sedim. Petrol.* **40,** 1226–40.

WILSON, A. A., FROST, D. V. 1964. Hydrogeology. 27–31 in Anon. (1964) q.v.

C. K. PATRICK. B.SC.

Department of Environmental Sciences, University of Lancaster, Lancaster LA1 4YQ

18

Environmental Geology

C. K. PATRICK

Environmental Geology is a fashionable term coined during the last decade to cover a wide range of applied geological and geomorphological topics. In common with other 'environmental' offshoots of the natural sciences no generally agreed definition exists, except that man is somehow involved. Three aspects of man's impact on the operation of geological processes in the Lake District are discussed in this chapter as an indication of the scope of environmental geology; construction in the intertidal area, water resources development and motorway construction. Each indicates a different degree of man's involvement in the environment but all indicate the inevitability of changes, even from the most subtle alterations, when natural systems are upset.

CONSTRUCTION IN THE INTERTIDAL AREA

The intertidal areas of Morecambe Bay and the Solway and their associated estuaries (Fig. 81), including the Duddon and Walney Channel, have been affected to varying extents by, and have influenced, construction activities since the early 19th century. Morecambe Bay has been altered more extensively than the Solway and provides excellent examples of the interplay between natural and induced changes. Structures in, and around the margin of, the Bay have altered the environment and caused permanent changes. Some of these merely accelerated natural processes but others have caused artificial alterations ranging from minor changes restricted to the area around the structure to major changes affecting large areas at considerable distances. In extreme cases these changes may damage, or negate the purpose of, the original structure. All changes result from interference with the delicate balances which exist in intertidal and estuary areas (Kestner 1966, 1972). Construction works alter tidal propagation, current patterns and velocities, and the accretion/erosion balances. The resulting changes may achieve a new balance rapidly, over a period of a century or more, or may upset a previously stable situation replacing it with an oscillatory one. Changes in Morecambe Bay and its estuaries have been investigated by Inglis and Kestner (1958), Kestner (1961, 1970), the Hydraulics Research Station (Anon. 1970b) and Patrick (1970, 1978).

Large-scale changes in estuary geometry have been investigated by Patrick (1970) using the prism-area method. O'Brien's (1931) original method, using one observation from each estuary, has been extended (Patrick 1970) to show that in the estuaries of the Kent and Lune estuary cross-sectional areas are related to the volumes of water stored between high and low water levels, the tidal prisms (Fig. 82). These prism-area relationships indicate the control exerted by tidal processes on the

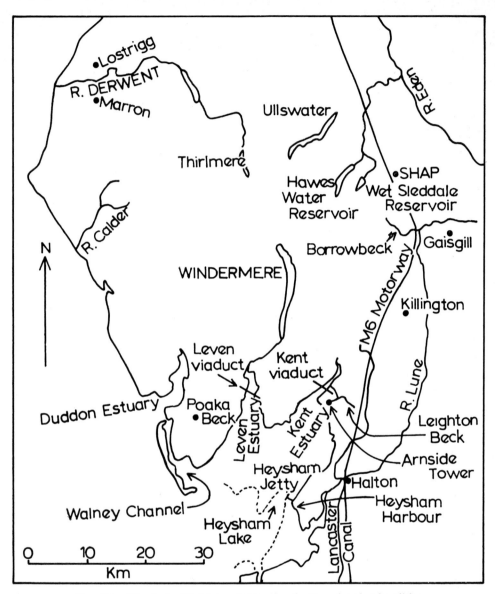

FIG. 81.   The Lake District and Morecambe Bay showing localities

gross geometry of the estuaries, and the unity of each. Data for the Lune estuary (Fig. 82) before and after the major reduction in prism which followed the construction of training walls demonstrate that a single prism-area relationship exists for each estuary which not only illustrates the existing state of the estuary but can be used to predict the nature and extent of changes which will result from changes in estuary prism. Reduction in prism normally follows reclamation or accelerated deposition caused by construction. Prediction of the new prism allows estimates of adjusted cross-section areas downstream to be made. In the Lune estuary a 50% reduction in prism was accompanied by reductions in cross-section areas of up to 30%. In the Kent and Leven estuaries reclamation and accretion associated with viaduct and embankment construction caused large reductions in prism which must have reduced depths in both estuaries and in the upper part of Morecambe Bay (Lockett 1976, Patrick 1978). Saltmarsh extensions probably reflect the resultant general shoaling of the area (Gray 1972). Future proposals may affect the prism in the area by reclamation or accretion, giving reduced prism, or dredging, giving increased prism. Patrick (1970) considered that accretion associated with construction of the proposed Kent viaduct would reduce prism by up to 45% giving a reduction in depth downstream between 0.2 m and 0.6 m. This could be accompanied by saltmarsh advances of up to 200 m. If existing rates of accretion and saltmarsh extension can be applied these changes could occur within 10 to 20 years. Construction of any of the barrages or poldered reservoirs for water storage in the inter-tidal area would reduce tidal prism with consequent reductions in water depths (Anon. 1970b, Kestner 1972). Proposals for extracting iron ore from the Duddon estuary (C. M. Bristow pers. comm.) would presumably involve polder construction, with similar results. These would not extend into Morecambe Bay but could alter conditions in the Walney Channel. Increases in prism could result from gravel and sand extraction in the inter-tidal area. This could be relatively large in the estuaries. In the Lune, for example, extraction of only part of the post-glacial gravel under the saltmarsh could cause a large increase in prism. In the Bay changes would be proportionately much smaller but could still affect local depositional systems controlled by prism-area considerations (e.g. Heysham Lake). Experience with the removal of over 0.5 million cubic metres of sand from Middleton Sands for fill at Heysham Nuclear Power Station (O'Connor 1975) suggests that replenishment takes place rapidly so that ill-effects would result from interruption of the sediment transport path rather than changes in prism.

Morecambe Bay and its estuaries are all in a state of dynamic equilibrium in which sediment carried in by the flood tide is balanced, over a period of months or years, by the amount of sediment removed by the ebb tide and the river flow. This is known as a regime condition. Regime conditions (Inglis and Kestner 1958) are only semi-permanent as estuaries eventually fill with sediment (Marker 1967) due to natural interferences with the regime balance. Three factors are involved in maintaining a regime condition, the flood and ebb tide velocities and the mobility of the low-water channel. If any one is disturbed then the system will adapt its geometry until new conditions which satisfy the regime condition are established. Tidal velocities can be altered by changes in estuary geometry causing changes in the propagation of the tidal wave. Decreased durations of the flood or ebb tides will increase current velocities and the amounts of sediment transported into or out of the estuary. Increased durations will cause decreases in velocities and sediment transport. Low-water channel mobility is critical to the maintenance of a regime condition as it provides the only persistent mechanism for removing the fine cohesive sediment normally deposited in the middle and upper reaches of estuaries. Low-water channels meander and migrate laterally undercutting sediment banks, placing sediment in suspension and carrying it out to sea. When a regime condition exists this removal balances the supply of sediment by the flood tide so that the total mass of sediment in the estuary is constant, although its distribution and the details of the estuary will vary. If the mobility of the low-water channel is impeded the regime balance will be upset in favour of accretion and the estuary will fill. Classic examples of all degrees of interference with the low-water channel, and the consequences, occur around Morecambe Bay. The Duddon represents one extreme as minimal alterations have left the wide intertidal area swept by a freely meandering low-water channel unaffected, and a completely natural regime condition has developed. At the other extreme the Lune estuary contains extensive training walls built to improve navigation to Lancaster by stabilising the main channel and increasing the depths. Instead, the walls interfered with the movement of the low-water channel, allowed extensive accretion (up to 5 m) behind the training walls and caused a large loss of prism (Fig. 82). The navigation channel has been stabilised but depths are similar to those before, as these are related to the normal river thalweg. In the Kent estuary (Fig. 83) a succession of small changes to the railway viaduct have been accompanied by changes in the low-water channel behaviour. These illustrate the very delicate balance which exists in these estuaries (Patrick 1970). The railway viaduct and embankments have disrupted the normal pattern of mutually evasive flood and ebb currents in a sinuous estuary, and their associated channels. The present stable low-water channel below the viaduct is due to the coincidence of the normal flood current channel with the ebb channel. This combined flood–ebb channel has anchored the low-water channel in the present position since 1953. Prior to that time four other sets of conditions related to different viaduct structures existed.

Intertidal areas are controlled by a complex set of processes conveniently summarised as regime behaviour, prism-area relationship, tidal current flow pattern, and low-water channel behaviour. These examples demonstrate, however, that no situation can be considered wholly in terms of any one alone as all will be inevitably involved. The intertidal area is a highly complex system which is always in a state of very delicate balance. Proposals to interfere with it must be based on the best available understanding of the way it operates, and the nature and extent of any changes which may be produced, so that minimal interference and harmful consequences will result. If possible structures should aim to conform with natural patterns and processes so that they blend easily into the environment (Kestner 1972).

WATER RESOURCES DEVELOPMENTS

Reservoirs and their associated aqueducts and tunnels have formed the most important single group of man-made features in the Lake District for a century. During this time they have become increasingly important as both sources of water and as sources of conflict between water supply and conservation interests. This conflict has emphasized the effects of reservoirs on amenity and has tended to overlook their equally important, and more far-reaching, effects on the environment as a whole. The influence of the environment on the choice of sites for reservoirs, and their construction, is well documented (Lewis 1921, Taylor 1951, Atkinson 1955, Walters 1962, Kennard and Knill 1969, Knill 1970, Morton 1973) but the effects of reservoirs and associated works on the environment have received little attention in the United Kingdom, and virtually none in the Lake District.

The choice of the central Lake District as a major source of water for Manchester was determined by the combination of high rainfall, impermeable rock and deep valleys (Walters 1936). The precipitation is the highest in England, the highly impermeable rocks give high surface runoff and excellent dam foundations, and the valley cross-sections give good reservoir area–storage relationships. Outside the central Lake

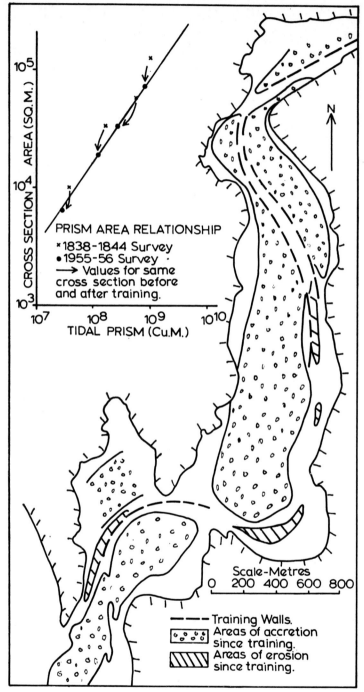

PRISM AREA RELATIONSHIP
× 1838-1844 Survey
• 1955-56 Survey ·
→ Values for same cross section before and after training.

Scale-Metres
0 200 400 600 800

– – – Training Walls.
Areas of accretion since training.
Areas of erosion since training.

FIG. 82. The estuary of the River Lune showing training walls, areas of subsequent accretion and erosion, and the prism-area relationship

District large reservoirs have been constructed on Boulder Clay at Poaka Beck, near Barrow, and Killington. Extensions to the latter were proposed but have now been rejected (Anon. 1971, 1976a) and new developments at Marron and Lostrigg Beck are still under consideration (Knill 1970, Anon. 1976b). Estuary barrages have also been proposed.

The environmental effects of water resource developments arise almost entirely from their disruption of the local hydrological cycle. The nature and scale of the effects depend on the type of scheme involved and its operation. Examples of all the normal types of scheme exist, or are proposed, for the Lake District area (Anon. 1970a). The three main reservoirs abstract water for direct supply. These impound the headwater portion of each catchment and normally discharge only compensation water, at a rate designed to maintain fishing and other riparian interests, and occasional higher discharges during overflow periods. The Haweswater catchment is extended to north and south by catchwaters which collect part or all of the flow of streams intercepted along the hillsides. These major sources are supplemented by large-scale abstractions from Ullswater and Windermere which are pumped into storage or directly into supply. Future developments may involve pumped storage, river regulation, water transfer or estuary abstraction either alone or in one of a variety of possible combinations (Anon. 1976a). Pumped storage schemes will abstract water from the lower reaches of rivers, take it into storage, and use it to augment the normal reservoir yield. River regulation will use natural river channels as aqueducts to transfer water from storage to downstream abstraction points and will aim to maintain river flow at or above a predetermined value by augmenting low flows and, if combined with pumped storage, reducing flood flows by taking them into storage. Regulation of flow in the Rivers Eden and Calder may also be achieved using groundwater abstracted from the Permo-Triassic aquifers (Reeves 1975, Monkhouse and Reeves 1977). Water transfers have been proposed to take water from the River Eden, via Haweswater into the River Lune and, after abstraction at Halton, into the River Wyre. Estuary abstraction may involve large-scale impounds or, more probably, abstraction within or at the head of the estuary with storage in nearby bunded reservoirs. Indications of the wide range of possible consequences of these schemes can be gained by consideration of three general sets of problems; the effects on river channels and estuary stability, reservoir sedimentation, and groundwater.

River channels are affected by water resource developments because the natural discharges, and particularly the flood–drought sequence, are severely modified or eliminated. Rivers receiving compensation water suffer a second alteration as they receive reduced suspended sediment loads, as these have been deposited within the reservoir. River channels adjust their geometry, cross-section and plan, to achieve equilibrium with the dominant discharge and sediment load (Harvey 1969, Gregory and Walling 1973). All rivers involved in water resource schemes will have dominant discharges differing from those which existed under natural conditions so that channel cross-sections and overall plan will change by erosion or deposition until a new equilibrium is established (Gregory and Park 1974). Cross-sectional changes will take place most readily if the bedload material is sandy and the mean discharge is relatively high (e.g. a regulated river) but with streams fed by compensation water, below abstraction points and with coarse bed material changes will be slow or non-existent. Changes in channel plan will also occur but will achieve a new equilibrium less rapidly than the changes in cross-section. In rivers fed by compensation water, the River Eden below the abstraction point for Haweswater and in the River Lune if the Borrowbeck

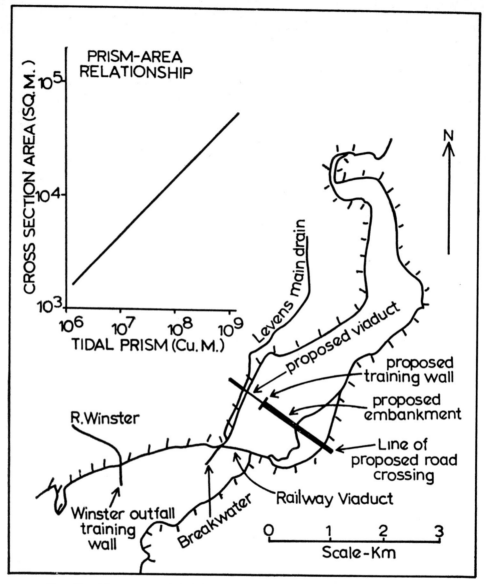

FIG. 83. The estuary of the River Kent showing the existing railway viaduct, details of the proposed road viaduct, and the prism-area relationship

reservoir is filled by pumped storage, the new dominant discharges will be less than the pre-existing natural ones, so that reductions in channel cross-section and changes in plan will result. In the River Eden above the abstraction point for Haweswater, the River Calder and the Lune, the new dominant discharges with flow regulation will be greater than the natural dominant discharge. This will increase cross-sectional areas and may alter the channel plan. Here flooding should not be a problem, if storm predictions are effective, but general erosion, loss of land and possible damage to structures may occur. The efficient operation of river regulation schemes can be adversely affected by changes in the channel as detailed information on the time of travel of water from reservoir to river intake, and a certainty that all the water released will reach the intake, are required. Continuous variations in channel geometry during the early life of a regulated river scheme will alter travel times introducing uncertainty into the operational procedures. In flood plain channels the passage of an artificial discharge pulse will be accompanied by temporary abstraction of water into bank storage, which will later be released gradually as part of the river baseflow. This will reduce the total volume of water reaching the abstraction point rapidly. In extensive flood plains channel adjustments will have little effect on bank storage behaviour as they will make only relatively small alterations to the volume of alluvial material but in valleys with limited alluvium reductions in channel cross-sections could cause significant changes with accompanying variations in apparent losses during the early stages of a scheme.

Estuaries will be drastically altered by water resources schemes involving construction in the intertidal area and less dramatically by changes in river discharge patterns. The elimination or reduction of flood discharges will alter the regime balance in the estuary, allow deposition to occur and reduce the tidal prism. The head of the estuary will eventually be colonized by vegetation and the transition to dry-land will be commenced. In the lower estuary reduced tidal scour will permit accretion with general decreases in mean depth. Alterations in the low-water channel pattern and associated additions to dry-land can have intriguing economic and political consequences.

Reservoirs and regulated lakes are similar to normal lakes in their abilities to act as sediment traps and to modify water chemistry. Their distinction from normal lakes lies in the different patterns of water levels caused by withdrawal of water from storage. Water levels in natural lakes fluctuate over only a small range during the year (e.g. Ullswater had a range of 2 m). In reservoirs the range can be up to 17 m (Donovan and Archer 1975) but is normally less (Gill and Bradshaw 1971). The median water level in a lake changes very little during the year but in reservoirs the summer level is considerably below that in winter. These differences result in marked differences in the sedimentary processes operating at the waters edge, and in the size of the beach developed. In lakes the consistent water level concentrates wave activity over a narrow vertical zone and allows an equilibrium beach profile, with minor seasonal variations, to develop. In contrast the wide variations in water level in reservoirs cause erosion and beach development over an extended zone, giving the wide beach strip to which conservation organizations frequently object, limit the development of an equilibrium beach and allow reworking of sediment into deeper parts of the reservoir (Donovan and Archer 1975). This encourages shoreline erosion, further increases in beach width and small-scale losses in reservoir storage. Beach and associated processes were studied in Haweswater during a period of very low water levels in 1973 by Donovan and Archer (1975), and Orrick (1976) has carried out a pilot study on part of the shoreline of Windermere. Both studies have important implications for successfully predicting and managing the evolution of reservoir and regulated-lake margins. Orrick's work is particularly important as it indicates, for the first time, the nature and extent of the damage which can be caused on lake beaches by visitors and their activities. It suggests that without careful management the recently won increased access to reservoir shores may be instrumental in increasing erosion, and the size of the objectionable beach.

Methods of camouflaging reservoir beaches using plants able to tolerate submergence have been investigated by Gill (1970) and Gill and Bradshaw (1971). Their primary aim has been to conceal the beach but any vegetation which could be established would help to protect lake margins from wave attack and limit beach development.

Groundwater will be affected to the greatest extent by schemes utilizing groundwater abstraction, and to lesser extents by reservoirs, regulated rivers, tunnels, aqueducts and canals. Groundwater abstraction in unconfined aquifers lowers the water table, reduces or cuts off spring flows, alters soil water levels and may induce infiltration from rivers (Monkhouse and Reeves 1977). In confined aquifers surface effects will be negligible but increased leakage through aquitards may introduce water with low quality into the aquifer and accelerate solution within the aquitard. This could cause progressive alterations in aquitard properties by the development of pipes in sandy clay, solution of mineral veins and evaporite horizons (R. J. Firman pers. comm.; cf. Burgess and Holliday 1973), and reductions in permeability if saline waters are induced to infiltrate through the aquitard (Hardcastle and Mitchell 1976). Reduced permeability in the aquitard could significantly affect the yield of the groundwater resource if leakage was expected to make a significant contribution. Reservoir construction has probably had only limited effects on the groundwater as all the major dams are built on impermeable foundations. Proposals for reservoirs at Marron, Lostrigg Beck (Knill 1970) and in Morecambe Bay, and an earlier site for Wet Sleddale, could have detrimental effects on the groundwater and the reservoir yields as the dams will, or would have been, built on permeable foundations. At Marron alluvial material may allow leakage to occur unless extensive precautionary measures are undertaken. The site is not ideal and was chosen largely because it lies outside the National Park. Knill (1970) comments that the boundaries of the Park were not drawn with regard to possible future reservoir sites. At Lostrigg Beck the reservoir will overlie old mine workings which have induced secondary fissure permeability in the Coal Measure rocks. This may cause leakage which, together with the load imposed by impounding water, may accentuate the permeability if subsidence is not yet complete. At Wet Sleddale possible leakage under the dam through the Carboniferous Limestone could have been prevented by grouting but previously plugged fissures, inherited from earlier periods of weathering (Ashmead 1974), and not invaded by grout, might have been flushed out after filling as hydrostatic pressure and seepage built up. In Morecambe Bay the poldered reservoir proposed on the saltmarsh near Silverdale (Anon. 1976a) could have a daily seepage loss of 5 to 20 Ml/d, depending on dimensions and waterproofing measures. Most of this will be discharged at the rising at Arnside Tower increasing its mean daily discharge by two to twenty times. The discharge from the rising will then enter Leighton Beck, be pumped over the sea wall into the barrage in the Kent estuary and then back into the polder. Regulated rivers will affect groundwater levels in alluvium because of the consistently high river levels. These may increase the need for land drainage and interfere with outfalls to the river.

MOTORWAYS

Modern high-speed roads and motorways were introduced into the Lake District area in 1970 with the opening of the M6 Motorway between Carnforth and Carlisle (Fig. 81). Since then several sections of new road have been constructed and major improvements have been made to, and are in progress on, other roads. When these developments are completed nearly 200 miles of high-speed road will have been built in the Lake District area. Two-thirds of these will be in the upland areas. High-speed roads are important new additions to the environment and must be distinguished from pre-existing roads because of their scale, both during and after construction, their unified design, the extent to which they interfere with the natural environment, and because they introduce a wholly new physico-chemical system into the area which interacts with adjacent established natural systems. Major roads are an example of a totally man-made system imposed on, and cross-cutting, existing natural ones. Their design makes little allowance for the resulting interactions, unless these are likely to affect the construction or use of the road. Problems arise when the motorway system interacts with, and interrupts, adjacent natural systems. These interactions have received only a limited amount of attention in the United Kingdom, partly because of their novelty but also because the official view has been that major roads present no new problems. This view is now widely disputed (Sartor and Boyd 1972, Anon. 1973, Solomon 1974, Shaheen 1975). Examination of the possible effects of road construction is particularly important in an area like the Lake District where much of the development affects marginal land characterized by relatively simple and unsophisticated geological and ecological systems. These are well adjusted to existing natural conditions but may be readily disrupted, or totally destroyed, if major roads introduce new processes, boundary conditions, or chemical species to which they cannot adapt.

Motorways interfere with the environment in different ways during and after construction. During construction excavation, embankment tipping, and realignment of minor roads and rivers produce obvious large-scale disturbances. On completion the motorway lies in a distinct corridor of disturbed ground which gradually readjusts to a new equilibrium. The final persistent effects of the motorway are less obvious and depend on the link between artificial and natural systems provided by the motorway drainage system. A minor, but highly significant, connexion is also made by the aerial dispersal of spray generated by vehicle tyres. The drainage system links together a set of hydrological, sedimentological and geochemical processes which are characteristic of the road system. The major effects of motorways on the environment can be discussed in terms of their effects on surface water, sediments, and chemical processes.

The surface water effects of motorway construction are caused by the need for rapid drainage of water from considerable lengths of carriageway. This produces a set of highly impermeable sub-catchments discharging steeply-rising flood waves with high peak discharges (Swinnerton et al. 1973). These enter streams which may normally have mean annual flood discharges similar to or smaller than the discharges received from the road drainage system. In addition the road drainage system may discharge similar high peak flows several times a year. As a result the hydraulic equilibrium of the receiving stream downstream from the outfall is disturbed. This is recognized by road designers in the limited protective measures taken at and below outfalls (Halls et al. 1969) but the effects of increased dominant discharge on the river channel further downstream are ignored. Bank erosion, channel realignment and flooding may, and do, take place. Flood peak discharges could be reduced with balancing ponds, which would also intercept sediment, but these have not been incorporated in the designs of road schemes in the Lake District.

Sediment is discharged from major roads during construction and use. Very high concentrations occur and can cause deposition at outfalls, affect fish, and produce unacceptable turbidity at water supply intakes. During construction major realignments of water courses, as at Gaisgill on the River Lune, will give extreme effects which will persist for some time as the river adapts to its new course. More common sources of sediment during construction are provided by newly excavated material and fill, vehicles driving on the road base, and erosion of unvegetated slopes. Measurements in the U.S.A. (Weber and Reed 1976) indicate sediment losses of 650 kg per metre of 30 m wide road during construction. Yields as extreme as this have not been experienced in the Lake District but very high levels of turbidity were experienced in the River Lune during construction of the M6 Motorway south of Tebay (Anon. 1974). After completion these sources of sediment cease and are replaced by new sources related to road and vehicle wear and spillages. Wear of bitumen carpet roads supplies rock particles and bitumen. Vehicle wear supplies particles of rubber and metal oxides and road-salting supplies large quantities of marl. The types of particles derived from the M6 north of Lancaster, their relative proportions, properties and behaviour have been studied by Patrick (1975). Sediment particles are flushed rapidly from the road surface by rainfall and pass through the drainage system into streams producing high concentrations of suspended sediments (Tucker 1975, Hedley and Lockley 1975) which may have adverse effects on fish and plant life. Harmful effects may also arise from the nature of the sediment particles. The mineral grains will be chemically inert, except for clays capable of absorbing heavy metal ions, and will behave in a predictable manner. Rubber and bitumen may degrade physically, releasing toxic inorganic or organic additives, and chemically and biochemically, producing toxic products, and exerting a biochemical oxygen demand on the water. The toxic compounds and oxygen demand may damage or inhibit the fauna in both the water and the bottom sediment. Organic sediment particles differ from normal inorganic ones as they can change their characteristics during transport or deposition so that prediction of their dispersal either initially or after re-erosion is difficult unless established procedures are modified. Sediment traps have been constructed on drainage outfalls to the Lancaster and Kendal Canal. These collect some sediment, which formed the basis of the investigation by Patrick, but are probably highly inefficient.

The chemical characteristics of road drainage water are probably the most harmful aspect of the interaction between the motorway and the natural environment. Road drainage water contains large amounts of common salt and other inorganic ions, moderate amounts of dissolved and immiscible organic compounds and small, but highly significant, quantities of trace metals (Sartor and Boyd 1972, Shaheen 1975, Tucker 1975, Hedley and Lockley 1975, Laxen and Harrison 1977). This mixture of chemical components leaves the road surface via the drainage system and as spray, which can be deposited in significant amounts at distances up to 100 m from the road (Laxen 1975). Salt applied for de-icing purposes has the most immediate, and possibly most drastic, effects as concentrations over 10,000 mg/l have been recorded (Laxen 1975). It also acts as an excellent tracer for investigating the complex pattern of pathways and stores along which motorway-derived water and other solutes pass. In the high concentrations discharged from the road surface, and the lower concentrations found in streams after dilution has taken place, salt may constitute a hazard to human and animal health, may damage vegetation or encourage the establishment of salt-tolerant vegetation (Anon. 1973, Roth and Wall 1976, Mathews and Davison 1976), damage the road structure, and remobilize heavy metal ions previously adsorbed on clay minerals. Saline water draining into lakes may induce density stratification which will inhibit or prevent the autumn and winter overturn, and damage the ecology of the lake. This may happen in Killington Reservoir. Saline water entering the soil or groundwater systems will cause water pollution and deflocculation of clay minerals. This deflocculation will reduce soil permeability, giving impaired drainage, increased soil moisture contents and possibly a loss of shear strength. On steep slopes this may eventually permit slumping and endanger the structure. Salt supplied to the soil as spray dissolves in the soil water and is then discharged as part of the normal baseflow to streams, causing increased salinity. Laxen (1975) demonstrated that at Shap salt is accumulating in the soil so that continuous increases in baseflow salinity must be expected. The 'philosophy' of road drainage design assumes that discharges of dissolved and suspended materials from motorways are unimportant as they will coincide with increased stream discharges produced by the same storm, providing adequate dilution. This is nearly always incorrect as the rapid drainage from roads delivers water to the receiving stream faster than any natural catchment can respond. Under conditions of high soil moisture deficit the natural catchment will not respond at all. In virtually all situations dilution will not occur unless the receiving stream is already in flood following an earlier storm.

The motorway system and its interactions with natural systems provides excellent, possibly unique, opportunities for applying the full range of geological techniques to a completely defined mini-catchment. Detailed systematic investigations will provide novel and valuable insights into the operation of sedimentological and geochemical processes in extreme yet controlled conditions in addition to providing the basic information needed to determine what physico-chemical effects motorways and other high-speed roads have on the environment.

Postscript

The three aspects of environmental geology discussed in this chapter illustrate the nature and scope of the subject. They show that none of the subject matter is unique since all the information and discussion could appear in chapters concerned with applications of hydrogeology, sedimentology, geochemistry or engineering geology. This is unimportant as the essential element of environmental geology, which justifies its existence as a separate branch of geology, is the synthesis of all information and lines of enquiry required to investigate situations in which man plays a significant geological role.

REFERENCES

ANON. 1970a. Water resources in the north. Northern Technical Working Party Report. *Wat. Resour. Bd. Publs.* No. 7, lx + 253 pp. H.M.S.O. London.
—— 1970b. Feasibility study of water conservation in Morecambe Bay. Prototype studies and tidal results from the fixed bed model. *Hydraulics Research Station, Wallingford. Ext. Rept.* EX 502, 4 vols.
—— 1973. Environmental degradation by de-icing chemicals and effective countermeasures. *Highway Res. Rec.* 425.
—— 1974. Ninth Annual Report of Lancashire River Authority for year ending March 1974. 118 pp.
—— 1976a. Water resources planning group report on alternative water resources developments. *North West Water Authority.* 32 pp.
—— 1976b. Water resources in West Cumbria. *North West Water Authority.* 13 pp.
ASHMEAD, P. 1974. The caves and karst of the Morecambe Bay area. pp. 201–26 *in* Waltham, A. C. (Ed.). *Limestones and caves of north-west England.*
ATKINSON, A. 1955. Manchester Waterworks. 1945–55. *J. Instn. Wat. Engrs.* **9**, 375–422.
BURGESS, I. C., HOLLIDAY, D. W. 1973. The Permo-Triassic rocks of the Hilton borehole, Westmorland. *Bull. geol. Surv.* No. 46, 1–34.
DONOVAN, R. N., ARCHER, R. 1975. Some sedimentological consequences of a fall in the level of Haweswater, Cumbria. *Proc. Yorks. geol. Soc.* **40**, 547–62.

GILL, C. J. 1970. The flooding tolerance of woody species — a review. *Forestry Abstr.* **31**, 671–87.
—— BRADSHAW, A. D. 1971. Some aspects of the colonisation of upland reservoir margins. *J. Instn. Wat. Engrs.* **25**, 165–73.
GRAY, A. J. 1972. The ecology of Morecambe Bay. v. The salt marshes of Morecambe Bay, *J. Appl. Ecol.* **9**, 207–20.
GREGORY, K. J., PARK, C. C. 1974. Adjustment of river channel capacity downstream from a reservoir. *Wat. Resour. Res.* **10**, 870–73.
—— WALLING, D. E. 1973. Drainage basin form and process. x + 456 pp Arnold, London.
HALLS, P. N., KNOWLES, P. D., McNEE, J. S. 1969. Design of the Lancaster–Penrith section of M6 motorway. *Roads and Road Construction (April)*, 109–15.
HARDCASTLE, J. H., MITCHELL, J. K. 1976. Water quality and aquitard permeability. *J. Irrign. Drainage Div., Amer. Soc. Civ. Engrs.*
HARVEY, A. M. 1969. Channel capacity and the adjustment of streams to hydrologic regime. *J. Hydrol.* **8**, 82–98.
HEDLEY, G., LOCKLEY, J. C. 1975. Quality of water discharged from an urban motorway. *Wat. Pollut. Control.* 659–74.
INGLIS, C. C., KESTNER, F. J. T. 1958. The long-term effects of training walls, reclamation, and dredging on estuaries. *Proc. Instn. Civ. Engrs.* **9**, 193–216.
KENNARD, M. F., KNILL, J. L. 1969. Reservoirs on limestone, with particular reference to the Cow Green scheme. *J. Instn. Wat. Engrs.* **23**, 87–136.
KESTNER, F. J. T. 1966. The effects of engineering works on tidal estuaries. pp. 226–38 *in* Thorn, R. B. (Ed.). River engineering and water conservation works, Butterworths, London.
—— 1961. Short-term changes in the distribution of fine sediments in estuaries. *Proc. Instn. Civ. Engrs.* **19**, 185–208.
—— 1970. Cyclic changes in Morecambe Bay. *Geog. Jour.* **136**, 85–97.
—— 1972. The effects of water conservation works on the regime of Morecambe Bay. *Geog. Jour.* **138**, 178–208.
KNILL, J. L. 1970. Environmental, economic and engineering factors in the selection of reservoir sites, with particular reference to Northern England. pp. 124–43 *in* Warren, P. T. (Ed.). *Geological aspects of development and planning in northern England. Yorks. geol. Soc.*
LAXEN, D. P. H. 1975. Water pollution and the highway: a framework for assessment. *Unpub. M.Sc. thesis, University of Lancaster.*
—— HARRISON, R. M. 1977. The highway as a source of water pollution: an appraisal with the heavy metal lead. *Wat. Res.*
LEWIS, L. H. 1921. Manchester Corporation Waterworks: Haweswater Scheme. *Trans. Instn. Wat. Engrs.* **26**, 130– .
LOCKETT, A. 1976. Ports and people of Morecambe Bay. *North Lonsdale Publications, Barrow-in-Furness.* 52 pp.
MARKER, M. E. 1967. The Dee Estuary, its progressive silting and salt marsh development. *Trans. Instn. Brit. Geog.* **41**, 65–72.
MATHEWS, P., DAVISON, A. N. 1976. Maritime species on roadside verges. *Watsonia.* **11**, 146–47.
MONKHOUSE, R. A., REEVES, M. J. 1977. A preliminary appraisal of the groundwater resources of the Vale of Eden. Tech. Note No. 11. *Cent. Wat. Planning Unit, Reading.*
MORTON, E. 1973. A review of the influence of geology on the design and construction of impounding dams. *J. Instn. Wat. Engrs.* **27**, 243–71.
O'BRIEN, M. P. 1931. Estuary tidal prisms related to entrance areas. *Civ. Engng. (May).*
O'CONNOR, K. 1975. Civil engineering aspects of Heysham nuclear power station. *J. Instn. Civ. Engrs.* **58**, 377–93.
ORRICK, M. J. 1976. Recreation and restriction on Brockhole Beach, Windermere. *Unpub. Report to the Director, National Park Visitor Centre, Brockhole.* 33 pp.
PATRICK, C. K. 1970. Proof of evidence presented to Public Inquiry into proposed Arnside Link Road. *Unpub. rep.*
—— 1975. The nature and significance of sediments derived from roads. *Proc. Ninth Int. Conf. Sedimentol.*, Nice, Theme 10, 99–104.
—— 1978 (in press). The evolution of the approaches to the port of Heysham. *Proc. 2nd Int. Conf. on Dredging Technol.*, Brit. Hydromech. Res. Assn.
REEVES, M. J. 1975. Proposals for the development of the Triassic sandstone aquifer in the Calder Valley, Cumbria. *Cent. Wat. Planning Unit, Reading.* Tech. Note No. 5, 15 pp.
ROTH, D., WALL, G. 1976. Environmental effects of highway de-icing salt. *Groundwater.* **14**, 286–89.
SARTOR, J. D., BOYD, G. B. 1972. Water Pollution aspects of street surface contaminants. *United States Env. Protn. Agency Rept.*, EPA–R2–72–081. 236 pp. Washington.
SHAHEEN, D. G. 1975. Contributions of urban roadway usage to water pollution. *United States Env. Protn. Agency Rept.*, EPA 600/2–75–004. 228 pp. Washington.
SOLOMON, D. 1974. Environmental research and highways. *Public Roads.* **37**, 297–305.
SWINNERTON, C. J., HALL, M. J., O'DONNELL, T. 1973. Conceptual model design for motorway stormwater drainage. *Civ. Engng.* **68**, 123–29.
TAYLOR, G. E. 1951. The Haweswater Reservoir. *J. Instn Wat. Engrs.* **5**, 355–9.

THORN, R. B. (Ed.). 1966. River engineering and water conservation works. xiv + 520 pp. Butterworths, London.

TUCKER, C. G. J. 1975. Stormwater pollution — sampling and measurement. *J. Instn Municipal Engrs.* **101,** 269–73.

WALTERS, R. C. S. 1936. The nation's water supply. xv + 244 pp. *Ivor Nicholson and Watson, London.*
—— 1962. Dam geology. *Butterworths, London.* viii + 335 pp.

WALTHAM, A. C. (Ed.). 1974. Limestones and caves of North West England. 477 pp.

WARREN, P. T. (Ed.). 1970. Geological aspects of development and planning in northern England. *Yorks. geol. Soc.* viii + 167 pp.

WEBER, W. G., REED, L. A. 1976. Sediment runoff during highway construction. *Civ. Engng – A.S.C.E.* (March) 76–79.

C. K. PATRICK, B.SC.

Department of Environmental Sciences, University of Lancaster, Lancaster LA1 4YQ

INDEX